Applied Econometrics Association Series

General Editors: **Jean H. P. Paelinck**, Emeritus Professor, Erasmus University, Rotterdam, and **Henri Serbat**, Chamber of Commerce, Paris

The vital importance of econometrics for understanding economic phenomena is increasingly recognised in every field of economics.

The discipline is based on 'scientific processes which aim to identify, explain and forecast economic phenomena using formalised tools to produce, measure, structure and model the information' (Gerard Duru and Henri Serbat, 1991).

The Applied Econometrics Association, established in 1974, seeks to develop the use of econometric tools by producing regular updates on the state of the art and the progress made in each specific field, and so to further the transformation of unconnected facts into pertinent information for the use in analysis and decision-making.

The series was conceived to form a set of working references for advanced students and researchers in each specific field, as well as a guide to development across the discipline more broadly.

This exchange of knowledge will be continued permanently by the opening of a debate-site on the Internet [http://www.lyonl.fr/aea].

Titles include:

Françoise Bourdon and Jean Bourdon (*editors*)
WAGE ECONOMETRICS AND MODELLING

Fabrizio Carlevaro and Jean-Baptiste Lesourt (*editors*)
MEASURING AND ACCOUNTING ENVIRONMENTAL NUISANCES AND SERVICES

Arthur Getis, Jesús Mur and Henry G. Zoller (*editors*)
SPATIAL ECONOMETRICS AND SPATIAL STATISTICS

Siv S. Gustafsson and Danièle E. Meulders (*editors*)
GENDER AND THE LABOUR MARKET
Econometric Evidence of Obstacles to Achieving Gender Equality

Hans Heijke and Joan Muysken (*editors*)
EDUCATION AND TRAINING IN A KNOWLEDGE-BASED ECONOMY

Pavlos Karadeloglou (*editor*)
EXCHANGE-RATE POLICY IN EUROPE

Emile Quinet and Roger Vickerman (*editors*)
THE ECONOMETRICS OF MAJOR TRANSPORT INFRASTRUCTURES

Philippe Thalmann and Milad Zarin-Nejadan (*editors*)
CONSTRUCTION AND REAL ESTATE DYNAMICS

Applied Econometrics Association
Series Standing Order ISBN 0–333–71460–1
(*outside North America only*)

You can receive future titles in this series as they are published by placing a standing order. Please contact your bookseller or, in case of difficulty, write to us at the address below with your name and address, the title of the series and the ISBN quoted above.

Customer Services Department, Macmillan Distribution Ltd, Houndmills, Basingstoke, Hampshire RG21 6XS, England

Spatial Econometrics and Spatial Statistics

Edited by

Arthur Getis
Stephen and Mary Birch Foundation Chair of Geographical Studies, Department of Geography, San Diego State University, USA

Jesús Mur
Associate Professor, Department of Economic Analysis, University of Zaragoza, Spain

and

Henry G. Zoller
Department of Geography, University of Louvain, Belgium

palgrave
macmillan

First published 2004 by
PALGRAVE MACMILLAN
Houndmills, Basingstoke, Hampshire RG21 6XS and
175 Fifth Avenue, New York, N. Y. 10010
Companies and representatives throughout the world

PALGRAVE MACMILLAN is the global academic imprint of the Palgrave Macmillan division of St. Martin's Press, LLC and of Palgrave Macmillan Ltd. Macmillan® is a registered trademark in the United States, United Kingdom and other countries. Palgrave is a registered trademark in the European Union and other countries.

ISBN 1–4039–0797–8

This book is printed on paper suitable for recycling and made from fully managed and sustained forest sources.

A catalogue record for this book is available from the British Library.

Library of Congress Cataloging-in-Publication Data
Spatial econometrics and spatial statistics / edited by Arthur Getis, Jesús Mur, and Henry G. Zoller
p. cm.
"All but one of the chapters in this volume were first presented as papers in the joint sessions of the Western Regional Science Association (WRSA) and the Applied Econometrics Association (AEA) on the occasion of the annual meeting of the WRSA held in Palm Springs, California, in February 2001" --P..
Includes bibliographical references and index.
ISBN 1–4039–0797–8
1. Geography--Statistical methods--Congresses. 2. Spatial analysis (Statistics)--Congresses. 3. Econometrics--Congresses. I. Getis, Arthur, 1934- II. Mur, Jesús. III. Zoller, Henry. IV. Western Regional Science Association. Meeting. (2001 : Palm Springs, Calif.) V. Applied Econometrics Association.
G70.3.S56 2003
910'.01'5195--dc21

2003042900

10 9 8 7 6 5 4 3
13 12 11 10 09 08 07 06 05 04

Printed and bound in Great Britain by
Antony Rowe Ltd, Chippenham and Eastbourne

Contents

	List of Tables	vii
	List of Figures	ix
	Preface	xii
	Notes on the Contributors	xiv
	Part I Introduction	**1**
1	Contributions to Spatial Econometrics and Spatial Statistics ARTHUR GETIS and JESÚS MUR	3
2	Theoretical Economic Geography and Spatial Econometrics: Bridging the Gap Between Theory and Evidence BERNARD FINGLETON	8
	Part II Spatial Econometrics Estimation	**29**
3	Spatial Autoregressive Local Estimation R. KELLEY PACE and JAMES P. LESAGE	31
4	Aggregation Bias in Maximum Likelihood Estimation of Spatial Autoregressive Processes TONY SMITH	53
5	Econometric Estimation of a Spatial Dynamic Model in Continuous Space and Continuous Time: An Empirical Demonstration KIERAN P. DONAGHY and MARIA PLOTNIKOVA	89
6	Using Matrix Exponentials to Estimate Spatial Probit/Tobit Models JAMES P. LESAGE and R. KELLEY PACE	105

Part III Spatial Dependence and Spatial Errors **133**

7 Properties of Tests for Spatial Error Components: A Further Analysis 135
HARRY H. KELEJIAN and YEVGENY YUZEFOVICH

8 Some Proposals for Discriminating Between Spatial Processes 150
F. JAVIER TRIVEZ and JESÚS MUR

9 Serial and Spatial Error Dependence in Space-Time Models 176
J. PAUL ELHORST

10 Spatial Dependence in the Evolution of Regional Income Distributions 194
SERGIO J. REY

Part IV Spatial Patterns and Local Spatial Analysis **215**

11 On Connectropy 217
JOHAN F. KAASHOEK, JEAN H.P. PAELINCK and HENRY G. ZOLLER

12 Locally Weighted Regression and Time-Varying Distance Gradients 232
DANIEL P. MCMILLEN

13 The Statistical Significance of the Maximum Local Statistic 250
PETER A. ROGERSON

14 Detecting Spatial Clusters: A Bayesian Approach with Application to Identifying Disability Patterns in Mississippi and Alabama 265
LI-AN XU and SHUMING BAO

Bibliography 283
Subject Index 296

List of Tables

2.1	Full (initial) model with groupwise heteroscedasticity	18
2.2	Summary statistics and diagnostics – Full model	18
2.3	Reduced model – groupwise heteroscedastic estimates	18
2.4	Reduced model – Summary statistics and diagnostics	19
2.5	Final model – groupwise heteroscedastic estimates	19
2.6	Summary statistics and diagnostics	19
2.7	Returns to scale estimates	20
2.8	Recursive parameter estimates	21
2.9	Recursive standard error estimates	21
5.1	Proportionate Root-Mean-Square Errors of In-Sample, One-Period Forecasts	98
5.2	Portmanteau (or Q) Test Statistics for Autocorrelations in Residuals of Variables	99
5.3	Quasi-FIML Estimates of Parameters and their Asymptotic Standard Errors Parameter Estimate	100
5.4	Augmented Engel-Granger Test of Noncointegration	103
6.1	A comparison of estimates based on 70% sample censoring	119
6.2	A comparison of estimates based on 30% sample censoring	120
6.3	A comparison of Probit estimates	126
7.1	Experimental values of σ_u^2 and σ_ψ^2	141
7.2	Rejection frequencies under H_0	143
7.3	Rejection frequencies under H_1	144
8.1	The Wald Test. Selected points of the estimated discrimatory function	160
8.2	The King Test. Pairs of parameters analysed	163
8.3	Specification errors and estimated variance. Comparison of results	175
9.1	Estimation results of the Phillips curve model	187
9.2	Estimation results of the labor force participation rate equation	187
10.1	Estimated Annual Transition Matrix, US 1929-99	199
10.2	Differences in Regional Mobility	212

11.1 Contiguity degrees of the Dutch provinces 221
11.2 Entries of table 11.1 222
11.3 GRP's of the Dutch provinces 224
11.4 Contiguity degrees of the Belgian provinces 225
11.5 Entries of table 11.4 227
11.6 GRP shares of the Belgian provinces 230

12.1 Descriptive Statistics and Base Hedonic Regression 240

13.1 Upper and Lower Bounds, and Comparison to Simulated 260

List of Figures

2.1	Recursive ML b_1 estimates	22
2.2	Recursive ML ρ estimates	23
3.1	Cross-validation MAPE for LWR and SALE vs. Subsample Size	42
3.2	LWR Income Estimates vs. Subsample Size	43
3.3	Equivalent Number of Variables for LWR vs. Sample Size	43
3.4	10th, 50th, and 90th Percentiles of *a* Estimates Across Sample Sizes	44
3.5	SALE Population Estimates by W-E Order	45
3.6	LWR Population Estimates by W-E Order	46
3.7	SALE Education Estimates by W-E Order	46
3.8	LWR Education Estimates by W-E Order	47
3.9	SALE Homeownership Estimates by W-E Order	47
3.10	LWR Homeownership Estimates by W-E Order	48
3.11	SALE Income Estimates by W-E Order	48
3.12	LWR Income Estimates by W-E Order	49
4.1	A Simple Spatial Aggregation Example	58
4.2a	Bimodal with Positive Global Maximum	59
4.2b	Bimodal with Negative Global Maximum	59
4.3a	Disaggregate Model Estimates of Rho (3x6)	62
4.3b	Aggregate Model Estimates of Rho (3x6)	62
4.4a	Regions and Subregions (16x100)	63
4.4b	Philadelphia Block Groups (43x312)	63
4.5a	Disaggregate Model Estimates of Rho (16x100)	64
4.5b	Aggregate Model Estimates of Rho (16x100)	64
4.6a	Disaggregate Model Estimates of Rho (43x312)	65
4.6b	Aggregate Model Estimates of Rho (43x312)	65
4.7a	Limiting Likelihood Function (3x6)	68
4.7b	Histogram for 100 Replications (3x6)	69
4.8a	Corrected Asymptotic Log-Likelihood (3x6)	76
4.8b	Corrected Limit Likelihood (3x6)	76
4.9a	Histogram of Corrected Estimates (3x6, N=100)	77
4.9b	Histogram of Corrected Estimates (16x100)	77
4.10	Histogram for $\rho_0 = 0$ (100 replications of 3x6)	79

6.1	The impact of censoring on β estimates	120
6.2	Posterior density for α from two models	121
6.3	Posterior density for ρ from two models	122
6.4	Posterior density for m (# neighbours) from two models	123
6.5	Imputed y values for 60 percent censored observations	124
6.6	Imputed y values for 90 percent censored observations	125
6.7	Predicted 0,1 values for non-spatial probit and MESS models	126
6.8	Predicted 0,1 values for spatial probit MESS models	127
8.1	Right Decision. SMA Processes. P=2	157
8.2	Right Decision. SAR Processes. P=2	158
8.3	Right Decision. SMA Processes: P=3	159
8.4	Right Decision. SAR Processes: P=3	162
8.5	The King Test. SMA Processes: Positive autocorrelation	164
8.6	The King Test. SAR Processes: Positive autocorrelation	165
8.7	The King Test. SMA Processes: Negative autocorrelation	166
8.8	The King Test. SAR Processes: Negative autocorrelation	167
8.9	The estimated variance method: SMA Processes	168
8.10	The estimated variance method: SAR Processes	169
9.1	The area on which the (concentrated) log-likelihood function is defined for τ and δ	183
9.2	The concave shape of the concentrated log-likelihood in function of τ and δ	189
10.1	US Regional Per-Capita Income Distribution 1929-99	196
10.2	Dispersion:US Regional Per-Capita Income Distribution 1929-99	198
10.3	Skewness and Kurtosis: US Regional Per-Capita Income Distribution 1929-99	198
10.4	Spatial Autocorrelation of Per-Capita Income Distribution 1929-99	207
10.5	Spatial Cohesion Index: 1-year interval	209
10.6	Spatial Rank Mobility Index: 1-year interval.	210
10.7	Spatial Rank Mobility Index: 5-year interval	210
10.8	Spatial Rank Mobility Index: 10-year interval	211
10.9	Spatial Rank Mobility Index: 25-year interval	211
11.1	Map of Dutch provinces	223
11.2	Map of Belgian provinces	229
12.1	Number of Sales by Quarter	241
12.2	Estimated Price Indexes	243
12.3	Estimated Distance Gradients	244
12.4	OLS 95% Confidence Intervals for Distance Gradients	245

12.5	LWR 95% Confidence Intervals for Distance Gradients	246
12.6	LWR Distance Gradients	247
14.1	An illustration of spatial mixture processes	267
14.2	Simulated Samples of the Potts model with different β	270
14.3	The disability rates with counties in Mississippi and Alabama States	273
14.4	The clusters calculated using K-Means methods	274
14.5	The result of the smoothing method	275
14.6	Tests of Local Moran's on the disability rate data in the Mississippi and Alabama counties	277
14.7	The results from spatial mixture model using Gibbs sampling (K=3)	278
14.8	The time series plots of the sampled p_0; p_1; p_2	279
14.9	The results from spatial mixture model using Gibbs sampling (K=2)	280
14.10	The time series plots of the sampled p_0; p_1	281

Preface

This volume contains recent contributions to the fields of spatial econometrics and spatial statistics. We make the distinction between these two compatible fields by the types of problems which practitioners attempt to solve. The field of spatial econometrics has come to include the methods and models that deal with estimation and testing problems encountered when attempting to implement regional economic models. The problems often are characterized by the difficulties associated with assessing the importance of spatial dependence and spatial heterogeneity. As Anselin (1988) puts it, the focus is on the *spatial effects*.

The field of spatial statistics is more broadly understood. In general, spatial statistics is concerned with the statistical and mathematical descriptors of spatial structure. While it may include some of the work of the spatial econometricians, it focuses less on the dependent-independent variable relationship in a regression setting and more on the nature of space and spatial data. In this volume, contributions to spatial statistics include work on spatial proximity, spatial patterning in general and, in particular, the spatial association (dependence) contained in local map patterns.

All but one of the chapters in this volume were first presented as papers in the joint sessions of the Western Regional Science Association (WRSA) and the Applied Econometrics Association (AEA) on the occasion of the annual meeting of the WRSA held in Palm Springs, California, in February 2001. The overall organizer of the AEA sessions was Henry Zoller. David Plane and Jean Paelinck represented the WRSA and the AEA, respectively. Jean Paelinck, as a managing editor of the present series, conceived of the idea for this book.

The papers were rigorously refereed blindly by the contributors to this volume. Referee reports were sent to each author and changes made accordingly. We supervised this process to insure that the authors received reviews that would be useful for the final construction of their chapters.

Fortunately, we were dealing with truly exemplary scholars, the most distinguished and sophisticated representatives of these fields of inquiry.

We would like to acknowledge the inspiration of Jean Paelinck for his stimulation, for his leadership, and for his good relations with the publisher, which made it possible to have this work entered into Palgrave's outstanding series on applied econometrics. We very much appreciate the cooperation of David Plane in his role as Executive Secretary of the WRSA. We thank the authors for their diligence, not only in providing extremely thoughtful and useful papers, but in meeting all deadlines in a timely manner and in following stringent editorial guidelines. In addition, we have benefited greatly from the editorial assistance of Laura Martin Makey, an instructor of geography at California State University, San Marcos. Her expertise in handling various word processing systems, formatting, and indexing, together with care and attention to detail helped immeasurably. Finally, we are grateful to Jared Aldstadt, a student assistant, for helping to keep the project well organized.

June 2002

ARTHUR GETIS, San Diego
JESUS MUR LACAMBRA, Zaragoza
HENRY G. ZOLLER, Louvain-la Neuve

Contributors

SHUMING BAO received his Ph.D. in applied economics from Clemson University in 1996. He was a research scientist at MathSoft from 1996 to 1997, and currently is a senior researcher at the China Data Center at the University of Michigan. His research interests are in spatial statistics, regional economics and geographic information systems. e-mail: sbao@umich.edu Address: Suite 3630; 1080 South University Avenue; Ann Arbor, Michigan 48109-1106; USA

KIERAN P. DONAGHY holds a Ph.D. in regional science from Cornell University. His recent research has concerned the coordination of macroeconomic and environmental policies, assessing regional impacts of climate change, and estimating and analyzing models of intertemporally optimizing agents. e-mail: donaghy@uiuc.edu Address: Department of Urban and Regional Planning, UIUC; 111 Buell Hall; 611 East Lorado Taft Drive; Champaign, Illinois 61820; USA

J. PAUL ELHORST graduated in econometrics at the University of Amsterdam in 1982. Since 1992 he has been associate professor of regional and labour economics at the University of Groningen. His current research topics are spatial econometrics, regional labour market analysis, and economic effects and social cost-benefit analysis of new transport infrastructure. e-mail: j.p.elhorst@eco.rug.nl Address: Department of Economics; University of Groningen; P.O. Box 800; NL-9700 AV Groningen; Netherlands

BERNARD FINGLETON holds the post of University Lecturer at Cambridge University, where he researches aspects of regional development, spatial econometrics and analytical economic geography. He has been a Jean Monnet Fellow in Economics at the European University Institute and a consultant for the European Commission, and is on the editorial boards of *Environmetrics*, *International Regional Science Review* and *Regional Studies*. e-mail: bf100@cam.ac.uk Address: Department of Land Economy; 19 Silver Street; Cambridge CB3 9EP; UK

ARTHUR GETIS holds the Stephen and Mary Birch Foundation Chair of Geographical Studies in the Department of Geography at San Diego State University. He has published widely in urban geography and spatial analysis, and co-edits the Journal of Geographic Systems. e-mail: arthur.getis@sdsu.edu Address: Department of Geography; San Diego State University; San Diego, California 92182; USA

JOHAN F. KAASHOEK is associate professor at the Econometric Institute of Erasmus University Rotterdam. He graduated in mathematics from Leiden University (1969), and received his Ph.D. from Erasmus University (1984). His main interest is in non-linear pattern formation. e-mail: kaashoek@wis.few.eur.nl

HARRY H. KELEJIAN is a Professor of Economics at the University of Maryland, College Park. He received his Ph.D. from the University of Wisconsin in 1968. Before coming to Maryland, he held academic positions at Princeton University and at New York University. He has published a wide variety of articles in leading journals relating to economic theory and its applications. e-mail: kelejian@econ.bsos.umd.edu Address: Department of Economics; University of Maryland; College Park, Maryland 20742; USA

JAMES P. LESAGE received his Ph.D. in economics from Boston College in 1983. His current research involves application of Bayseian methods to spatial econometric estimation, and he maintains a public domain collection of algorithms to implement these methods at www.spatial-econometrics.com. e-mail: jlesage@spatial-econometrics.com Address: University of Toledo; Department of Economics; 2801 West Bancroft Street; Toledo, Ohio 43606; USA

DANIEL P. MCMILLEN is a Professor of Economics and Director of the Center for Urban Real Estate at the University of Illinois at Chicago. He received his Ph.D. in economics from Northwestern University in 1987, and has since served on the faculties of University of Oregon, Santa Clara University, and Tulane University. e-mail: mcmillen@uic.edu Address: Department of Economics (MC144); University of Illinois at Chicago; 601 South Morgan Street; Chicago, Illinois 60607;USA

JESÚS MUR has been an associate professor at the University of Zaragoza in the department of Economic Analysis, since , since 1986, and received his Ph. D. there in 1990. His current research topics are spatial econometrics, mainly autocorrelation approaches, location tendencies, regional forecast techniques and short-term economic analysis. e-mail:jmur@posta.unizar.es Address: Department of Economic Analysis; University of Zaragoza; Gran Via 2-4; 50005 Zaragoza; Spain

R. KELLEY PACE holds the Louisiana Real Estate Commission Chair of Real Estate at Louisiana State University and has published more than 20 papers developing spatial statistical methods and applying these to economic data. She concentrates on augmenting the richness of spatial models while maintaining computational feasibility for large spatial data sets. e-mail: kelley@pace.am Address: 2164 CEBA; Department of Finance; E.J. Ourso College of Business; Louisiana State University; Baton Rouge, Louisiana 70803-6308; USA

JEAN H.P. PAELINCK is emeritus professor of Erasmus University Rotterdam and honorary research fellow of the Tinbergen Institute. He is presently a distinguished visiting professor at George Mason University's School of Public Policy. As a (co)-author or (co)-editor he has published over fifty books and three hundred and fifty articles, mostly in the field of theoretical spatial economics and spatial econometrics. e-mail: j.paelinck@planet.nl Address: Oranjelaan 36; NL-3062 BT Rotterdam; Netherlands

MARIA PLOTNIKOVA is a Ph.D. student in Regional Planning at the University of Illinois at Urbana-Champaign. E-mail: plotniko@uiuc.edu Address: Department of Urban and Regional Planning, UIUC; 111 Buell Hall; 611 East Lorado Taft Drive; Champaign, Illinois 61820; USA

SERGIO J. REY's research interests include regional economic growth and income inequality, spatial econometrics, open source geocomputation, integrated multiregional modeling, and industry cluster analysis. He has published widely in such journals as *Geographical Analysis*, *Regional Studies*, *Growth and Change*, *Environment and Planning A*, and *The International Regional Science Review*. e-mail: serge@rohan.sdsu.edu Address: Department of Geography; San Diego State University, San Diego, California 92182, USA

PETER A. ROGERSON is Professor of Geography and a member of the National Center for Geographic Information and Analysis. He has interests in spatial modeling and spatial statistics, and their applications in epidemiology and crime analysis. e-mail: rogerson@acsu.buffalo.edu Address: Department of Geography; University of Buffalo; Wilkeson Hall; Buffalo, New York 14261; USA

TONY SMITH is Professor of Regional Science and Systems Engineering at the University of Pennsylvania. His current research interests are concerned with the integration of micro spatial behavior into macro statistical models. e-mail: tesmith@ssc.upenn.edu Address: Department of Systems Engineering; University of Pennsylvania; Philadelphia, Pennsylvania, 19104; USA

F. JAVIER TRIVEZ has been an associate professor in the Department of Economic Analysis of the University of Zaragoza since 1981. He received his Ph. D. from the University of Zaragoza in 1986. He has worked in different econometrics fields but his main interests are in regional analysis, time series analysis, outliers and short-term economic analysis. e-mail:fjtrivez@posta.unizar.es Address: Department of Economic Analysis; University of Zaragoza; Gran Via 2-4; 50005 Zaragoza; Spain

LI-AN XU received his Ph.D. from the Department of Statistics at the University of Michigan in 2000 and is currently a bio-statistician at the Pharmaceutical Research Institute of the Bristol-Myers Squibb Company. e-mail: li-an.xu@bms.com Address: Bristol-Myers Squibb; Princeton, New Jersey 08543; USA

YEVGENY YUZEFOVICH is a Ph.D. student in the Economics Department at the University of Maryland. He received his M.Sc. in applied mathematics from Moscow State University in 1994 and an M.A. in economics from the New Economic School in Moscow in 1996. His dissertation focuses on estimation of spatial models. e-mail: yezefovich@econ.umd.edu Address: Department of Economics; University of Maryland; College Park, Maryland, 20742, USA

HENRY G. ZOLLER is professor in the Department of Geography of Louvain University (Louvain-la-Neuve, Belgium). Over the past twenty years he has worked on various topics in the fields of spatial economic analysis, multicriteria modelling and policy evaluation in urban planning. e-mail: zoller@geog.ucl.ac.be Address: 30 Avenue Penelope; B-1190 Brussells; Belgium

Part I

Introduction

1 Contributions to Spatial Econometrics and Spatial Statistics

Arthur Getis
San Diego State University

Jesús Mur
Universidad de Zaragoza

The word *spatial*, of little use and concern a mere 50 years ago, is now part of the working vocabulary of most social and physical scientists. Concern for the spatial manifestation of key social, economic, and physical variables has led to the development of various schools of thought dedicated to particular methodological approaches to the use of spatial data. If spatial data were no different in form than temporal data we might expect that the well-developed time series methodologies would be transferred directly into the world of space series. Such is not the case, however. Spatial data is more complex in structure than temporal data. Location relative to other locations (the spatial view) is a much more difficult structure to model than time relative to other times. Locations are multi-dimensional in a multi-directional sense; that is, they are related to one another over varying distances, horizontal and vertical, and varying directions, while time is uni-dimensional. Locations are represented as points, lines of any shape, and areas of many sizes and shapes, while time is only understood in well defined units such as day, week, month, year. This great difference has only recently been acknowledged as requiring specially devised techniques for modeling.

One of the schools of thought created to aid in the modeling of spatial data is called spatial econometrics. The characteristics of the field were first outlined in Paelinck and Klaassen's 1979 book *Spatial Econometrics*. From its modest beginnings there has grown a small, but sophisticated number of practitioners that have broadened and deepened the field, and solved some of the problems inherent in the peculiarities of taking space into account. This book represents the latest set of advances in this field.

We have divided these contributions into four parts: this introduction and a chapter on *new economic geography* theory, four chapters each on spatial econometrics estimation, the analysis of spatial dependence and spatial errors, and the detection and analysis of spatial patterns and spatial errors. Chapter 2, on new economic geography theory, serves as the introduction to the kinds of issues and problems addressed in these fields and sets the tone for the volume. The chapter, by BERNARD FINGLETON, is more than an introduction, however. Fingleton addresses the spatial econometric issues that have arisen due to the recent writing in the "new" economic geography. The author demands a greater symbiosis between economic geography theories and spatial econometric methods as a way of guaranteeing more balanced scientific progress in the spatially sensitive field of economic geography. The application discussed in the paper is enlightening. The topic analyzed is the Verdoorn law on productivity growth, which is linked to urban and economic geography principles such as externalities and increasing returns to scale. The specification proposed is estimated using a broad set of European regions. The results obtained tend to corroborate the hypotheses scrutinized, although there is evidence of a certain structural instability in the time dimension.

The problem of estimation in spatial econometric modeling is tied not only to the usual assumptions of the various regression formulations but also to special characteristics such as spatial dependence and spatial heterogeneity. Spatial econometricians are acutely aware that when dealing with georeferenced data, residuals that are specified as normally distributed must be spatially randomly distributed. In Part II, Spatial Econometrics Estimation, four chapters range over critical methodological issues. Chapter 3, by R. KELLEY PACE and JAMES P. LESAGE, comes to grips with estimation of parameters of local linear models such as Brunsdon, Fotheringham, and Charlton's geographically weighted regression model (GWR) and McMillen's linear weighted regression model (LWR). The local models often suffer from their inability to deal with the assumption of spatial heterogeneity. The use of a single distance (bandwidth) parameter constrains these models from accounting for a variety of distance effects. The single distance parameter may not eliminate the spatial dependence inherent in the spatial data. In addition, parameter estimates for both of these models, which by their nature are estimated using a small set of data, are affected by outliers. In order to avoid these problems Pace and LeSage propose a new estimation procedure called SALE, spatially autoregressive local estimations. They provide a computationally feasible approach for estimating LWR-type models that allow for parameter variation and spatial dependence in moderate sample sizes.

In Chapter 4, TONY SMITH develops a procedure for reducing aggregation bias in a spatial context. As Smith explains, there is "often a mismatch between the scale at which data are available and the scale at which key spatial dependencies are known to occur." He addresses the question of how to combine fine grain information about spatial dependencies with typically aggregated data such that there is an improvement in model fit. By means of demonstration he selects a simple spatial autoregressive model to show how the variance-minimizing tendency of maximum likelihood estimation often sends correlations into negative territory when they should be positive. His attempt at correcting this problem leads to a rescaling of the spatial autogressive parameter. The method used is much like a piecewise linear transformation.

KIERAN P. DONAGHY and MARIA PLOTNIKOVA, in Chapter 5, say "that spatial econometricians have made so much progress in understanding and coping with these conditions [spatial dependence and spatial heterogeneity] over the last two decades is surely a remarkable achievement." They point out that less well developed than spatial models are spatial dynamic models mainly because of the difficulty in identifying and controlling for the influence of spatio-temporal dependence. In earlier work, Donaghy had estimated a macroeconomic model of optimal space-time development originally put forward by Isard and Liossatos. In this chapter the authors extend this work by application to temporal data of spatially contiguous western European countries. They show that their general approach is feasible which provides an opening for future analyses of spatio-temporal autocorrelation.

The final chapter in this section by JAMES P. LESAGE and R. KELLEY PACE considers spatial data samples that contain binary dependent variables or are truncated due to censoring, usually for confidentiality requirements. They show how their Bayesian approach, introduced in 2000, can be used to produce spatial probit and tobit models. One of a number of advantages of the approach is the computational speed relative to more traditional specifications. They point out that for probit models their method can produce a generalization of traditional logit models. For the tobit model, their method produces the usual model parameters and imputations for the censored observations. In addition, they introduce a flexible spatial weight structure rather than using the usual fixed spatial weights matrix

While Part II concerned itself more generally with spatial econometric modeling, Part III focuses more directly on spatial dependence and the assumptions concerning spatial errors in spatial econometric models. A variety of tests are used to test for spatial error correlation. Among these are the Lagrangian multiplier and the Kelejian-Robinson tests. In a recent study by Anselin and Moreno, it is shown that the Lagrangian test has more power than the Kelejian-Robinson test. In Chapter 7, by HARRY H. KELEJIAN and YEVGENY YUZEFOVICH, the authors show that by the nature of the Monte Carlo procedure used by Anselin and Moreno to perform their analyses one would expect low power. Thus, they set about performing what may be called untainted Monte Carlo experiments and show that the power of the Kelejian-Robinson test is much higher but still not as high as the Lagrangian test. Further, they show that the basic structure of these tests is considerably different in that assumptions about heteroskedasticity and the spatial weights matrix are different. They conclude that tests on spatial errors are "not entirely straightforward."

In Chapter 8, F. JAVIER TRIVEZ and JESUS MUR discuss the problem of the selection of the most adequate dependence structure for a given spatial series. The alternatives they consider are limited to autoregressive or moving average processes for series that must be centered on zero. In the chapter different possibilities are studied, including the King point optimal test. The authors end by proposing a discrimination method called the minimum sampling variance, which behaves relatively well in medium-sized samples.

J. PAUL ELHORST addresses the question of bidimensional dependencies in Chapter 9, proposing an econometric model that simultaneously incorporates a temporal and a spatial autoregressive structure in the disturbance term. There is a detailed analysis of this model's maximum likelihood estimation, which faces severe problems of non-linearity which in turn requires the use of numerical methods. The author suggests an iterative solution based on the traditional approach of feasible generalized least squares. The consistency of this proposal seems guaranteed in relatively weak conditions.

SERGIO J. REY, in Chapter 10, examines the role of spatial dependence in a class of models for regional income convergence over time. He introduces three new measures for testing for spatial dependence. The first deals with regional conditioning, that is, an evaluation of the extent to which the rgeional income distributions and nationally conditioned distributions depart from spatial randomness. It is derived as a trace statistic from a modified Markov transition. Two statistics are intended for use in a dynamic context and measure the degree of spatial clustering and regional cohesion in income rank mobility.

The final section of the book, Part IV, Spatial Patterns and Local Spatial Analysis, explores characteristics of map patterns, in particular, the focus is on the relationship between nearby data points. This leads to formal definitions of spatial clustering and to what is called a local approach to discerning pattern. A local approach is usually designed to focus on a particular site where tests usually concern the site's relationship with other sites.

In Chapter 11, JOHAN F. KAASHOEK, JEAN H.P. PAELINCK, and HENRY G. ZOLLER, propose an interesting reflection on the concept of connectivity in a regional system. The authors first carry out an axiomatic approach, introducing a nearness metric, which can also be understood as a measure of entropy. The core of the paper is dedicated to the exploitation of the information contained in a spatial matrix of entropy or connectropy. The objective is to identify circuits of regions with the maximum interaction potential. The technique proposed by the authors is relatively simple and comes down to a problem of linear or quadratic assignment.

In Chapter 12, DANIEL P. MCMILLEN points out that "the non-parametric estimator used in spatial analysis is a version of locally weighted regression [LWR] in which observations closer to a target point receive greater weight." He claims that non-parametric estimation procedures are useful alternatives to standard regression analysis when functional forms are unknown. Non-parametric estimators produce a full set of coefficients for each target point. McMillen shows how locally weighted regression can be used to construct an index of distance gradients that vary over time. He demonstrates the procedures by using a kernel weighting function that places more weight on recent events.

PETER A. ROGERSON, in Chapter 13, notes that "in many applications, there is often interest in testing the statistical significance of more than one local spatial statistic." Multiple testing requires a Bonferroni-type adjustment to the significance level of each test so that the overall significance level can be maintained at a desired level. The commonly employed Bonferroni adjustment is conservative, since it assumes that individual tests are independent; the correlation between the local test statistics for nearby areas requires a less conservative adjustment. In this chapter, Rogerson shows how the Bonferroni upper bound for the critical value of a local statistic may be improved. He accomplishes this by taking into account the correlation between tests.

The final chapter is by LI-AN XU and SHUMING BAO. They introduce a new method for the identification of spatial clusters. Their methodology is based on a Markov random field designed to model the latent process that creates spatial clusters. They call this a spatial mixture process. The resultant spatial mixture model is estimated using a Bayesian approach. Finally, a Gibbs sampling algorithm is used to make inferences about the model for binomial processes. The model is built around a parameter that describes the degree of local spatial dependence. Their general approach is applicable to a wide variety of spatial clustering problems.

2 Theoretical Economic Geography and Spatial Econometrics: Bridging the Gap Between Theory and Evidence

Bernard Fingleton
Cambridge University

Abstract

Contemporary urban economic and new economic geography theory (Huriot and Thisse, 2000; Fujita et al, 1999) is characterized by an assumption of increasing returns to scale, thus introducing an element of realism which was absent from earlier, neoclassically-based, models, although other assumptions such as the absence of technological externalities mean that none the less the theory remains very much in conflict with the real world. On the other hand the more eclectic and disconnected nature of theory in economic geography 'proper' means that it is weak from the point of view of theoretical coherence. This chapter represents an example of the so-called 'third way', which strives for both empirical and theoretical consistency. The chapter breaks new ground by estimating spatial econometric model parameters recursively. The estimates support the assumption of increasing returns, showing that for manufacturing across the EU returns to scale fall through the 1980s, and increase in the 1990s. The chapter suggests estimation of the fundamental determinants of returns to scale according to the underlying theory as a way of understanding the reasons for this apparent structural break.

2.1 Introduction

Recent advances in urban economics and 'new economic geography' (Huriot and Thisse, 2000; Fujita et al, 1999) have reinforced the theoretical basis of the concept of increasing returns, which has long been held as realistic and appropriate by many in the regional science and economic geography community, moving it more to the centre stage of mainstream economic thought. However new economic geography theory is highly abstract and offers few concessions to empirical reality, although it does recognize the importance in the real world of monopolistic competition and the lack of realism of perfect competition assumptions in the context of urban and regional economies (Krugman, 1991; Fujita, Krugman, and Venables, 1999). The problem with the current theory is that it provides only a partial basis for those seeking to build empirically consistent econometric models. Some of the key assumptions on which it is based are acknowledged to be tendentious ('funny but useful' is the phrase used by Fujita, Krugman, and Venables, 1999, to describe them) and this has led some (notably Martin, 1999) to be sceptical about the whole exercise, advancing the case of economic geography 'proper'. However, economic geography 'proper' is characterized by eclectic theories, imprecise concepts and a qualitative emphasis, which make it hard to see how scientific progress can be made. This has led others, notably Fingleton (2000) (see also Plummer and Taylor, 2001), to suggest a so-called 'third way' which strives for both theoretical and empirical consistency by advocating economic geography in the context of spatial econometrics. In this chapter, while the whole gamut of urban economics and new economic geography theory provides a somewhat indigestible menu for applied econometric analysis, elements of the theory are used which can be reconciled with a reduced form which has a reasonable empirical track record. This, it is argued, provides a way forward towards a more scientific economic geography in which data confronts theory. Section 2 of the chapter therefore sets out the theory leading to this reduced form (see also Fingleton, 2001a; 2001b). Section 3 sets out the arguments for incorporating technological externalities and spatially differentiated technical progress. Knowledge spillover has, for reasons to do with the need for a clean cut theoretical exposition, been disregarded in new economic geography theory, although Marshall (1920) emphasized it as one of the triad of forces (along with labour pooling and intermediate goods supply) leading to geographic concentration. Technological externalities are evidently a necessary component of the model on empirical grounds, as we show subsequently. Section 4 is concerned with fitting models to real-world data, in this case manufacturing productivity growth over the period 1975-95 for 178 regions of the EU. Various specifications are considered, the outcome being a final model, incorporating an endogenous spatial lag and groupwise heteroscedasticity, which reaffirms the existence of increasing returns. Section 5 focuses on the robustness of the model across time by obtaining recursive maximum likelihood parameter estimates, which is an easy-to-implement methodological innovation for spatial econometrics. Finally, attention is focused on the fundamental determinants of returns to scale according to the foregoing theory, as a way towards understanding the returns to scale parameter dynamics. Section 6 concludes.

2.2 Theoretical Basis

The empirical model which is the main focus of the chapter has at its core the relationship between the exponential growth rates of manufacturing productivity (p) and manufacturing output (q) known as the dynamic Verdoorn law, thus

$$p = \lambda + b_1 q \qquad (2.1)$$

in which λ is the rate of technical progress or the growth of total factor productivity. This equation is familiar to regional economists working in a non-neoclassical framework and was popularized by Kaldor (1957) in particular. It lies at the heart of some models of cumulative causation (see for example Dixon and Thirlwall, 1975) and is associated with disequilibrium dynamics engendered by increasing returns to scale. The estimate of the coefficient b_l is normally significantly different from zero, with point estimates consistently in the vicinity of 0.5 across a wide range of data sets. A coefficient significantly greater than 0 indicates the presence of increasing returns, which can be shown for example if one assumes that the Verdoorn Law is derived from a Cobb-Douglas production function (see Fingleton and McCombie 1998; Fingleton 2001a). In this literature, increasing returns are normally attributed to learning by doing and the efficiency enhancing effects of the division of labour as workers become more specialized as the scale of the economy expands. This has echoes in the theory advanced below.

Although it has mainly been associated with the post-Keynesian tradition, the Verdoorn Law is a strong empirical regularity which is consistent with different theoretical positions, as indicated in Fingleton (2001a). In the chapter emphasis is placed on its consistency with some basic tenets of urban economics and new economic geography, namely aggregate increasing returns to scale derived from realistic market structure assumptions. In this context the coefficient b_l can be expressed as a function of some fundamental parameters and this leads to new insights regarding the causes of increasing returns. The theory, which is outlined below, originates from the urban economics literature, for example Abdel-Rahman and Fujita (1990) and Fujita and Thisse (1996), but in addition we assume the existence of diminishing returns per unit area due to the effects of congestion, as in Ciccone and Hall (1996). This latter assumption has the consequence that returns to scale may be either increasing or decreasing depending on the parameters of the model, although as the subsequent empirical analysis shows that the net effect remains one of increasing returns to scale.

Assume, as in urban economic theory, that the regional economy is divided into two sectors, which we shall call manufacturing and services. Services are non-traded and immobile and characterized by a large number of firms within a monopolistic competition market structure. Manufacturing is competitive, but as we show later, it incurs increasing returns in the number of specialized service firms due to the link between the scale of activity and variety of services provided, so that manufacturing productivity is enhanced. A variation on this theme is set out in new economic geography theory by Fujita et al. (1999) with two sectors consisting of manufacturers and farmers, and in this case it is manufacturers under monopolistic competition while agriculture is perfectly competitive. Also, transport costs have an explicit effect on wage rates, with real wage differences driving worker migration and the dynamics of the system of regions. However the critical role of transport costs and relative wage rates in the EU regional context is questionable, and the preferred approach is to commence with the urban economics model and build an economic geography model incorporating spatial effects on a different basis.

Accordingly, assume a Cobb-Douglas production function (equation 2.2) for the level of output produced by manufacturers. In equation (2.2), the level of manufacturing output (*Q*) on a fixed area of land is determined by two inputs, manufacturing labour (*M*) and composite immobile intermediate services (*I*), produced under constant returns technology. In addition, with a fixed area of land, increasing congestion with increasing output results in diminishing returns per unit area as given by $\alpha < 1$.

$$Q = (M^{\beta} I^{1-\beta})^{\alpha} \tag{2.2}$$

In contrast intermediate services production involves internal increasing returns to scale and with preference for variety each firm specializes in its own variety so that the number of firms equals the number of varieties. This is consistent with equation (2.3) which defines the constant elasticity of substitution subproduction function derived from the product variety theory of Dixit and Stiglitz (Fujita, Krugman, and Venables, 1999; Dixit and Stiglitz, 1977).

$$I = \left[\sum_{t=0}^{t=x} i_t^{1/\mu} \right]^{\mu} \tag{2.3}$$

In equation 2.3 x is the number of varieties of intermediate good, i_t is the amount of variety type t and μ is the average cost to marginal cost ratio, with elasticity of substitution equal to $\mu/(\mu\text{-}1)$. A low elasticity of substitution among the differentiated varieties implies high scale economies and gives the producers of intermediate products a high level of monopolistic power. As noted by Fujita, Krugman, and Venables (1999), the Dixit-Stiglitz model ensures that the intermediate goods sector responds to a larger market by increasing the variety of goods rather than the scale of production of existing varieties. As shown below (see equation 2.5), the theory implies that $I=x^{\mu}\, i$ in which i is the 'average' output per intermediate producer, and x is the number of varieties or firms, so that as μ shrinks to 1 then the existence of varieties x provides no added enhancement to I, but otherwise I is a power function of x.

The equilibrium output level of each intermediate firm (i) is the consequence of profit maximizing behaviour. Under monopolistic competition, firms enter while profits are available, but new entrants dilute the market share and profits of each firm as the quantity demanded at any given price falls. Equilibrium is the break-even point with zero profit. Below this point firms leave and above this point firms enter. It is possible to calculate i since the zero profit point is when revenues ($wa\mu i$) equal costs $w(ai+\zeta)$ and thus the equilibrium level of output i is $\varsigma / a(\mu\text{-}1)$. In this, w is the wage rate, ς is the fixed cost of production in terms of labour requirement (entailing internal increasing returns), a is the marginal labour requirement, and μ is the mark-up on marginal costs that typifies equilibrium under monopolistic competition, so that price equals average cost and the mark-up reflecting monopoly power or the elasticity of substitution of the service varieties. Costs at equilibrium are based on an assumption that, for any individual variety of intermediate good, it takes $(ai+\varsigma)$ labour units in order to produce i, so that given wages w, production costs equal $w(ai + \varsigma)$. Equilibrium revenues are equal to price charged $(wa\mu)$ multiplied by output (i). Note that since the units of output are an arbitrary choice, they can be chosen so that $a = 1$, in which case $I = \varsigma / (\mu-1)$.

Denote the total amount of labour by N. In order to obtain x, divide the total amount of labour employed producing intermediate goods, $(1-\beta)N$, by the common amount of labour used to produce a variety, hence

$$x = (1-\beta)N/(i+\varsigma) \tag{2.4}$$

The total labour employed producing intermediate goods is $(1-\beta)N$ because $\beta N=M$ is directly employed in manufacturing production. This is a result which follows from standard theory about the equilibrium allocation of labour under competitive conditions with workers paid wages equal to the value of their marginal product.

Given x and i which is a constant across intermediate producers it follows from equation (2.3) that

$$I = x^{\mu}i = \left[(1-\beta)N/(i+\varsigma)\right]^{\mu} i \tag{2.5}$$

Thus we see that with $\mu > 1$, I is an increasing function of the number of varieties, so that high density areas have disproportionately high levels of services I. Substituting equation (2.5) into equation (2.2) gives

$$Q = \phi N^{\gamma} \tag{2.6}$$

in which γ, the elasticity we wish to estimate, is a function of exogenously determined constants, thus

$$\gamma = \alpha\left[1 + (1-\beta)(\mu-1)\right] \tag{2.7}$$

This is also true of ϕ, but because it is a more complex function of constants and is of limited interest, it is not explicitly defined here.

In order to estimate the elasticity γ, we rearrange equation (2.6) to obtain manufacturing output per unit of total labour, given by

$$\ln(Q/N) = \ln(\phi)/\gamma + \left[(\gamma-1)/\gamma\right]\ln(Q) \tag{2.8}$$

and from this it is easy to obtain the relationship between manufacturing productivity and manufacturing output, which is the subject of the next section.

2.3 Incorporating Technological Externalities and Spatially Differentiated Technical Progress

In order to obtain unbiased estimates of returns to scale, it is also necessary to extend the specification to adjust for other, so far omitted, variables. Instead of equation (2.2), we adopt the production function technology

$$Q_t = ((A_0 e^{\lambda t} M_t)^{\beta} I_t^{1-\beta})^{\alpha} \tag{2.9}$$

in which technology augments labour. The level of technology is denoted by $A_0 e^{\lambda t}$ and indicates that there is growth in technology from a base level equal to A_0 depending on exogenous technical change or the growth of total factor productivity λ. On this basis, we obtain a model for the level of manufacturing productivity at time t,

$$\ln(Q/M)_t = \ln(\phi)/\gamma + [(\gamma-1)/\gamma]\ln(Q_t) - \ln(\beta) + \ln(A_0) + \lambda t \tag{2.10}$$

The dynamic specification is obtained by differentiating equation (2.10) with respect to time, thus giving

$$p = \lambda + b_1 q \tag{2.11}$$

in which p is the exponential growth rate of manufacturing productivity, q is the exponential growth rate of manufacturing output, and $b_1=(\gamma-1)/\gamma$ is a constant.

Assume that the rate of technical progress λ is a variable which is a function of regional characteristics and their location with respect to other regions. These characteristics all relate to the access to, and flow of, information and knowledge. Developing this hypothesis further, it is assumed that part of the reason for regional differences in the rate of technical progress is technological externalities involving knowledge or information transmission between firms. Assume that the amount of knowledge spilling over between firms is partly determined by their respective locations in physical or economic space. We focus here on the determination of flows across region boundaries, holding constant temporarily the causes of knowledge spillover intensity within regions. The idea that knowledge can easily cross the boundary into a neighbouring region is a reasonable proposition in the EU context, since workers embodying knowledge who change jobs are likely to do this not by changing residence but by migrating between firms, and the new place of employment, while in the same commuting zone as the previous firm, may well be in a different regions. In the EU, household migration tends to be prohibitive for a number of reasons, such as language, culture and regional differences in housing costs. The crossing of regional boundaries with job migration is quite inevitable under the NUTS 2 region system, which is one of formal, administrative regions, such as counties or groups of counties in the UK, rather than a system of functional urban regions which would be more self-contained and tend more to coincide with commuting zones.

Another potential knowledge transmission mechanism involves direct firm to firm interaction. Firms cooperate and compete with other firms in their local economy, but again the boundaries of local economies will not necessarily coincide with those of NUTS 2 regions. Thus technical progress in a region is likely to be a stimulus to technical progress in nearby regions, as might result for example from the exchange of information to create a coordinated just-in-time supply chain within a local economy that straddles region boundaries. Likewise, firms will be quick to adopt new knowledge if it enhances the productivity of their competitors in the local economy.

Distance is likely to be a major factor constraining information flow, but it will, to some extent, be offset by the 'size' of the economy. For instance one would anticipate a greater information flow from a large conurbation than from a rural region at the same distance. The combined effects of size and distance are represented in the square matrix **W**

$$W_{ij} = A_{tj}^{\kappa} d_{ij}^{-\delta} / \sum_j A_{tj}^{\kappa} d_{ij}^{-\delta} \quad (2.12)$$

in which d_{ij} is a measure of distance between regions i and j, and A_{tj} is a measure of the size of region j at some point in time t, and κ and δ are parameters with suitably chosen values[1]. The presence of the denominator in equation (2.12) means that **W** is scaled to row totals equal to 1, which has certain computational advantages (although as pointed out by Anselin, 1988, the existence of a consistent decay with distance embodied in the function prior to scaling may be lost). Applying this to the technical progress rate produces the spatial lag variable $\mathbf{W}\lambda$, which is the vector resulting from the matrix product of **W** and λ, which has typical cell i equal to the weighted average of technical progress in regions interacting with region i as determined by **W**. The assumption is that, for reasons such as cited above, if region i interacts principally with regions with comparatively slow technical progress, then this slows i's technical progress compared to the rate it would have seen if it interacted principally with fast growing regions.

The discussion so far has highlighted how technical progress rates might differ between regions due to spillovers across region boundaries and so depend on each region's spatial context, but there has been no consideration of the influence of within-region characteristics. We now extend our thesis regarding the causes of different technical progress rates to incorporate the differential adoption rate of innovations, which is assumed to depend on the inherent socio-economic character of each region. Assume that there exists knowledge that is available for adoption irrespective of the location of the firm. However, it is not of the same value to every potential adopter, instead we assume that the value declines as the technology level increases. Firms in high technology regions are likely to be, on average, close to the technological frontier and will find minimal benefit for productivity growth from the knowledge available for adoption. Firms in low technology regions will benefit more from innovation adoption and the acquisition of new knowledge. I choose to measure the level of technology of a region by the 'technology gap' defined as $G=1-P_0/P_{*0}$ in which P_0 is the level of manufacturing productivity at the start of period (time 0), and P_{*0} is the time 0 level in the leading technology region. In this way the technology gap is *exogenized*, since technical progress over the subsequent period is assumed to be a positive function of the initial level of technology gap.

There are other within-region variables which also come into consideration as additional causes of technical progress. For instance, there has been no mention thus far for the differing extent to which knowledge will originate from, and circulate among, firms in the same region, but it seems reasonable to hypothesize that the creation and assimilation of innovations is likely to be greater in regions characterized by a greater density of human capital. Human capital is concentrated mainly in urban regions with high population densities, which have higher educational attainment rates (see below). This urban effect is denoted by u. In addition, however, in the European context, it is assumed that human capital is not simply differentiated according to whether regions are an urban or rural, but is also related to whether a region is located in the core or periphery of the continent. Peripheral regions are characterized by lower educational attainment rates. In fact, it appears that the main contrast is between regions of the Low Countries, Denmark, Germany, Britain and France which have on the whole high educational attainment levels, and Portugal, Greece, Italy and parts of Spain in which the proportion of the population attaining the highest level is less, and so we proxy the core/periphery human capital contrast by a crude distance from Luxembourg measure (l). In fact educational attainment data are available for EU regions, but we have opted not to use this directly since it is prone to measurement error and appears likely to be an endogenous response to productivity growth rather than being an exogenous variable[2]. Both the proxies can be justified since using these recent educational attainment data as the dependent variable in a regression with u (in fact the 1995 population density) and l as regressors, the coefficient to standard error ratios are 3.21 for u and -2.66 for l, both of which are correctly signed and are highly significant[3]. Thus the exogenous variables u and l enter into the empirical model, and this facilitates maximum likelihood estimation.

Combining the assumed determinants of the technical progress rate, namely information flows across region boundaries ($\mathbf{W}\lambda$), the technology gap effect (G) and human capital effects (u,l) gives

$$\lambda = b_0 + b_2 G + \rho \mathbf{W}\lambda + b_3 u + b_4 l + \xi \qquad (2.13)$$

in which $\xi \sim N(0,\sigma^2)$ represents exogenous shocks. Using equations (2.13)[4] and (2.11), we obtain

$$p = \rho \mathbf{W}\lambda + b_0 + b_1 q + b_2 G + b_3 u + b_4 l + \xi \qquad (2.14)$$

One problem with equation (2.14) is that $\mathbf{W}\lambda$, the weighted average of the technical progress rate in 'surrounding' regions, is 'invisible'. We therefore make it visible as a residual, since

$$\lambda = p - b_1 q \tag{2.15}$$

and therefore

$$p = \rho \mathbf{W}(p - b_1 q) + b_0 + b_1 q + b_2 G + b_3 u + b_4 l + \xi \tag{2.16}$$

hence

$$p = \rho \mathbf{W} p + b_0 + b_1 q + b_2 G + b_3 u + b_4 l - b_5 \mathbf{W} q + \xi \tag{2.17}$$

2.4 Model Estimation

Having attempted to give a theoretical rationale for the reduced form, I now proceed to confront this model with data. The data are manufacturing productivity and output growth over the period 1975-95 for 178 NUTS 2 regions of the EU. Of course the specification is a space lag model typical of the spatial econometric literature and with the endogenous variable $\mathbf{W}p$ it is necessary to use maximum likelihood or two stage least squares (2sls) for consistent estimation. There are a number of problems related to the use of 2sls which are discussed in Fingleton (2000a) and references therein. These relate to the difficulties of obtaining appropriate instruments and to the possibility of an unstable model resulting from the unbounded range for ρ in 2sls estimation. In contrast, with $\mathbf{W}$ standardized to row sums equal to unity, maximum likelihood produces estimates with $1/e_L < \rho < 1/e_U = 1$, where e_L and e_U are the maximum and minimum eigenvalues of the $\mathbf{W}$ matrix, thus avoiding the instability associated with estimates outside this range. In this chapter we therefore choose to avoid problems of 2sls and obtain maximum likelihood estimates, thus assuming that other than $\mathbf{W}p$ the variables are exogenous. This produces an estimate not dissimilar to that obtained by treating q as an endogenous variable in a system of equations estimated via full information maximum likelihood (Fingleton, 2000b). Likewise, estimation of a spatial SUR model via three stage least squares (Fingleton, 2001b) gives a similar outcome for appropriately specified $\mathbf{W}$ matrices.

The strategy for model selection (see Florax and Folmer, 1992; Florax, Folmer and Rey, 1998) is to commence with a full model containing all (first order) lags (ie $\mathbf{W}l$, $\mathbf{W}u$ and $\mathbf{W}G$ in addition to $\mathbf{W}q$), and then seeking to reduce this model. This idea here is to improve the basis for inference by attempting to purge systematic variation from the error, leaving nothing but random noise, since that is what is assumed in our significance tests. One disadvantage is that by increasing the number of variables, there is an increased chance of problems associated with multicollinearity, a problem that we seemingly do encounter below.

With the additional lags, the model becomes

$$p = \rho \mathbf{W}p + b_0 + b_1 q + b_2 G + b_3 u + b_4 l + b_5 \mathbf{W}q + b_6 \mathbf{W}G + b_7 \mathbf{W}u + b_8 \mathbf{W}l + \xi \qquad (2.18)$$

However, we further augment the model by modeling error heterogeneity, which has been shown previously to be a feature of these data[5]. Previous evidence suggests smaller residual variance for the core regions compared with the peripheral regions, with the periphery containing both some of the fastest growing regions (Ireland) and some of the slowest. The basic cause of this heterogeneity may relate to the comparatively dispersed geography and the presence of smaller, more volatile, economies in the peripheral regions. For simplicity this is modeled by groupwise heterogeneity with groups defined by a core/periphery (0/1) dummy variable with the group of core regions identified as those that are less than 500km distance from Luxembourg. Therefore the initial (full) model is given by equation (2.18) with error distribution $\xi \sim N(0, \sigma_i^2)$ in which σ_i^2 denotes error variance conditional on whether region i is a core or a periphery region.

The resulting maximum likelihood estimates of the full model with groupwise heterogeneity are given in Table 2.1, and the Likelihood Ratio test in Table 2.2 confirms the presence of error heterogeneity[6]. Table 2.2 indicates the degree of concordance between model and data, as shown by the R^2 value of 0.5146 and by the similar value for the square of the correlation between observed and fitted values (sq. corrn.). While this means that approximately 50% of variation is unaccounted for and attributed to random disturbances, none the less, the R^2 values reported for this and the subsequent models are considered to be reasonably acceptable in the context of cross-sectional growth data, particularly as there is no (informal) evidence of spatially autocorrelated residuals as might result from significant omitted variables. The parameter to standard error ratios show that the variables *Wp, q, G,l, Wq* and *u* are, as anticipated, significant[7]. Regarding the spatial lag $\mathbf{W}q$, the sign is as suggested by the earlier theoretical discussion, although the parameter value is somewhat larger than the product of ρ and b_1. The other exogenous spatial lags are insignificant. The hypothesis that these are jointly null can be tested via the likelihood ratio test. The log likelihood for the full model is 528.902, compared with 525.527 for the nested reduced model with parameters on $\mathbf{W}u$, $\mathbf{W}l$ and $\mathbf{W}G$ are set to zero. The likelihood ratio test statistic (equal to twice the difference in log likelihoods) is therefore 6.75 which is not significant when referred to the upper 5% point of the χ^2 distribution with 3 degrees of freedom. We therefore eliminate the other exogenous lags, but retain $\mathbf{W}q$.

Table 2.1. Full (initial) model with groupwise heteroscedasticity

	Estimate	st. err.	ratio	p-value
W*p*	*0.64279*	*0.158402*	*4.057971*	*0.000050*
constant	*-0.0228913*	*0.0126461*	*-1.810143*	*0.070274*
q	*0.442366*	*0.0627129*	*7.053835*	*0.000000*
G	*0.0621823*	*0.00782821*	*7.943369*	*0.000000*
l	*-1.84748*	*0.39989*	*-4.619963*	*0.000004*
u	*2.42262*	*1.13913*	*2.126731*	*0.033442*
W*q*	*-0.564792*	*0.249053*	*-2.267754*	*0.023344*
W*G*	*0.0151991*	*0.051527*	*0.294973*	*0.768014*
W*l*	*2.63183*	*1.78431*	*1.474986*	*0.140216*
W*u*	*-3.30669*	*2.81565*	*-1.174399*	*0.240235*

Table 2.2. Summary statistics and diagnostics – Full model

R^2	*0.5146*	sq. corrn.	*0.5619*
Likelihood	*528.902*	Convergence	4 iterations
σ^2 (core)	*0.000104958*	σ^2 (periphery)	*0.000192215*
Test	df	statistic	p-value
LR groupwise hetero.	1	*6.625244*	*0.010054*

The resulting, reduced, model, summarized by Tables 2.3 and 2.4, produces a more acceptable estimate for the coefficient on $\mathbf{W}q$, which is now very close to the value one would anticipate (ρb_1) from theory. The change in the $\mathbf{W}q$ coefficient estimate compared to the estimate of Table 2.1 indicates that it was distorted in the presence of the other exogenous lags, presumably due to multicollinearity. In fact since this coefficient is no longer significantly different from zero, we chose for simplicity to eliminate it from the model. There is no (informal) evidence of systematically organized residuals which might suggest the inclusion of another spatial variable.

Table 2.3. Reduced model – groupwise heteroscedastic estimates

	Estimate	st. err.	ratio	p-value
W*p*	*0.693049*	*0.121379*	*5.709809*	*0.000000*
constant	*-0.0174691*	*0.00389128*	*-4.489293*	*0.000007*
q	*0.4524*	*0.0634033*	*7.135280*	*0.000000*
G	*0.0625151*	*0.00787027*	*7.943192*	*0.000000*
l	*-1.18866*	*0.285876*	*-4.157971*	*0.000032*
u	*2.70251*	*1.12983*	*2.391955*	*0.016759*
W*q*	*-0.288323*	*0.198891*	*-1.449653*	*0.147155*

Table 2.4. Reduced model – Summary statistics and diagnostics

R^2	*0.4839*	sq. corrn.	*0.5461*
Likelihood	*525.527*	Convergence	4 iterations
σ^2 (core)	*0.000107094*	σ^2 (periphery)	*0.000200814*
Test	df	statistic	p-value
LR groupwise hetero.	1	*7.112465*	*0.007655*

This leads to the final model summarized by Tables 2.5 and 2.6, which produces very similar estimates to those in Table 2.3. It is apparent that we can approximate the (lagged) rate of technical progress by the endogenous productivity growth rate lag (as has been the practice in earlier work), that there are increasing returns as shown by the significant positive coefficient on q, that there is a significant positive link between the technology gap G and productivity growth, attributed to innovation diffusion, and that technical progress is also increased by human capital proxied[8] by population density u and peripherality l. Finally, the likelihood ratio test of groupwise heteroscedasticity is also significant, with the estimated error variance for peripheral regions about twice as large as that for core regions.

Table 2.5. Final model – groupwise heteroscedastic estimates

	Estimate	st. err.	ratio	p-value
Wp	*0.631316*	*0.123497*	*5.112010*	*0.000000*
constant	*-0.01844*	*0.00386927*	*-4.765771*	*0.000002*
q	*0.429095*	*0.0615165*	*6.975287*	*0.000000*
G	*0.0644585*	*0.00781893*	*8.243909*	*0.000000*
l	*-1.3627*	*0.256974*	*-5.302887*	*0.000000*
u	*2.81814*	*1.13278*	*2.487801*	*0.012854*

Table 2.6. Summary statistics and diagnostics

R^2	*0.4816*	sq. corrn.	*0.5387*
Likelihood	*524.584*	Convergence	4 iterations
σ^2 (core)	*0.000108465*	σ^2 (periphery)	*0.000204128*
test	Df	statistic	p-value
LR groupwise hetero.	1	*7.249735*	*0.007091*

It is evident that the estimated regression coefficients on the variable q are quite similar and significantly different from zero, indicating increasing returns to scale, across all three specifications. Table 2.7 gives the estimates, together with the corresponding estimates of $\gamma=-1/(b_1-1)$ and the 95% confidence interval, which consistently excludes $\gamma=1$, which is the value which should be exceeded for increasing returns[9]. It is apparent from Table 2.7 that while the point estimate for returns to scale from the final model is below that of the initial model, it remains within the 95% confidence envelope[10] for the initial model.

Table 2.7. Returns to scale estimates

Model	b_l	95% lb	95% ub	γ	95% lb	95% ub
Full	0.4424	0.3194	0.5653	1.7933	1.4694	2.3003
Reduced	0.4524	0.3281	0.5676	1.8262	1.4883	2.3127
Final	0.4291	0.3085	0.5497	1.7516	1.4462	2.2207

2.5 Recursive ML Estimation

The suggestion from the analysis carried out so far is that the returns to scale are rather robust to different model specifications. However, it would be rather surprising if returns to scale were exactly the same for the entire period 1975-95, since this covers many changes in the EU economy, so that returns to scale may not be robust to time period analyzed. Of course there will be some natural variation over time, but the question is whether there is evidence of systematic variation, or sudden shifts, and how might this be interpreted. The approach I adopt is very simple, I carry out recursive maximum likelihood estimation of the final model parameters commencing with data for the period 1975-81. The method thus simply involves re-estimation of the model parameters as each additional year is added to the series, and plotting parameter paths and standard errors to highlight possible trends or regime shifts[11]. The resulting estimates are presented in Table 2.8 and the standard errors are given in Table 2.9.

These recursive estimates contain some intriguing information, probably the most salient trend being the increase in the strength of autoregressive spatial interaction (as measured by the estimated ρ). This is made apparent via Figure 2.1, which shows the progression of the point estimates and the 95% confidence interval. The lower bound for the 1995 end year estimates is about equal to the upper bound for the 1981 end year estimate, indicating that there has been a real significant increase over time in the importance of technological externalities which result from spatial interaction transmitting knowledge or information across region boundaries. This would be consistent with the progressive lowering of boundaries between the EU regions, the more extensive commuting field that have developed, and the strengthening of cross-border interaction between firms.

Figure 2.2 gives the recursive estimates of $b_l=(\gamma-1)/\gamma$ and the 95% confidence envelope[12]. It is apparent from this that there were increasing returns throughout the period of analysis, although there is also an indication of an absence of parameter constancy and a suggestion of a structural break at about 1990. The more or less consistent fall in returns to scale from $b_l=0.7588$ ($\gamma=4.146$) estimating with data up to 1983, to $b_l=0.4184$ ($\gamma=1.719$) using data up to 1991, is followed by a change to an upward-sloping path for the 1990s.

Table 2.8. Recursive parameter estimates

year from 1975	**W***p*	*q*	G	*l*	*u*
1981	*0.1585*	*0.7070*	*0.10744*	*-2.007*	*4.947*
1982	*0.1622*	*0.7194*	*0.09578*	*-1.754*	*4.911*
1983	*0.1868*	*0.7588*	*0.08841*	*-1.656*	*5.301*
1984	*0.2654*	*0.6872*	*0.08567*	*-1.493*	*4.952*
1985	*0.4198*	*0.6764*	*0.08314*	*-1.647*	*4.952*
1986	*0.5470*	*0.5366*	*0.09095*	*-1.683*	*3.994*
1987	*0.4600*	*0.5382*	*0.08936*	*-1.546*	*3.890*
1988	*0.5165*	*0.5242*	*0.08366*	*-1.493*	*3.801*
1989	*0.4927*	*0.4829*	*0.08400*	*-1.363*	*3.380*
1990	*0.6134*	*0.4242*	*0.08160*	*-1.413*	*3.642*
1991	*0.7136*	*0.4184*	*0.07842*	*-1.612*	*3.689*
1992	*0.7092*	*0.4210*	*0.06678*	*-1.626*	*3.910*
1993	*0.7519*	*0.4257*	*0.06623*	*-1.751*	*2.950*
1994	*0.6587*	*0.4275*	*0.06713*	*-1.404*	*3.009*
1995	*0.6313*	*0.4291*	*0.06445*	*-1.363*	*2.818*

Table 2.9. Recursive standard error estimates

year from 1975	**W***p*	*q*	G	*l*	*u*
1981	*0.1354*	*0.04568*	*0.011321*	*0.3559*	*1.607*
1982	*0.1445*	*0.04527*	*0.010360*	*0.3404*	*1.485*
1983	*0.1516*	*0.05005*	*0.010220*	*0.3433*	*1.455*
1984	*0.1628*	*0.04810*	*0.010184*	*0.3418*	*1.436*
1985	*0.1504*	*0.04885*	*0.010106*	*0.3279*	*1.422*
1986	*0.1390*	*0.05538*	*0.010127*	*0.3207*	*1.433*
1987	*0.1443*	*0.05589*	*0.010232*	*0.3228*	*1.451*
1988	*0.1370*	*0.05996*	*0.009937*	*0.3143*	*1.420*
1989	*0.1404*	*0.06085*	*0.009550*	*0.3078*	*1.365*
1990	*0.1233*	*0.06189*	*0.008558*	*0.2796*	*1.225*
1991	*0.1111*	*0.06110*	*0.008157*	*0.2639*	*1.175*
1992	*0.1183*	*0.06397*	*0.008073*	*0.2673*	*1.163*
1993	*0.1052*	*0.06483*	*0.007983*	*0.2671*	*1.142*
1994	*0.1187*	*0.06321*	*0.007925*	*0.2632*	*1.142*
1995	*0.1235*	*0.06152*	*0.007819*	*0.2570*	*1.133*

It might be argued that Figure 2.2 reflects putative business-cycle effects, with output falling faster than labour on the downswing and vice-versa on the upswing (the so-called Okun effect), so that as the end-year for estimation moves forward in time, this might itself be responsible for the changes in the path of b_1. However, it is inconceivable that all the different economies will be exactly at the same

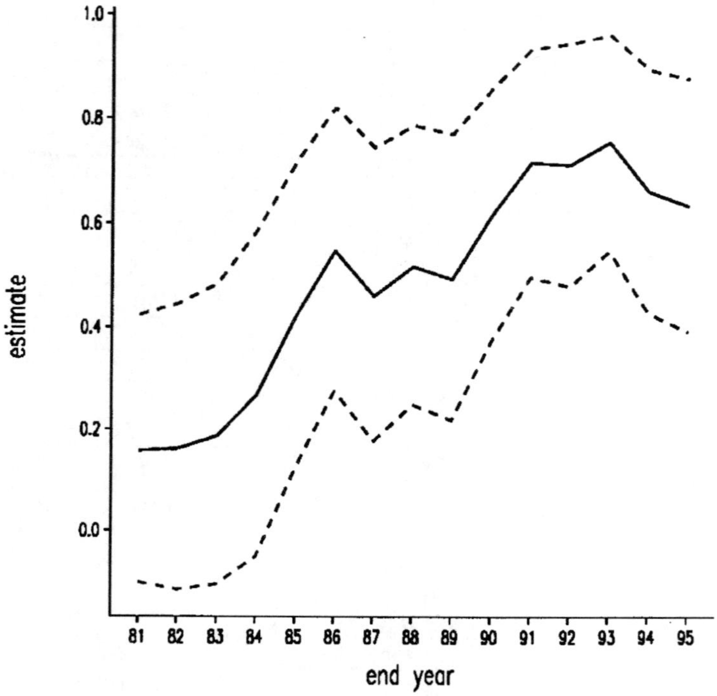

Figure 2.1. Recursive ML ρ estimates

stage of the business cycle at any one point in time, so it is possible that the b_l path represents the net effect of different causes, or the same causes but with different intensity, in different regions. The change from steeply downward to slightly upward path in 1990 coincides with some momentous decisions in the EU. It was the year agreed in the Madrid council of 1989 for the commencement of the first stage of European monetary union. Also the Maastricht treaty of 1991 reinforced the movement towards monetary union followed by the establishment of the European Monetary Institute and the European system of central banks. These and possibly other changes in the wider socio-economic framework, leading to closer socio-economic integration, may have been the reason why the decline in returns to scale was arrested in the 1990s, although this is speculation.

Some additional insight as to the significance of these changes is provided by the companion paper to this Chapter (Fingleton, 2001b), using a panel analysis approach, namely spatial seemingly unrelated regression (SUR) with spatially lagged dependent variables (Anselin, 1988). Using break points of 1981 and 1989 to define three periods, a model is estimated for the whole 1975-95 period which differentiates

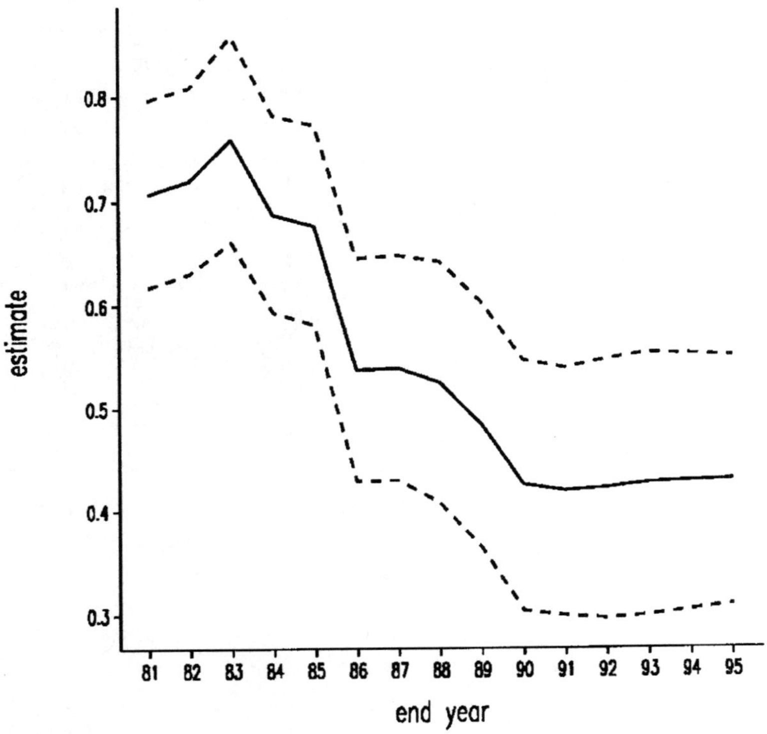

Figure 2.2. Recursive ML b_1 estimates

parameters according to period. With regard to the returns to scale parameters, the unrestricted estimates fall then rise, a pattern that is also apparent from the current recursive analysis. Likewise, the panel analysis shows a significant and inexorable rise in the ρ estimates, a trend that is echoed by Figure 2.1. Also, in the companion paper, formal hypothesis testing shows a significant loss of fit for the overall model with parameters restricted to equality across periods. While in contrast the recursive analysis of this Chapter is informal and does not provide a single simultaneous equation model for the entire period, it does provide a useful picture of parameter dynamics and points to a possible structural break.

In order to identify more precisely the mechanism causing these dynamics, it is edifying to examine changes in the parameters α, β and μ which our theory tells us are the fundamental determinants of returns to scale. Given that $\gamma=\alpha[1+(1-\beta)(\mu-1)]$, in which $\alpha<1$ represents diminishing returns due to the effects of congestion, holding the other parameters constant, as α goes to 0 congestion effects are maximized and returns to scale minimized. The β parameter reflects the relevance of intermediate services to production, so that as β goes to 1 the relevance of intermediate services is minimized, so as β goes to 1 returns to scale (which ultimately derive from the service sector) are minimized. As μ reduces to 1 the number of varieties becomes less important as a determinant of composite intermediate services available in a region since there is more substitution between the differentiated intermediate goods and less monopoly power available to intermediate producers. Hence, again holding constant the other parameters, shrinking μ towards 1 causes returns to scale to fall. It is possible that the initially downward path traced by the recursive returns to scale estimates simply reflects increasing congestion through the 1980s, assuming β and μ remained constant, or is the net effect of changes in all three parameters.

The question is, can we estimate these parameters and move beyond the realm of speculation? The answer is yes, given appropriate data. Commence with the production function (2.9), take natural logs, and (for simplicity) differentiate with respect to time, hence

$$q = \lambda + \alpha\beta m + \alpha(1-\beta)i \qquad (2.19)$$

in which m is the growth of manufacturing labour, i is the growth of intermediate products and λ is the technical progress rate. For simplicity, assume that λ is constant across regions and the equation is re-expressed so that

$$q = c_0 + c_1 m + c_2 i \qquad (2.20)$$

It follows that

$$\beta = \frac{c_1}{c_1 + c_2} \qquad (2.21)$$

and

$$\alpha = c_1 + c_2 \qquad (2.22)$$

and given γ, α and β,

$$\mu = \frac{\gamma - \alpha}{\alpha(1-\beta)} + 1 \qquad (2.23)$$

The recursive estimation of the fundamental parameters seems an obvious next step, together with further exploration of the robustness of estimates using Bootstrap methods which do not rely on specific error distribution assumptions. In addition, further work might involve obtaining parameter estimates that are demonstrably purged of any simultaneity bias that might be due to assuming as exogenous variables that might be endogenous. The variable that initially comes to mind in this context is output growth

(q) which has been modeled in (some) previous research as an endogenous variable, although previous work (Fingleton and McCombie, 1998, among others) has shown that regardless of whether q is modeled as endogenous or exogenous, the estimated returns to scale parameter are more often than not in the vicinity of 0.5. Human capital is another variable that poses some problems because, in the form of educational attainment, it evidently embodies measurement error and endogeneity. In the chapter it is represented by proxy exogenous variables. While it might appear from the above remarks that 2sls would be a step forward methodologically, there are additional issues which are absent from maximum likelihood, such as the choice of appropriate instrumental variables, and the avoidance of ρ estimates outside the stable parameter space.

2.6 Conclusion

The main contributions of the chapter are three fold. First, the chapter is an example of the 'third way', positioned between new economic geography and economic geography 'proper', which strives for both theoretical coherence and empirical consistency by applying spatial econometric methods. The model developed in the chapter stands apart from new economic geography (as currently practised) by revisiting the common ground in urban economic theory, which provides a more obvious route to empirically estimable relationships and the incorporation of technological externalities. Neither of course is it a typical example of what we might call economic geography 'proper' (meaning economic geography as typically practiced by geographers rather than economists) because of the role given to scientific hypothesis testing and mainstream economic concepts. Second, by forging a link with fundamental parameters of production functions, there is a new insightful interpretation of the increasing returns parameter of the Verdoorn law. Thus we see increasing returns as the (probable) net outcome of diminishing returns due to congestion effects (α), of the market structure and division of labour of the service sector in which the variety of services (μ) feeds manufacturing productivity growth, and of the relative importance of these service inputs to manufacturing production (β). Third, the chapter is evidently the first to explore the application of recursive estimation in the spatial econometrics context. The recursive estimates show that increasing returns appear not to be a product of the estimation period chosen, but is a consistent feature over a range of estimation periods. Some changes are apparent over time, with estimates falling sharply then rise marginally as the period extends beyond 1990. Speculative reasons are suggested for what appears to be a structural break, but it is anticipated that there is more understanding to be gained from estimating the fundamental drivers α, β and μ.

To summarize, the chapter suggests a new approach which seems to be a promising way forward for both theory and analytical method, although there is clearly further work needed before the estimates and techniques outlined in this chapter can be considered to be established facts or an easily accessible part of the spatial econometrics toolkit.

Acknowledgements

This chapter was first presented in the Applied Econometric Association conference held in conjunction with the Western Regional Science Association Fortieth Anniversary Annual meeting at Palm Springs, California, February 2001. I am grateful to Giuseppe Arbia and to participants at that meeting for their comments.

Endnotes

[1]Following Fingleton (2001b), the assumption is that κ=1, δ=2, thus giving weight to both size and distance. As argued below, this is considered more reasonable than, for example, nullifying the influence of either by assuming either κ=0 or δ=0.
[2]Data for 1999 giving the share of population aged 25-59 with higher educational attainment levels by EU region have been provided by the European Commission. However, cross-national comparisons based on these data may be unreliable. Also, the available data post-date the period 1975-95, and so may be a response to productivity growth over that period rather than an exogenous determinant of it.
[3]This is ignoring the fact that the dependent variable is limited and assuming normality, but allowing for nuisance error dependence.
[4]We use the coefficient ρ to reflect its special status below as the coefficient on an endogenous variable.
[5]Normality is assumed ab initio by choice of the likelihood function, and judging from graphical plots of the residuals this assumption appears to be reasonable.
[6]Because of the units of measurement of u and l, SPACESTAT gives estimates that differ by a factor of 100000 or 1000000 from those reported in the subsequent tables. For purposes of clarity we omit these factors. The parameter to standard error ratios given here are of course identical to those produced by the software.
[7]Some caution should normally be exercised since the true standard errors will tend to be larger than those reported by SPACESTAT, which are based on asymptotic theory, although in this case the use of asymptotics is unlikely to make any difference.
[8]One suggested interpretation of the significance of u and l is that they reflect variables other than human capital, such as the composition of regional production. However the preferred hypothesis is that human capital is the fundamental causal variable, and that this may determine other variables that are collinear with u and l.
[9]Note that the γ distribution is not normal, hence the asymmetry of the upper and lower bounds. Using nonlinear regression as an analogy, the estimated variance of γ would be approximately equal to the square of the first derivative of γ with respect to b_l multiplied by the asymptotic variance of b_l. Denoting this by $[1/(1-b_l)^4]$ var(b_l) the estimated standard error of γ from the final model is 0.19.
[10]It should however be mentioned that this is based on asymptotic theory.
[11]This is a standard feature of contemporary time series analysis, for example via PcFIML.

[12]Unlike the time series analogue, the width of the 95% confidence interval does not automatically diminish as the estimation period lengthens. This is because the sample size remains the same for each estimate, equal to the number of regions.

Part II

Spatial Econometrics Estimation

3 Spatial Autoregressive Local Estimation

R. Kelley Pace
Louisiana State University

James P. LeSage
University of Toledo

Abstract

We introduce a recursive approach for maximum likelihood estimation of spatial autoregressive models. This allows consideration of a series of estimates based on sub-samples of varying size, similar to the nonparametric locally linear estimators described in McMillen (1996, 1997) and Brunsdon et al. (1996). One important difference between the approach introduced here and other methods is that our model includes a spatial autoregressive term reflecting spatial dependence (similar to Brunsdon et al. (1998)). Relative to the traditional approach, this improves prediction, the stability of the resulting parameter estimates, and decreases the sensitivity of the performance to the bandwidth parameter. The method is illustrated using a sample of 3107 US counties with a model for voter participation in the 1980 election.

3.1 Introduction

One branch of spatial econometrics uses distance weighted sub-samples of the data in conjunction with OLS to produce parameter estimates at various points in space. McMillen (1996, 1997) introduced this form of nonparametric locally linear weighted regression (LWR) which Brunsdon, Fotheringham and Charlton (1996) term geographically weighted regressions (GWR). By estimating separate models using data near each observation these intuitively appealing methods attempt to overcome the problem of spatial heterogeneity. If spatial dependence arises due to inadequately modeled spatial heterogeneity, LWR can potentially eliminate this problem and these models often rely on the estimated parameters to detect systematic patterns over space.

LeSage (2001) points out two problems that plague these locally linear estimators. First, most locally linear schemes provide a single bandwidth determined using cross-validation which does not provide the user much information on the sensitivity of the resulting estimates to other bandwidths. Insofar as the bandwidth parameter exerts a major impact on the estimates, the inability to observe this sensitivity may reduce the utility of local linear estimation. Second, aberrant observations due to spatial enclave effects, shifts in regime, or outliers can exert a large impact on the locally linear estimates which by necessity rely heavily upon a small number of observations. As possible evidence of these impacts, practitioners of LWR methods often encounter estimates that exhibit high variability across the spatial sample[1].

In addition, LWR methods may be subject to a third problem. Spatial dependence may not vanish at the optimal bandwidth, although practitioners usually assume that spatial dependence is eliminated. This issue can be examined using the method proposed here, that can produce less biased estimates even in the presence of spatial dependence in the locally linear samples.

To address some of the problems of LWR, we introduce spatial autoregressive local estimation (SALE). Our approach estimates a sequence of spatial autoregressions for each observation in the data set using a range of sub-sample sizes. This resembles the approach taken by Brunsdon et al. (1998) who integrated LWR and the autoregressive error model. Despite working with only a small number of observations (118), they report that 'one notable characteristic of the calibration of the Ord model with spatially reweighted maximum likelihood is the high computational overhead'. This chapter provides a computationally feasible approach for estimating LWR models that allows for both parameter variation and spatial dependence in moderate sample sizes. We illustrate the method using a data sample containing 3107 observations on US counties.

A notable feature of spatial autoregressive local estimation is that the sequence of estimates arising from increasing sample sizes converges to the global spatial autoregressive estimates in the limit (which could also include OLS when the spatial autoregressive parameter equals 0). In contrast, locally linear estimates converge to OLS in the limit, a model which ignores spatial dependence in the data.

Under ideal conditions locally linear least squares estimates based on sub-samples reflecting local information may eliminate spatial dependence. Since SALE nests a version of LWR (when the spatial autoregressive parameter is restricted to 0), one can examine whether LWR eliminates spatial dependence in practice. Of course, one could test for spatial autocorrelation using the residuals from an LWR model. In cases where the test rejects the hypothesis of spatial independence, one would need to turn to an estimation method such as SALE to model the spatial dependence. Use of the SALE method at the outset eliminates pretest complications that might plague inference in the former case. In addition, the more general maximum likelihood SALE method we propose has the potential to produce more efficient estimates while at the same time providing a test for spatial dependence.

Use of spatial autoregressive estimates in place of least squares has other advantages. For example, inclusion of the spatial lags tends to improve predictive performance without resorting to estimates based on very small sample sizes that are highly volatile. Hence, the reduced variability of the spatial autoregressive local estimates aids the exploratory or diagnostic use of SALE. Estimates based on sub-samples can be used to explore parameter variation over regions of space as well as the strength of spatial dependence in sub-regions.

A final advantage of our approach is that the problem of cross-validation to select an optimal sub-sample size is overcome by allowing the user to examine a sequence of sub-sample sizes ranging from small to large. This applies to both SALE and LWR methods. As already noted, SALE estimates from the sub-samples are comparable to those from the entire sample, so the impact of changing the sub-sample size on estimates and inferences can be viewed in a continuum[2].

On the surface, SALE appears computationally infeasible since a single spatial autoregression may require as many as $O(n^3)$ operations and local estimation methods can involve as many as $O(n^2)$ regressions, for a potential total operation count of $O(n^5)$. What we show is that a recursive computational aspect of matrix decompositions required for computing the log-determinant along with use of sparse matrix techniques (see Pace and Barry, 1997) can be used in conjunction with recursive least squares techniques to produce dramatic simplifications in this otherwise difficult problem. Essentially, one can perform SALE using as few as $O(n)$ computations. In comparison, LWR methods generally require anywhere from $O(n)$ to $O(n^3)$ operations depending upon the implementation.

Section 3.2 of the chapter sets forth our recursive estimation scheme for spatial autoregressive models. We also compare and contrast the resulting estimation approach to locally linear least squares estimation methods. An illustration of the method is undertaken in section 3.3 where we use a sample of 3107 US counties to estimate a model of voter participation in the 1980 presidential election. Conclusions are drawn in section 3.4.

3.2 Recursive Spatial Maximum Likelihood Estimates

A recursive approach to maximum likelihood estimation of spatial autoregressive models is taken up in this section. Section 3.2.1 discusses aspects of the spatial weight matrix needed to motivate the recursive estimation scheme. The spatial mixed model that we rely on is examined in section 3.2.2 and a recursive estimation approach is set forth in section 3.2.3. Section 3.2.4 places our approach in the context of other locally linear approaches to spatial estimation based on least squares methods.

3.2.1 The Spatial Weight Matrix

Spatial statistics requires some means of specifying the spatial dependence among observations. The spatial weight matrix fulfils this role for lattice models (Besag, 1974, 1975). One specification of the spatial weight matrix begins by forming the binary adjacency matrix **N** where $N_{ij} = 1$ when observation j is a neighbour to observation i ($i \neq j$). These elements may be further weighted to give closer neighbours higher weights and incorporate whatever spatial information the user desires. By itself, **N** is usually asymmetric. To insure symmetry, we can rely on the transformation $\mathbf{N_S} = (\mathbf{N} + \mathbf{N}')=2$.

At this stage, users often re-weight the adjacency matrix to create a row-stochastic matrix or a matrix similar to a row-stochastic matrix. This can be accomplished in the following way. Let N_{Sri} represent the row sum of the ith row of $\mathbf{N_S}$. Define a diagonal matrix **V** such that $V_{ii} = N_{Sri}^{-1/2}$. The matrix $\mathbf{D_r} = \mathbf{V}\ \mathbf{VN_S}$ is row-stochastic, its eigenvalues do not exceed 1 in absolute value, and the maximum eigenvalue equals 1 via the properties of row-stochastic matrices (see section 5.13.3 of Marcus and Minc, 1993, p. 133).

Despite the symmetry of $\mathbf{N_S}$, the matrix $\mathbf{D_r}$ will be asymmetric. One can however invoke a similarity transformation:

$$\mathbf{D_S} = \mathbf{V}^{-1}\,\mathbf{D_r}\,(\mathbf{V}^{-1})^{-1} = \mathbf{V}\,\mathbf{N_S}\,\mathbf{V} \tag{3.1}$$

with the resulting $\mathbf{D_s}$ having eigenvalues (which we denote as λ) equal to those of $\mathbf{D_r}$ (Ord, 1975, p. 125). Note, the eigenvalues of $\mathbf{D_r}$ do not exceed 1 in absolute value via the properties of row-stochastic matrices (5.13.3 Marcus and Minc, 1993, p. 133). Because $\mathbf{D_s}$ is similar to $\mathbf{D_r}$, $-1 \leq \lambda_i^{D_s} \leq 1$ due to the equivalent eigenvalues.

From a statistical perspective, one can view $\mathbf{D_r}$ as a spatial averaging operator. Given an arbitrary n-vector υ, the row-stochastic normalization $\mathbf{D_r}\upsilon$ results in a form of local average or smoothing of υ. In this context, one can view elements in the rows of $\mathbf{D_r}$ as coefficients of a linear filter. From a numerical standpoint symmetry of $\mathbf{D_s}$ simplifies computing the log-determinant and has theoretical advantages as well. See Bavaud (1998) for more information on spatial weight matrices.

3.2.2 The Mixed Spatial Model

The spatially autoregressive model in (3.2) provides a flexible way of modeling the dependent variable,

$$y = \mathbf{X}\boldsymbol{\beta} + \alpha\mathbf{D}y + \varepsilon \tag{3.2}$$

where y is an n element vector of observations on the dependent variable, **X** is a n by k matrix of n observations on k independent variables, **D** is a n by n spatial weight matrix, and ε is an n element vector of disturbances. For added flexibility, **X** can include spatially lagged independent variables as shown in (3.3),

$$\mathbf{X} = [\iota\ \mathbf{X_{sub}}\ \mathbf{DX_{sub}}] \tag{3.3}$$

where ι is a n element vector of ones and $\mathbf{X}_{sub}$ represents a matrix of n observations on p independent variables and $\mathbf{X}$ is an n x k matrix with $k = 2p + 1$. The specification of $\mathbf{X}$ in (3.3) is termed a mixed regressive spatially autoregressive model and it subsumes the corresponding error autoregression (see Anselin, 1988, p. 225-230).

One can transform the model as $(\mathbf{I} - \alpha\mathbf{D})y = \mathbf{X}\beta + \varepsilon$ and solve for the errors using the least squares idempotent matrices $\mathbf{P}_{ols} = \mathbf{X}(\mathbf{X}'\mathbf{X})^{-1}\mathbf{X}'$ and $\mathbf{M} = \mathbf{I} - \mathbf{P}_{ols}$.

$$e_o = \mathbf{My}$$
$$e_d = \mathbf{MDy}$$
$$e_{mix} = e_o - \alpha e_d$$

Maximum likelihood estimation of the mixed model proceeds by maximization of the profile loglikelihood $L(\alpha) = \ln|\mathbf{I} - \alpha\mathbf{D}| - (n/2)\ln(SSE(\alpha))$ where $SSE(\alpha) = e_{mix}' e_{mix}$ (Pace and Barry, 1997, p. 235).

3.2.3 Recursive Estimation

Since the mixed model estimates rely on least squares estimates and the log-determinant, recursive spatial estimation is possible. Efficient methods exist for computing recursive least squares (Björck, 1996, p. 130-131), and the recursive nature of matrix decompositions used to compute log-determinants (Strang, 1976, p. 166-167) can be combined with these to yield recursive spatial estimation.

Specifically, one approach to computing log-determinants decomposes $(\mathbf{I}-\alpha\mathbf{D})$ into two triangular matrices $\mathbf{L}$ and $\mathbf{U}$, that is $(\mathbf{I}-\alpha\mathbf{D}) = \mathbf{LU}$. The overall log-determinant of $(\mathbf{I}-\alpha\mathbf{D})$ is the log-determinant of $\mathbf{L}$ plus the log-determinant of $\mathbf{U}$. Note that the log-determinant of a triangular matrix is the sum of the logs of the diagonal elements or pivots. Since the matrix $\mathbf{L}$ has ones along its diagonal for positive-definite $(\mathbf{I}-\alpha\mathbf{D})$, its log-determinant is 0. The matrix $\mathbf{U}$ has positive pivots on the diagonal and the log of the first pivot is actually the log-determinant of the first element that we denote $(1-D_{11})$. The sum of the logs of the first and second pivots are the log-determinant of expression (3.4).

$$\begin{pmatrix} 1-\alpha D_{11} & -\alpha D_{12} \\ -\alpha D_{21} & 1-\alpha D_{22} \end{pmatrix} \tag{3.4}$$

This pattern continues indefinitely for larger and larger submatrices (e.g., theorem 4D in Strang, 1976, p. 166). These results apply to the Cholesky decomposition as well, since this is a special variant of the LU decomposition. Intuitively, Gaussian elimination uses substitution to eliminate one variable at a time, relying only on prior equations during the process. For our purposes, note that stopping the decomposition at any point results in a system that has the correct relations among variables based on information already processed.

A small numerical example may make this clear. Consider the arbitrary example matrix **Z** and its associated **LU** decomposition such that $\mathbf{L_Z U_Z} = \mathbf{Z}$;

$$\mathbf{Z} = \begin{pmatrix} 1.612 & 0.163 & 0.266 \\ 0.163 & 1.714 & 0.890 \\ 0.266 & 0.890 & 2.127 \end{pmatrix} \qquad (3.5)$$

$$\mathbf{L_Z} = \begin{pmatrix} 1.000 & 0.000 & 0.000 \\ 0.101 & 1.000 & 0.000 \\ 0.165 & 0.508 & 1.000 \end{pmatrix} \mathbf{U_Z} = \begin{pmatrix} 1.612 & 0.163 & 0.266 \\ 0.000 & 1.698 & 0.863 \\ 0.000 & 0.000 & 1.644 \end{pmatrix} \qquad (3.6)$$

If we independently compute the sequence of determinants for the sub-matrices of **Z**, we find the determinant of the first 1 by 1 sub-matrix is 1.612, the determinant of the second 2 by 2 sub-matrix equals 2.7370, and the determinant of the 3 by 3 matrix equals 4.5004. The cumulative product of the diagonal of $\mathbf{U_Z}$ equals 1.612, 1.612x1.698=2.7370, and 1.612x1.698x1.644 = 4.5004. These cumulative products of the diagonal $\mathbf{U_Z}$ exactly match the independent computations of the determinants of the sub-matrices. Naturally, the log-determinants match as well.

To formalize this further, let s represent the number of observations and t represent an observation between 1 and s. The log-determinant for observations 1 through t represents the cumulative sum of the logs of the pivots,

$$ln|I - \alpha_t D_{1...t,1...t}| = \Sigma_{j=1}^{t} ln(U_{jj})$$

where $t \leq s$. Note, there are no additional costs for computing the sequence of log-determinants over those incurred in computing the overall log-determinant (conditional upon this ordering of the rows and columns in the spatial weight matrix). In addition, the use of sparse **D** and a sample of size s which is smaller than the global sample lead to increased speed when computing the log-determinants. One could use eigenvalues to compute the log-determinants, which once calculated give the log-determinant for any value of α. However, the eigenvalues require recalculation for each sub-sample size. Computing the log-determinant using 15 iterations of the Cholesky decomposition (and interpolating since the log-determinant function is very smooth) in conjunction with the recursive nature of the pivots saves substantial time relative to the eigenvalue method. As a simple illustration, we computed the eigenvalues for a sequence of dense matrices from size 20 by 20 to size 500 by 500 and compared the time required to that needed to compute 15 Cholesky decompositions on the same 500 by 500 matrix. The ratio of speeds was in excess of a factor of 50 in favour of the Cholesky decomposition. The use of sparse matrices would further favour the Cholesky decomposition, since this decomposition performs better than eigenvalue computations for sparse matrices. Of course, one could decrease the number of sub-sample sizes and make the eigenvalue approach feasible. However, the cross-validation MAPE is not a smooth function over sample sizes, ruling out interpolation when using the eigenvalue approach.

The log-determinant of the entire sample ($\ln|\mathbf{I}-\alpha_s\mathbf{D}_s|$) takes into account relations among all observations. In contrast, the log-determinant for a sub-sample based on t observations, ($\ln|\mathbf{I}-\alpha_t\mathbf{D}_{1...t,1...t}|$), only takes into account relations among observations in the sub-sample, ignoring relations between sub-sample observations and those outside the sub-sample. If an observation in any particular sub-sample has no relation to other observations in this sub-sample, it will not affect the log-determinant.

For brevity, let $\mathbf{D(t)} = \mathbf{D}_{1...t,1...t}$ and let $SSE(t;\alpha)$ represent the sum of squared errors for a regression of the first t observations out of s possible observations. We can examine the recursive log-likelihood,

$$L(t,\alpha_t) = ln|\mathbf{I} - \alpha_t \mathbf{D(t)}| - (t/2)ln(SSE(t,\alpha_t))$$

Following Pace and Barry (1997), the recursive log-likelihood $L(t;\alpha_t)$ and its constituent parts across a grid of q values of α beginning with observation k+1 takes the form[3].

$$\mathbf{SSE(t,\alpha_t)} = \begin{pmatrix} SSE(k+1,\alpha_1) & SSE(k+1,\alpha_2) & \cdots & SSE(k+1,\alpha_q) \\ SSE(k+2,\alpha_1) & SSE(k+2,\alpha_2) & \cdots & SSE(k+2,\alpha_q) \\ \cdots & \cdots & \cdots & \cdots \\ SSE(t,\alpha_1) & SSE(t,\alpha_2) & \cdots & SSE(t,\alpha_q) \end{pmatrix}$$

$$ln|\mathbf{I} - \alpha_t \mathbf{D(t)}| = \begin{pmatrix} ln|\mathbf{I} - \alpha_1 \mathbf{D(k+1)}| & ln|\mathbf{I} - \alpha_2 \mathbf{D(k+1)}| & \cdots & ln|\mathbf{I} - \alpha_q \mathbf{D(k+1)}| \\ ln|\mathbf{I} - \alpha_1 \mathbf{D(k+2)}| & ln|\mathbf{I} - \alpha_2 \mathbf{D(k+2)}| & \cdots & ln|\mathbf{I} - \alpha_q \mathbf{D(k+2)}| \\ \cdots & \cdots & \cdots & \cdots \\ ln|\mathbf{I} - \alpha_1 \mathbf{D(t)}| & ln|\mathbf{I} - \alpha_2 \mathbf{D(t)}| & \cdots & ln|v - \alpha_q \mathbf{D(t)}| \end{pmatrix}$$

$$\mathbf{L(t,\alpha_t)} = \begin{pmatrix} L(k+1,\alpha_1) & L(k+1,\alpha_2) & \ldots & L(k+1,\alpha_q) \\ L(k+2,\alpha_1) & L(k+2,\alpha_2) & \ldots & L(k+2,\alpha_q) \\ \ldots & \ldots & \ldots & \ldots \\ L(t,\alpha_1) & L(t,\alpha_2) & \ldots & L(t,\alpha_{q1}) \end{pmatrix}$$

By choosing the column with the largest likelihood for each row, we arrive at a sequence of maximum likelihood estimates for α denoted by $\alpha^*(t)$ and the corresponding sequences of errors and estimates associated with this sub-sample.

3.2.4 A Comparison with Other Approaches

The non-parametric locally linear least squares models rely on a sequence of locally linear regressions to produce estimates at various points in space (possibly all n points) based on a sub-sample of data information from nearby observations. (We will refer to these models as LWR, following McMillen, 1996, 1997). Using the previously introduced notation and letting **W(i)** represent an n x n diagonal matrix containing distance based weights for observation i that reflect the distance (cardinal or ordinal) between observation i and all other observations, we can write the LWR model as:

$$\mathbf{W(i)}^{1/2} y = \mathbf{W(i)}^{1/2} \mathbf{X}\beta(i) + \mathbf{W(i)}^{1/2} \varepsilon_i \tag{3.7}$$

The notation $\beta(i)$ indicates that this kx1 parameter vector is associated with observation i. The LWR model produces n such vectors of parameter estimates, one for each observation. These estimates are calculated using:

$$\hat{\beta}(i) = (\mathbf{X'W(i)X})^{-1}(\mathbf{X'W(i)}y) \tag{3.8}$$

A number of alternative approaches have been proposed to construct the distance based weights for each observation i contained in the vector on the diagonal of **W(i)**.

McMillen (1998) suggests a tri-cube weighting function shown in (3.9) that has support only for some of the observations similar in spirit to the approach we take.

$$diag(\mathbf{W(i)}) = \left(1 - \left(d_i^{(j)} / d_i^{(n_{sub})}\right)^3\right)^3 I\left(d_i^{(j)} < d_i^{(n_{sub})}\right) \tag{3.9}$$

Where $d_i^{(j)}$ represents the distance between observation j and observation i, $d_i^{(n_{sub})}$ represents the distance between the n_{sub}th nearest neighbour and observation i; and $I()$ is an indicator function that equals one when the condition is true and zero otherwise. Least squares estimates for a variant of LWR that we compare to our approach are based on the n observations arranged in order of increasing distance from observation i (including i). Weighted least squares estimates can then be computed using a weighting matrix where the ith diagonal entry is zero. If we let **W(i)** represent such a matrix with the condition $W_{ii} = 0$, this effectively leaves out observation i and permits computation of the cross-validation prediction error at observation i.

We follow McMillen (1998) by using limited support, but deviate in using constant weights. Specifically, we assume **W(i)** contains constant weights for the n_{sub} nearest neighbours to observation i, except for the ith weight which equals 0. This represents a nearest neighbour form of geographic weighting with a bandwidth of n_{sub}, referred to as a uniform kernel in the nonparametric literature. This matches the weighting function used by Pavlov (2000). This could be generalized to other kernels, but at some computational cost.

By examining all of the cross-validation prediction errors, one could choose an optimal n_{sub}. Implemented with efficient recursive least squares techniques, this form of LWR would require $O(n_{sub}n)$ operations to produce estimates. If we were to compute optimal weights for each sub-sample size and estimates for all possible sub-samples from 1 to n, $O(n^3)$ operations would be required, which represents an upper bound on the operation count. Not surprisingly LWR estimation without employing computational accelerations can require substantial computational time. Pavlov reported that his estimation effort involving around 3000 observations took less than 24 hours using a fast PC. In its favour, LWR does lend itself to parallel processing.

Returning to the problem of quantifying the flexibility of LWR methods, one can view LWR methods in the context of linear smoothers. Buja, Hastie and Tibshirani (1989) developed a general theory of linear smoothers in a well-known paper. Linear smoothers take the form $\hat{y} = \mathbf{P}y$, where **P** is an nxn smoother matrix. In the case of least-squares based on the entire sample, $\mathbf{P_{ols}} = \mathbf{X(X'X)^{-1}X'}$, where P_{ols} represents an idempotent 'hat' matrix that projects (in a least-squares sense) the vector y onto the space spanned by the matrix **X**. Buja, et al. (1989) develop a measure of 'equivalent number of variables' required by a standard technique such as least-squares to achieve the same flexibility in estimation as a smoother such as LWR. Based on the hat matrix, one could employ any of the following expressions for the equivalent number of variables in least-squares is $tr(\mathbf{P_{ols}}) = tr(\mathbf{P_{ols}^2}) = tr(\mathbf{P_{ols}'P_{ols}}) = k$. Cleveland and Devlin (1998, p. 602) proposed this last measure of the equivalent number of variables.

We can apply this same measure of equivalent variables to LWR using its linear smoother. For each observation $i = 1,.....,n$, the LWR technique computes,

$$\mathbf{P(i) = W(i)^{1/2} X (X'W(i)X)^{-1} X'W(i)^{1/2}}$$

and thus makes the prediction for all sample observations of **P(i)**y. LWR discards the predictions for the observations other than the ith observation. Hence, the prediction for the ith observation is the dot product between the ith row of **P(i)** and y. Let $\mathbf{P}_{\mathbf{lwr}}$ represent the matrix constructed from the stacking the ith row of **P(i)** for all i and $\hat{y}_{lwr}\mathbf{P}_{\mathbf{lwr}}\,y$. The smoother matrix $\mathbf{P}_{\mathbf{lwr}}$ represents the least-squares projection matrix adjusted for the leave-one-out cross-validated prediction process and for the different sub-samples used during estimation. Noting that $tr(\mathbf{P}_{\mathbf{lwr}}) = 0$ by construction since $W_{ii} = 0$, we arrive at the implausible result that cross-validation completely avoids the issue of model dimension irrespective of the sub-sample size. This implies that $tr(\mathbf{P}_{\mathbf{lwr}}^2)$ or $tr(\mathbf{P}_{\mathbf{lwr}}'\mathbf{P}_{\mathbf{lwr}})$ should be used as a measure of the effective number of variables for the LWR smoother. We calculate these proxies for the number of variables in section 3.3 for an applied example. The calculation of the effective number of variables allows users to obtain some idea of the tradeoff between model flexibility and estimate variability.

Summarizing, we set forth a recursive approach to maximum likelihood spatial estimation that allows us to generalize past least-squares-based locally linear methods for dealing with spatial heterogeneity. One can follow a procedure similar to those used with past approaches, but replace the least-squares estimates with our recursive spatial maximum likelihood estimates. This results in flexible spatial modeling that accommodates spatially varying parameters alongside spatial dependence incorporated in traditional spatial autoregressive models.

3.3 An Applied Example: Voting Across Counties

In this section we illustrate the techniques presented in the previous section using data on the votes cast in the 1980 presidential election across U.S. counties originally used in Pace and Barry (1997). Section 3.3.1 discusses the data employed, section 3.3.2 sets forth the model specification, and section 3.3.3 presents estimation results.

3.3.1 Voting Data

To locate our sample of contiguous US counties in space, we relied on the geographic centroids of all counties (or their equivalents) from the 1990 Census[4]. We collected data on the total number of recorded votes in the 1980 presidential election (all parties), the population 18 years of age or older (POP) in each county, the population in each county with a 12th grade or higher education (EDUCATION), the number of owner-occupied housing units (HOUSES), and aggregate county-level income (INCOME).

3.3.2 Models

We fitted the model using leasts-quares and our SALE method. The dependent variable was transformed to reflect the proportion of votes cast for both candidates in the 1980 presidential election. That is, our dependent variable is ln(PRVOTES) = ln(VOTES/POP) = ln(VOTES)-ln(POP). The complete model is shown in (3.10), where **D** represents a spatial weight matrix.

$$\ln(PRVOTES) = \beta_0 + \ln(POP)\beta_1 + \ln(EDUCATION)\beta_2 + \ln(HOUSES)\beta_3 + \ln(INCOME)\beta_4 + \mathbf{D}\ln(POP)\beta_5 + \mathbf{D}\ln(EDUCATION)\beta_6 + \mathbf{D}\ln(HOUSES)\beta_7 + \mathbf{D}\ln(INCOME)\beta_8 + \mathbf{D}\ln(PRVOTES)\alpha + \varepsilon \quad (3.10)$$

If $\alpha = 0$ in (3.10), the model represents a non-spatial autoregressive variant similar in spirit to LWR that can be estimated using least-squares. For simplicity, we selected a row-stochastic spatial weight matrix **D** based on 15 nearest neighbours where the rate of geometric decline in the weights is 0.8 (e.g., the second nearest neighbour has 0.8 the weight of the first nearest neighbour, the third nearest neighbour has 0.8 the weight of the second nearest neighbour, and so forth). This choice of neighbours and associated weights was based on previous analyses of these data. The use of spatially lagged explanatory variables differs from the usual LWR methods, but facilitates comparison with our SALE estimates since only the single parameter α distinguishes the two models. This specification provides our version of LWR with both spatially varying β as well as spatially lagged explanatory variables to model space. The SALE model inherits these two approaches to capturing spatial effects, and augments them with the spatially lagged dependent variable. Hence, SALE completely nests the LWR model with spatially lagged independent variables. That is, a value of α significantly different from 0 would reject the LWR model with spatially lagged independent variables in favour of SALE. This hypothesis can be tested using likelihood ratio statistics based on a chi-squared density with one degree-of-freedom under the null hypothesis of $\alpha = 0$.

3.3.3 Estimation Results

The cross-validated prediction errors across different sample sizes provides a means of estimating an optimal sample size traditionally used to implement LWR. In the framework of our SALE method we can also examine the gains that arise from using sub-samples relative to a global spatial autoregressive model. Figure 3.1 provides a plot comparing LWR and SALE mean absolute prediction errors (MAPE) for various sample sizes. The symbol 'o' is used to denote LWR MAPE errors and '.' represents SALE errors. In addition, the figure shows global MAPE errors for least-squares using the symbol '-' and global SALE MAPE errors designated with the '-.' symbol.

From the figure we see that LWR fulfils its promise over global least-squares, producing a MAPE of 7.34% based on a cross-validated sub-sample size of 58 observations, compared to errors of 10.28% for global least-squares based on the entire

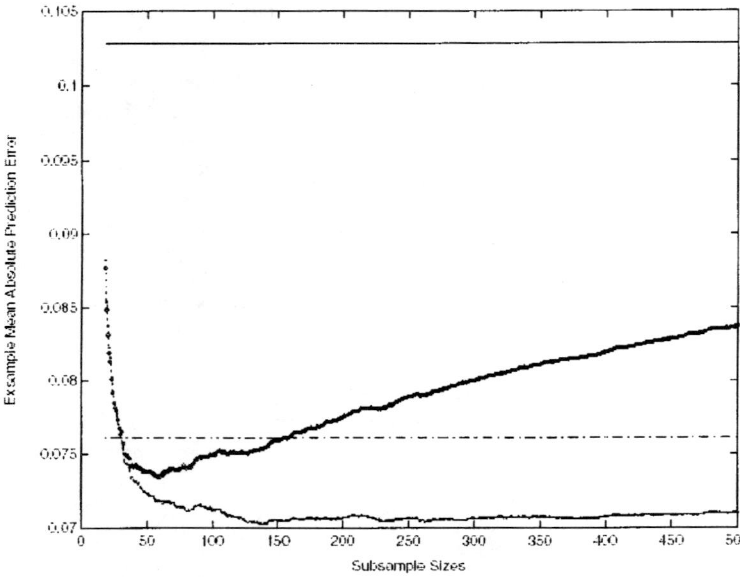

Figure 3.1. Cross-validation MAPE for LWR and SALE vs. Subsample Size

sample of 3107 observations. Our SALE method achieves a minimum MAPE of 7.02% using a sub-sample of 139 observations which compares quite differently to global spatial autoregressive model estimates that produced MAPE errors of 7.61%. An intuitive motivation for this result is that the global spatial autoregressive model estimates already incorporate space using both spatially lagged independent and dependent variables, so the additional gain from allowing the parameters to varying spatially is far less than obtained in the case of LWR versus least-squares.

Another point worthy of note is that the LWR sub-sample size exerts a sharp impact on the MAPE, whereas the SALE errors display much less sensitivity to sample size. Specifically, we see that the LWR MAPE rises to 8.37% when sub-samples of 500 observations are used, while SALE MAPE rises to only 7.10%. Intuitively, the addition of a single spatially lagged dependent variable reduces the SALE prediction errors relative to LWR. The magnitude of this reduction is such that the SALE errors associated with a sub-sample size of 500 observations are lower than the optimal LWR MAPE errors based on 58 observations determined using cross-validation.

The LWR method produces estimates that exhibit high variability around the optimal prediction sub-sample size. Figure 3.2 shows a plot of the coefficient on the income variable obtained from various sub-sample sizes using the local regression centered around the location of a selected county. Using small sub-sample sizes enhances the LWR model flexibility at the cost of highly variable estimates and inferences. It is interesting to note that the estimate at the optimum sub-sample size of 58 observations does not differ greatly from that based on a sub-sample size of 500.

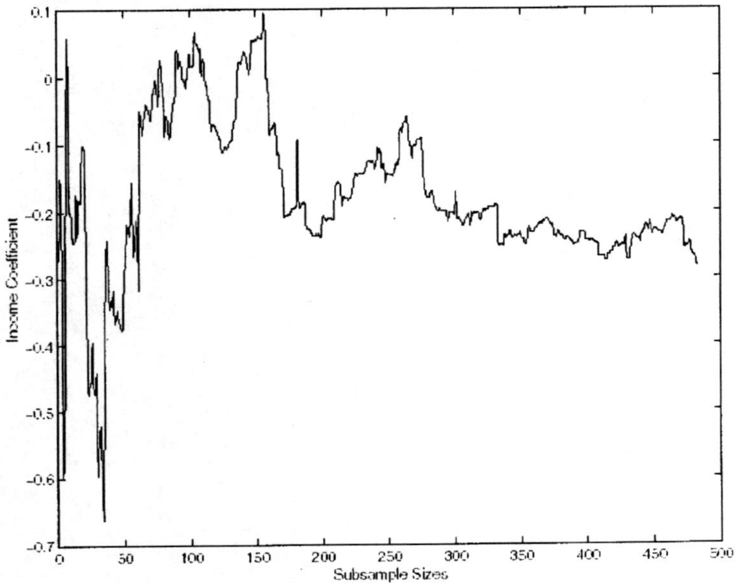

Figure 3.2. LWR Income Estimates vs. Subsample Size

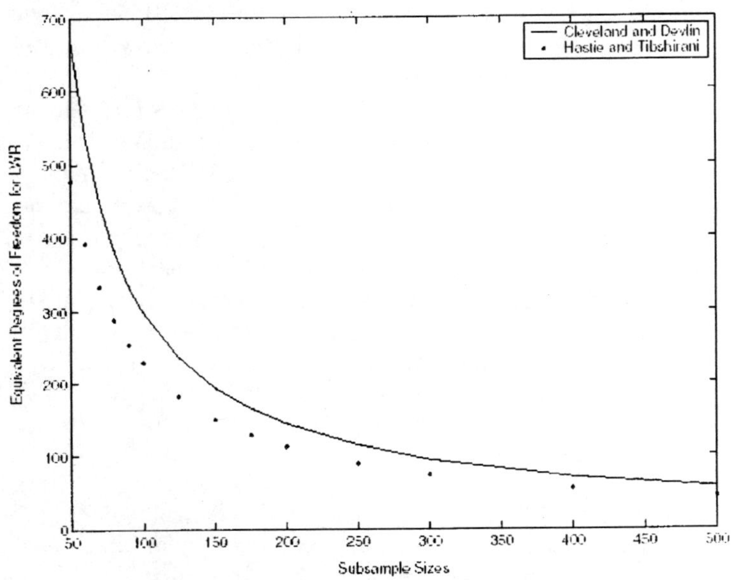

Figure 3.3. Equivalent Number of Variables for LWR vs. Sample Size

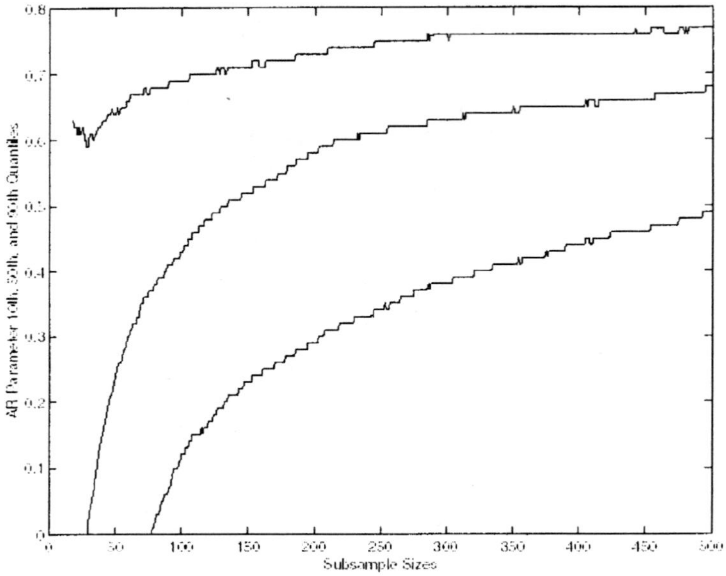

Figure 3.4: 10th, 50th, and 90th Percentiles of *a* Estimates Across Sample Sizes

Nonetheless, choosing other sub-sample sizes would result in coefficient estimates with very different magnitudes. Further, the coefficient magnitudes show little trend in the face of considerable variation in magnitude.

Figure 3.3 shows the effective number of variables across different sub-sample sizes for the election data measured using the approach of Buja, et al. (1989) and Cleveland and Devlin (1988, p. 602) as discussed in section 3.2.4. The figure shows that the effective number of variables employed by LWR to achieve low prediction errors is quite large. At the optimum sample size of 58, LWR uses the equivalent of over 400 variables based on the Buja et al. measure and over 550 variables using the Cleveland and Devlin approach. This suggests that LWR requires a choice between highly variable parameter estimates accompanied by low prediction errors and less variable parameter estimates associated with higher prediction errors. In contrast, our SALE estimates display relatively low prediction errors across all sub-sample sizes shown in Figure 3.3. This suggests that the SALE method can produce low prediction errors without the accompanying noise in parameter estimates.

Figure 3.4 provides some insight into the relation between LWR and SALE methods by plotting the 10th; 50th, and 90th quantiles for the optimal values of the spatial dependence parameter α across the n weighted maximum likelihood spatial autoregressions. LWR does appear to live up to the expectation that one can often eliminate spatial autocorrelation when sub-sample sizes are very small. Note, the 90th percentile curve of α remains quite high even for the small sample sizes. This suggests that one can not necessarily rely upon small sample sizes alone to eradicate spatial dependence in every case.

One of the motivations for use of LWR methods is the ability to examine parameter instability over sub-regions in space. To illustrate that this can be achieved with our SALE method, Figures 3.5 through 12 present a plot of the SALE and LWR parameter estimates sorted by east to west longitudes. The LWR estimates were based on the cross-validated sample size of n_{sub}=58 observations, while the SALE parameters in the graphs are for n_{sub}=500. This sample size was selected because cross-validation showed almost no difference in error between the optimum (n_{sub}=139) and selected n_{sub} (n_{sub}=500), while the smaller sub-sample size obviously equates to the equivalent of adding additional variables to the model, or a more flexible functional form. The difference in parameter variability between the LWR and SALE estimates is quite apparent in the figures.

In figure 3.5, SALE shows that voting participation declines more substantially with population at the coasts than in the center of the country, while LWR in figure 3.6 shows less systematic variation. This systematic variation could be viewed as depicting an actual phenomenon which LWR missed due to its high volatility, or as evidence of some more fundamental misspecification. In figures 3.7 and 3.8, both SALE and LWR show voter participation rates are more sensitive to education in the west than in other parts of the country.

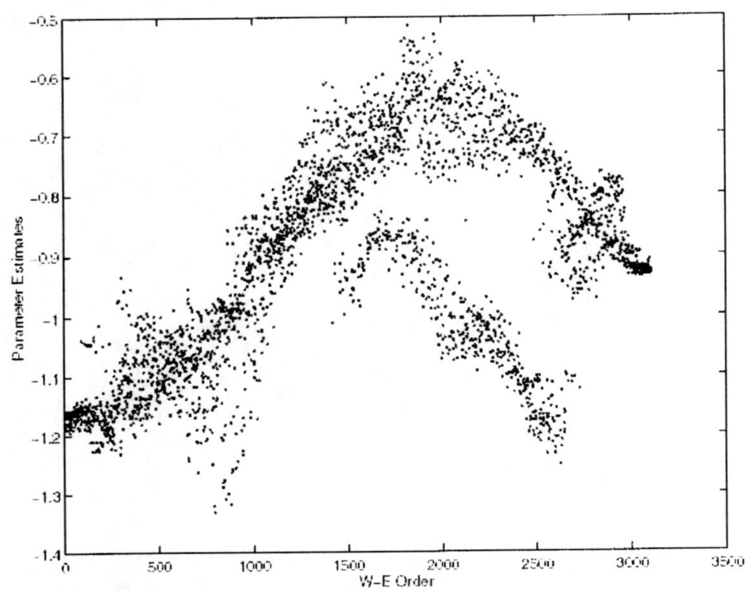

Figure 3.5: SALE Population Estimates by W-E Order

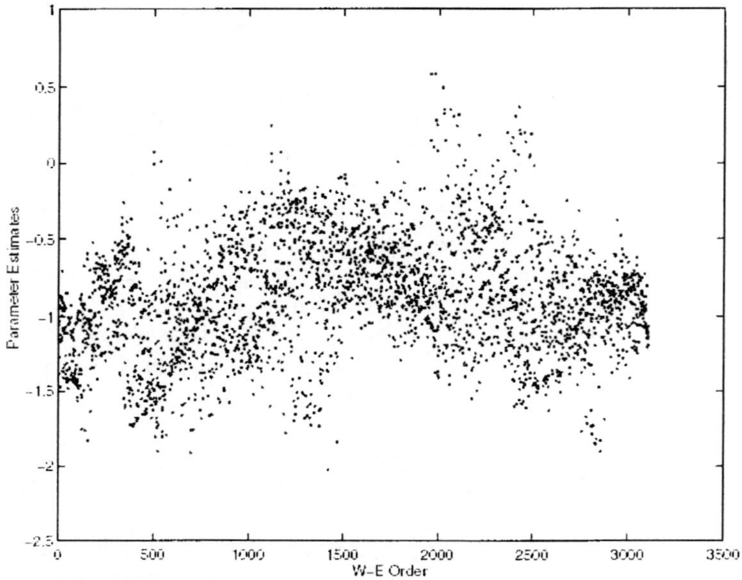

Figure 3.6: LWR Population Estimates by W-E Order

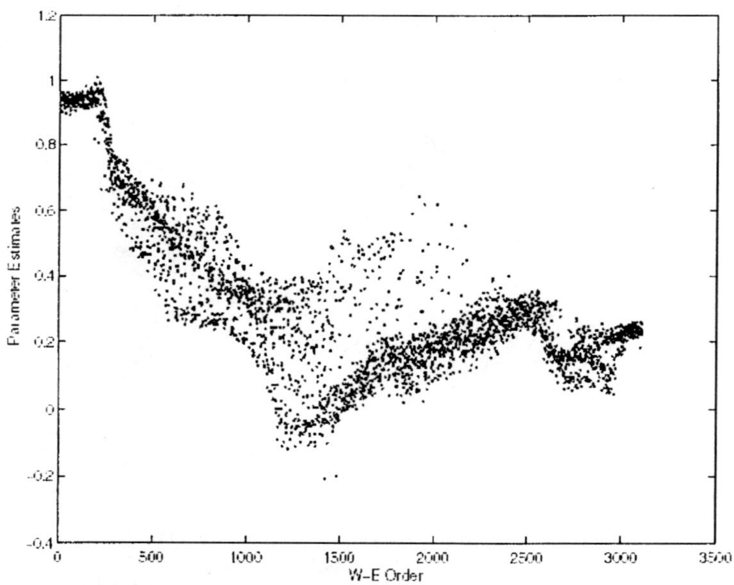

Figure 3.7: SALE Education Estimates by W-E Order

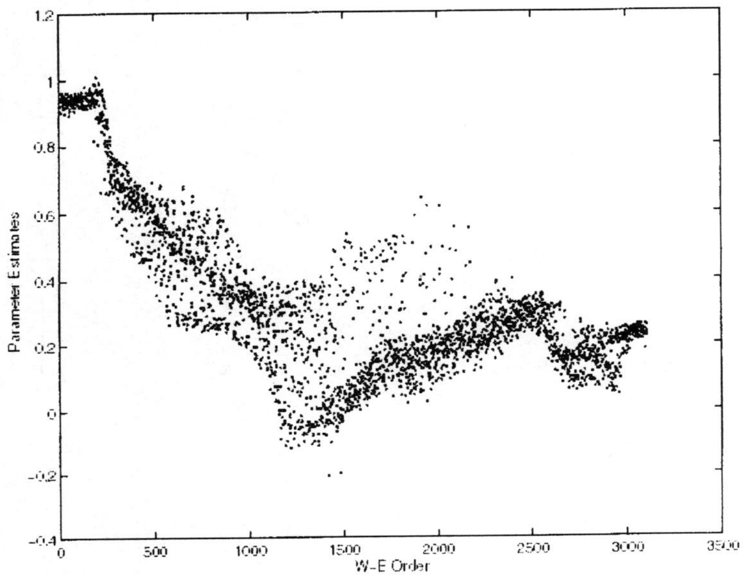

Figure 3.8: LWR Education Estimates by W-E Order

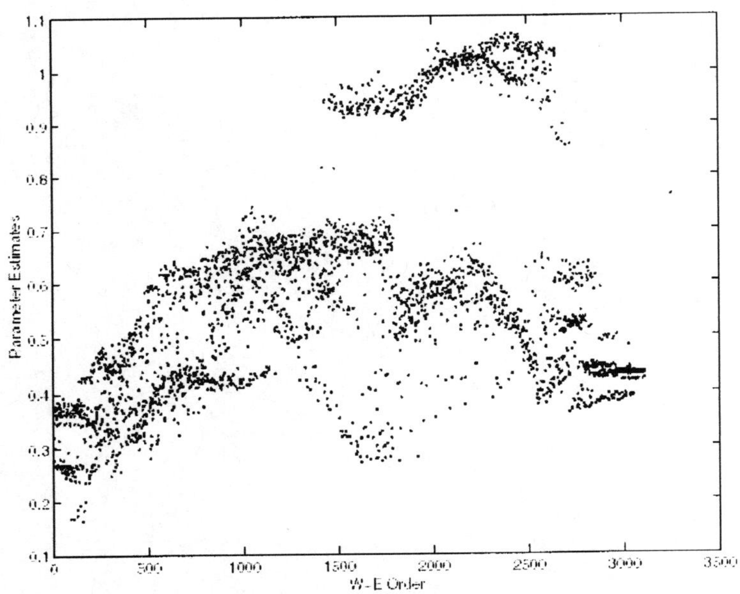

Figure 3.9: SALE Homeownership Estimates by W-E Order

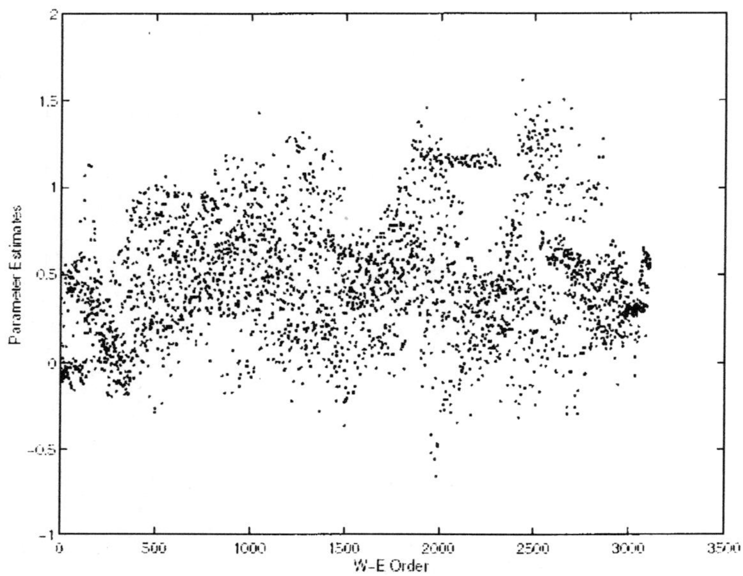

Figure 3.10: LWR Homeownership Estimates by W-E Order

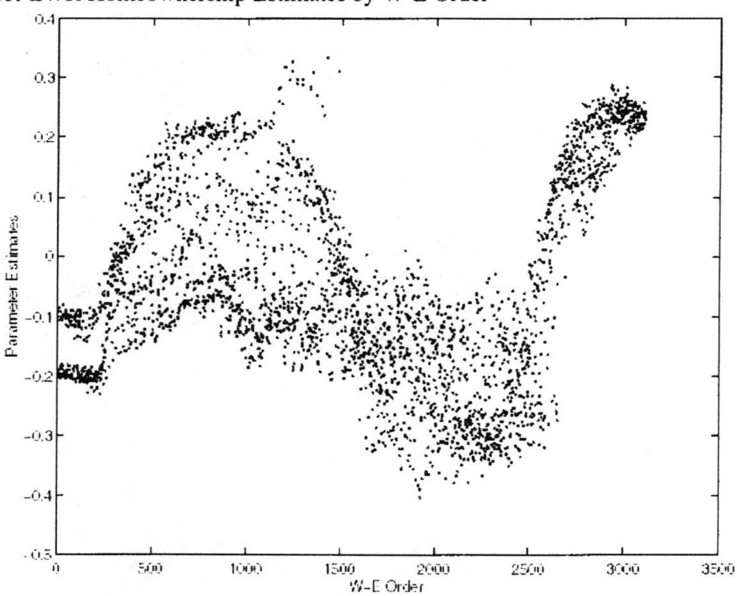

Figure 3.11: SALE Income Estimates by W-E Order

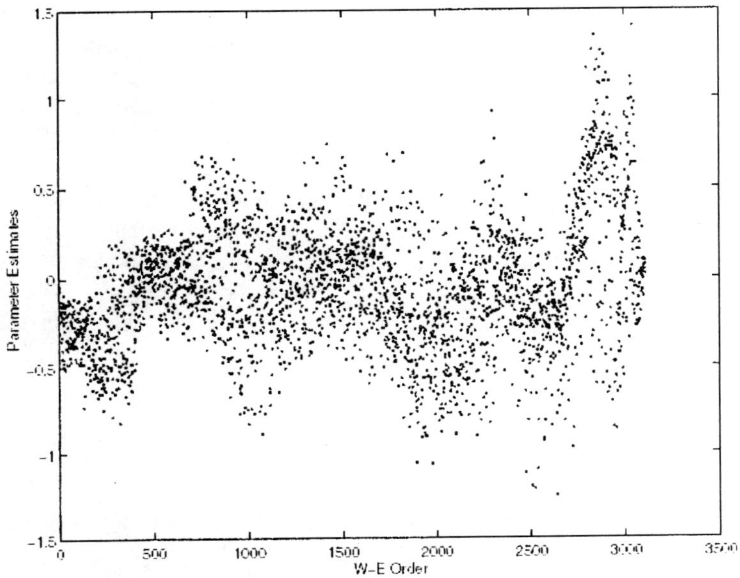

Figure 3.12: LWR Income Estimates by W-E Order

A benefit of the additional smoothing produced by the SALE estimation methodology is the simplicity of discerning patterns of instability over space. These patterns might be obscured by the highly volatile nature of more traditional LWR estimates.

Of course, both LWR and SALE may suffer from misspecification in this example, as voting behaviour may depend on more than the four independent variables chosen. In addition, the double-log specification may not match the true functional form. Our results indicating that the spatial autoregressive term is significant would be interpreted by some as evidence of misspecification. Both functional form misspecification and missing variables can give rise to spatial dependence. It seems likely that a more complete model might reduce the differences between SALE and LWR if the magnitude of spatial dependence were lowered. However, imperfect information and models tend to predominate in practice, making the SALE method a useful tool. One might also use SALE as a way to validate the assumption of zero spatial dependence made by LWR methods, by drawing an inference from the SALE parameter estimate of α. Use of SALE in this fashion would require that one produce SALE estimates, which might have other desirable features relative to those from LWR methods.

A final point is that exploring spatial heterogeneity via LWR or SALE methods remains computationally expensive for large sample sizes. It required just under 8 hours time to compute LWR and SALE estimates for the 3107 observations using Matlab 6.0 under Windows NT 4.0 with a 600 Mhz Pentium III machine. One of our contributions is to show that SALE estimation is computationally competitive with LWR methods, while subsuming the latter as a special case. Nonetheless, the relative computational intensity of these methods can be seen by noting that global spatial autoregressive estimation using maximum likelihood (including finding neighbours and creating the spatial weight matrices) required less than 10 seconds. As a positive note, SALE should prove amenable to parallel processing. Finally, for large sample sizes it may not be necessary to run a separate regression for each observation. For many purposes SALE or LWR estimates based on observations centered around selected spatially separated locations may prove adequate for exploring spatial heterogeneity. In fact, if the observations have enough separation, this may permit inference concerning differences between coefficients from individual regions when separate regressions do not share common observations. In this case, one could employ Chow-like tests to examine parameter stability over space.

3.4 Conclusion

We demonstrate the computational feasibility of recursive spatial autoregressive model estimation by employing properties of the matrix decomposition often used to calculate the log-determinant term in the likelihood function for these models. The cumulative sum of the logs of sequential pivots from the **LU** matrix decomposition of ($\mathbf{I}$-$\alpha\mathbf{D}$) yield a sequence of the exact log-determinant terms for any sub-sample of observations. Combining this with information from least-squares computations allows us to produce recursive maximum likelihood spatial autoregressive estimates for sub-samples of observations. We show that for moderate sample sizes involving all US counties, the computational intensity of our estimation methodology competes with those of conventional locally linear methods.

The computational insights noted above were used to extend traditional locally linear models based on least-squares estimates, often employed to explore spatial heterogeneity. Replacing least-squares estimates with our recursive spatial autoregressive estimates results in spatial autoregressive locally linear estimates. This approach subsumes the more traditional locally linear least-squares models as a special case, and we argue that this approach has advantages. The advantages are: improved prediction, increased stability of the spatially varying estimates, and less sensitivity to sample size choice.

We argue that traditional LWR methods exhibit a trade-off between increasing the sample size to produce less volatile estimates that contain increasing spatial dependence. Selecting a smaller sample size reduces the spatial dependence, but at the cost of increased parameter variability that impedes detection of patterns of parameter variation over space. The SALE method introduced here does not suffer from this problem because spatial dependence is modeled directly during estimation.

Acknowledgements

Kelley Pace would like to gratefully acknowledge the research support received from Louisiana State University. In addition, she would like to thank Ming-long Lee and Carlos Slawson for their remarks.

Endnotes

[1]LeSage (2001) proposed a Bayesian variant of the locally linear models that overcomes the second problem by introducing explicit stochastic spatial parameter transition relationships as prior information in the model. Since prior uncertainty regarding parameter variability is under control of the user, a continuum of estimates ranging from highly volatile to relatively constant over space can be produced.

[2]This assumes that the SALE estimates rely on the same sub-sample size for all observations, but it is possible to vary the sub-sample size for each observation. Note, that using a fixed number of neighbours (sub-sample size) effectively increases the geographic area used by SALE or LWR when the spatial density of observations is low and decreases the effective geographic area when the spatial density of observations is high.

[3]One cannot compute the likelihood for fewer observations, although one could begin with more observations.

[4]Three continental U.S. counties contributed zero votes in the election. These included Yellowstone National Park and other unusual entities. We employed a Lambert conformal conic projection of the latitude and longitude of the county centroids to construct the locational coordinates used in computing distances. The projection minimized the error for the 27.5 and 43.5 degree parallels.

4 Aggregation Bias in Maximum Likelihood Estimation of Spatial Autoregressive Processes

Tony Smith
University of Pennsylvania

Abstract

In statistical models of spatial behaviour, there is often a mismatch between the scale at which data is available and the scale at which key spatial dependencies are known to occur. However, in attempting to incorporate finer grain information about spatial dependencies, certain estimation problems arise. Here it is shown that maximum likelihood procedures can produce significantly negative estimates of positive autocorrelation. This problem is analyzed in the context of a simple spatial autoregressive process, and possible correction procedure is proposed for reducing aggregation bias.

4.1 Introduction

In statistical models of spatial behaviour, there is often a mismatch between the scale at which data are available and the scale at which key spatial dependencies are known to occur. For example, spatial dependencies in housing values may be strongest between adjacent houses or houses on the same block face, while available data may only be available at the blockgroup or even census tract level. Hence when modeling such dependencies in terms of spatial lag or spatial autocorrelation effects, the corresponding proximity weight matrices are necessarily limited to the same degree of aggregation as the given data. However, in many cases it is possible to obtain *finer grain* weight matrices in term of existing map data. In the illustration above, for example, it may be possible to determine proximity relations at a more appropriate scale, such as shared block faces rather than shared blockgroup boundaries.

In this context, the central question of interest here is how to combine such finer grain information about spatial dependencies with aggregated data in a manner which improves the overall goodness of fit. A number of initial efforts in this direction (including various attempts at lower-dimensional approximations to weight matrices for regression models with spatial autoregressive errors and/or spatial lags) all produced disappointing results. However, further investigation revealed that the poor fit of these models was partly a consequence of the *maximum likelihood estimation* procedure itself. In particular, when spatial interaction effects are at a finer scale than the basic data, the standard maximum likelihood procedure has a strong tendency to *underestimate the degree of spatial autocorrelation.* At first glance, this would appear to be consequence of the well-known fact that correlations among aggregates tend to diminish (shrink toward zero) as aggregation size increases (see for example Chapter 5 of Arbia, 1989). However, the present situation is quite different, and appears to be more a consequence of the *variance minimizing* tendency of maximum likelihood estimation which, in the presence of aggregation, tends to favor *negative* autocorrelation. In many instances, a substantial portion of the estimates are thus negative when actual autocorrelation is *positive*. Moreover, this is not simply a 'small sample' problem. While such estimates are theoretically consistent, examples show that even for very large sample sizes these problems need not disappear. Hence they serve to illustrate some of the practical limitations of consistency itself.

For purposes of analysis, we focus in the present chapter on the simplest possible case of a *pure spatial autoregressive model* involving only a single variable. While this model is generally of interest only as one component of a larger model, in the present context it has the advantage of exhibiting all of the key difficulties above while removing many extraneous factors. This model is formalized in Section 4.2 below. In addition, both the small sample and large sample properties of spatial autocorrelation estimates are developed for selected examples. In Section 4.3 the nature of this aggregation bias problem is explored in more depth. Here it is shown by means of a small example that this bias is at least in part due to the variance minimizing tendency of maximum likelihood estimation. In addition, a possible method is proposed for reducing this bias.

4.2 An Aggregated Spatial Autoregressive Model

Consider the n-vector *spatial autoregressive process,*

$$y = \rho \mathbf{W} y + \varepsilon \tag{4.1}$$

with nonnegative (nonzero) nxn *weight matrix*, $\mathbf{W} = (w_{ij})$, satisfying $w_{ii} = 0$ for all $i = 1, .., n$, together with *influence parameter*, ρ, and n x 1 *disturbance vector*, ε, normally distributed as

$$\varepsilon \sim \mathrm{N}(0, \sigma^2 \mathbf{I}_n) \tag{4.2}$$

In the analysis to follow, **W** will be assumed to be either symmetric or the row normalization of a symmetric matrix, both of which are known to possess real eigenvalues. If we let

$$\mathbf{B} = \mathbf{I}_n - \rho\mathbf{W} \tag{4.3}$$

and denote the minimum and maximum eigenvalues of **W** by λ_{min} and λ_{max}, respectively, then it is well known that the inverse matrix

$$\mathbf{D} = \mathbf{B}^{-1} = (\mathbf{I}_n - \rho\mathbf{W})^{-1} \tag{4.4}$$

exists for all ρ in the open interval $(1/\lambda_{min}, 1/\lambda_{max})$, and allows y in (4.1) to be expressed in terms of ε as [1]

$$y = \mathbf{D}\varepsilon \tag{4.5}$$

In this context, it is assumed that the vector y is not directly observable. Only an aggregated form of y is observable, here denoted by

$$x = \mathbf{A}y \tag{4.6}$$

where **A** is a nonnegative *kxn aggregation matrix* ($1 \leq k < n$) satisfying the 'partition' condition that for each $j = 1, .., n$, $a_{ij} > 0$ for exactly one $i = 1, .., k$, i.e., that each *subregion* (micro zone), j, belongs to exactly one *region* (macro zone), i. In the introductory example, if each component y_j were to represent the average housing value in block j, and a_{ij} were to represent the fraction of housing units of block group i belonging to block j, then each component, $x_i = \Sigma_j a_{ij}y_j$, would represent the average housing value in block group i[2].

By combining (4.5) and (4.6) we obtain the following *aggregate model* corresponding to the *disaggregate model* in (4.5):

$$x = \mathbf{AD}\varepsilon \tag{4.7}$$

To analyze this model, note first that by definition the rows of the aggregation matrix **A** are *orthogonal* (since the partition condition above implies that each column j has exactly one nonzero coefficient a_{ij}). Hence **A** is of full row rank, which together with the nonsingularity of **D** implies (from standard matrix results) that **AD** is of full row rank, and that **ADD'A'** is nonsingular. It thus follows at once from (4.7) that the aggregate data x is normally distributed as $N(0, \sigma^2\mathbf{ADD'A'})$, with k-dimensional density:

$$\phi_k(x;\rho,\sigma^2) = \left(2\pi\sigma^2\right)^{-k/2} \left|\mathbf{ADD'A'}\right|^{-1/2} exp\left\{-\frac{1}{2\sigma^2}x'\left(\mathbf{ADD'A'}\right)^{-1}x\right\}$$

(4.8)

So even though y is not directly observable, it is still possible to obtain *maximum-likelihood estimates* of ρ and σ^2 in this aggregate model by employing the log likelihood function,

$$L_k(\rho,\sigma^2;x) = const. - \frac{k}{2}ln(\sigma^2) - \frac{1}{2}ln\left|\mathbf{ADD'A'}\right| - \frac{1}{2\sigma^2}x'\left(\mathbf{ADD'A'}\right)^{-1}x \tag{4.9}$$

4.2.1 Estimation of ρ

Since our main concern here is the estimation of ρ, it is convenient to solve for the *mle* of σ^2 as

$$\sigma^2 = \frac{1}{k}x'\left(\mathbf{ADD'A'}\right)^{-1}x \tag{4.10}$$

and substitute (4.10) in (4.9) to obtain the *concentrated likelihood* of ρ,

$$l_k(\rho;x) = -\frac{1}{2}ln(x'\left(\mathbf{ADD'A'}\right)^{-1}x) - \frac{1}{2k}ln\left|\mathbf{ADD'A'}\right| \tag{4.11}$$

where for convenience we have multiplied through by $1/k$ and suppressed the constant, -(1+ln k)/2. To maximize this function, we begin with the first derivative,

$$\frac{\partial}{\partial\rho}l_k = -\frac{1}{2}\frac{x'\left[(\partial/\partial\rho)\left(\mathbf{ADD'A'}\right)^{-1}\right]x}{x'\left(\mathbf{ADD'A'}\right)^{-1}x} - \frac{1}{2k}\frac{\partial}{\partial\rho}ln\left|\mathbf{ADD'A'}\right| \tag{4.12}$$

where

$$\frac{\partial}{\partial\rho}(\mathbf{ADD'A'})^{-1} = -(\mathbf{ADD'A'})^{-1}\mathbf{A}\left[\frac{\partial}{\partial\rho}\mathbf{DD'}\right]\mathbf{A'}(\mathbf{ADD'A'})^{-1}$$
$$= -(\mathbf{ADD'A'})^{-1}\mathbf{AD}[\mathbf{WD}+\mathbf{D'W'}]\mathbf{D'A'}(\mathbf{ADD'A'})^{-1} \tag{4.13}$$

and

$$\frac{\partial}{\partial\rho}\ln|\mathbf{ADDA'}| = tr\left[(\mathbf{ADD'A'})^{-1}\mathbf{A}\left(\frac{\partial}{\partial\rho}\mathbf{DD'}\right)\mathbf{A'}\right]$$
$$= 2\cdot tr\left[(\mathbf{ADD'A'})^{-1}\mathbf{ADWDD'A'}\right] \tag{4.14}$$

By substituting (4.13) and (4.14) into (4.12) and employing the trace identity, $tr(\mathbf{M}_1\mathbf{M}_2) = tr(\mathbf{M}_2\mathbf{M}_1)$, one obtains the first-order condition:

$$\frac{\partial}{\partial\rho}l_k = \frac{x'(\mathbf{ADD'A'})^{-1}\mathbf{ADWDD'A'}(\mathbf{ADD'A'})^{-1}x}{x'(\mathbf{ADD'A'})^{-1}x}$$
$$-\frac{1}{k}tr\left[\mathbf{D'A'}x'(\mathbf{ADD'A'})^{-1}\mathbf{ADWD}\right] = 0 \tag{4.15}$$

In the case of no aggregation ($\mathbf{A} = \mathbf{I_n}$) it follows it follows that (4.15) reduces to the simpler condition that

$$\frac{\partial}{\partial\rho}l_k = \frac{y'\mathbf{B'WB}y}{y'\mathbf{B'B}y} - \frac{1}{k}tr(\mathbf{WD}) = 0 \tag{4.16}$$

where **B** is given by (4.3) above. It should be clear from a comparison of (4.15) and (4.16) that the case of aggregation involves a much more complex first-order condition. In particular, while it is readily shown (see Section 1 of the Appendix) that (4.16) always has a unique (maximal) root, this is not true of (4.15).

This is seen most easily by constructing a simple spatial example, as shown schematically in Figure 4.1. Here there are $n = 6$ subregions all connected by a simple {0,1}-contiguity relation, yielding the symmetric binary weight matrix **W** shown in the Figure[3]. These subregions are aggregated into $k = 3$ regions (where again the values, a_{ij}, of the aggregation matrix **A** in Figure 4.1 might reflect the fraction of housing units or

Figure 4.1: A Simple Spatial Aggregation Example

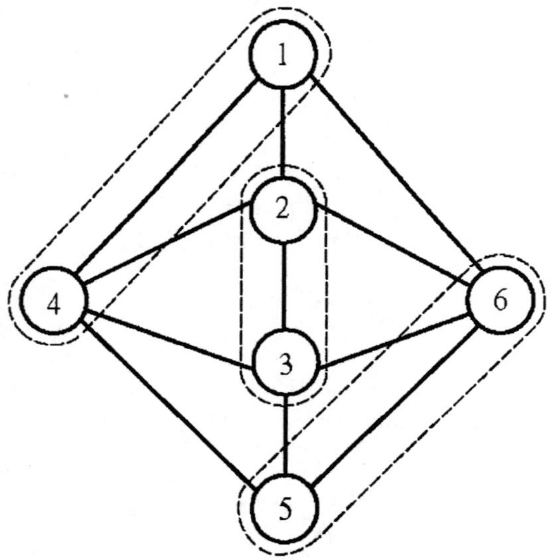

$$W = \begin{bmatrix} 0 & 0 & 0 & 1 & 0 & 1 \\ 0 & 0 & 1 & 1 & 0 & 1 \\ 0 & 1 & 0 & 1 & 0 & 1 \\ 1 & 1 & 1 & 0 & 1 & 0 \\ 0 & 0 & 0 & 1 & 0 & 1 \\ 1 & 1 & 1 & 0 & 1 & 0 \end{bmatrix}$$

$$A = \begin{bmatrix} 0 & .35 & .65 & 0 & 0 & 0 \\ 0 & 0 & 0 & .15 & .85 & 0 \\ .80 & 0 & 0 & 0 & 0 & .20 \end{bmatrix}$$

Figure 4.2a: Bimodal with Positive Global Maximum

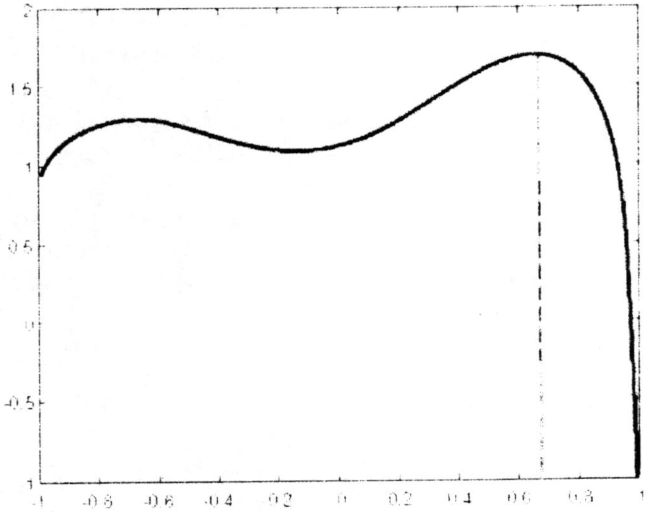

Figure 4.2b: Bimodal with Negative Global Maximum

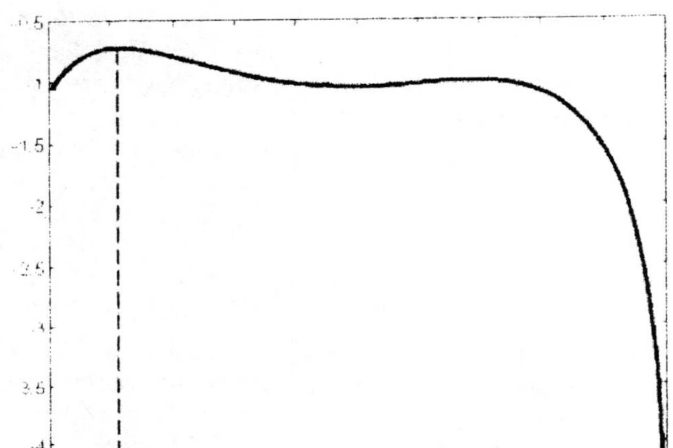

population of region i located in subregion j). The resulting aggregate model (4.7) was then simulated using many choices of ρ, and the concentrated likelihood functions were plotted. The two examples shown in Figure 4.2 are both for $\rho = .5$ and $\sigma = 1$, and illustrate multiple maxima with Figure 4.2a showing a *positive* global maximum at $\rho = .69$ and Figure 4.2b showing a *negative* global maximum at $\rho = .77$[4].

This example shows that local search procedures (such as gradient methods) can be very misleading in estimating ρ for the present class of models. But since the maximization is only over a one-dimensional bounded interval of possible ρ values, it should be equally clear that global maximization presents no real problem in this case. Hence it will be assumed throughout that global maximization procedures are used. In this context, the real questions of interest relate to the statistical properties of these global maxima.

4.2.2 Sampling Distributions of ρ

We begin with 1000 simulations of global maximum estimates for simple example above with ρ=.5 and σ=1. Figure 3a shows the results for the standard disaggregate model (5) with no aggregation. Here we see that the distribution is somewhat negatively skewed with sample mean, $\bar{\rho} = .335$, indicating a tendency to underestimate ρ in the standard model. Such biases have been well documented for a wide range of maximum likelihood estimates involving small sample sizes, such as the present case of n=6. However, for the results of the aggregate model (4.7) the situation is decidedly worse, as seen in Figure 4.3b. Here the sample mean, $\bar{\rho} = .02$, shows a pronounced bias. Much more important however is the erratic multimodal nature of the histogram. The strong mode near ρ=.1 is particularly troublesome. But since there are effectively only three samples here, namely the values of $x = \mathbf{A}y$ for the three aggregate regions, it can be argued that this may be simply an artifact of sample size.

To show that this is not the case, a number of larger models were constructed and simulated as well. The first is a square 10×10 grid of subregions aggregated into 16 regions, as shown in Figure 4.4a. Here the weight matrix **W** is again based on simple contiguities (with row normalized values w_{ij} now interpretable as the fraction of shared boundary for subregion i which is contiguous with subregion j). The aggregation matrix, **A**, in this case gives equal weight to the members of each region i (so that if n_i denotes the number of subregions in i then $a_{ij} = 1/n_i$ for all $j = 1, .., n_i$). The results of 1000 simulations for the disaggregate model (5) with ρ=.4 and σ=1 are shown for this case in Figure 4.5. Here it is seen that estimates for this model (with $n = 100$) are now behaving very well with a bell-shaped histogram centered at $\bar{\rho} = .39$. However, the results for the aggregate model (4.7) continue to exhibit the same difficulties: a low sample mean, $\bar{\rho} = .097$, and a multimodal histogram with a strong mode near ρ=.1.

A final example involving both a larger number of samples and a more realistic regional scheme is the set of blocks and block groups from West Philadelphia shown in Figure 4.4b. In this case **W** is again based on boundary shares as above, and **A** is based on the fraction of housing units in each block of a block group. There are 312 blocks within the 43 block groups shown. The results for 1000 simulations of the disaggregate model (5) with ρ=.4 and σ=1 are shown for this case in Figure 4.6a. Here again that estimates for this model (with n = 312) are right on target, with a sample mean $\bar{\rho}$ = .395. While the results for the aggregate model (4.7) are somewhat better than the example above, with a mean of $\bar{\rho}$ = .252, there continues to be a secondary mode near ρ= .1. Hence while there is some clear improvement in this case, there is still a significant fraction of *negative* estimates $\hat{\rho}$ (more than 22%) even though the degree of positive spatial autocorrelation (ρ=.4) is considerable. Moreover, while the effective sample size (k = 43) is not overwhelming, it should certainly be adequate to obtain estimates in such a simple two-parameter model.

4.2.3 Asymptotic Properties of ρ Estimates

Before proceeding to a more detailed investigation of the possible causes of this undesirable behaviour, it is of interest to ask whether such behaviour persists in the limit. The examples above suggest that this might simply be a case where relatively large samples are required in order to achieve reasonable sampling distributions of the maximum likelihood estimates. However, the following extension of the example in Figure 4.1 provides an informative counterexample.

While asymptotic results for spatial models are somewhat more tenuous than for temporal models, it is nonetheless possible to imagine that a given spatial pat- tern can be extended to the infinite plane by some form of expansion scheme (designated as 'increasing-domain asymptotics' by Cressie, 1993). In this context, Mardia and Marshall (1984) developed a set of 'growth, convergence, and continuity' conditions (based on the more general results of Sweeting, 1980) for both consistency and asymptotic normality of maximum likelihood estimates. In principle, this result can be applied to establish a comparable consistency result for the present aggregated case, under appropriate conditions[5]. However it is far too general to allow any conclusions to be drawn about the finite-sample behaviour of such estimators[6].

But there is one case in which such analysis is possible. In particular, if we simply *replicate* a given system of regions an arbitrarily large number of times, and treat each replicate as an 'island' then the (block diagonal) covariance structure of this composite system not only satisfies all conditions for convergence, but actually allows the limiting behaviour of estimates to be studied. In this case, a set of N replicates can be viewed as a sequence of N independent random sample from the k-dimensional aggregate model in (4.7), so that all classical results for maximum-likelihood estimation

Figure 4.3a: Disaggregate Model Estimates of ρ (3x6)

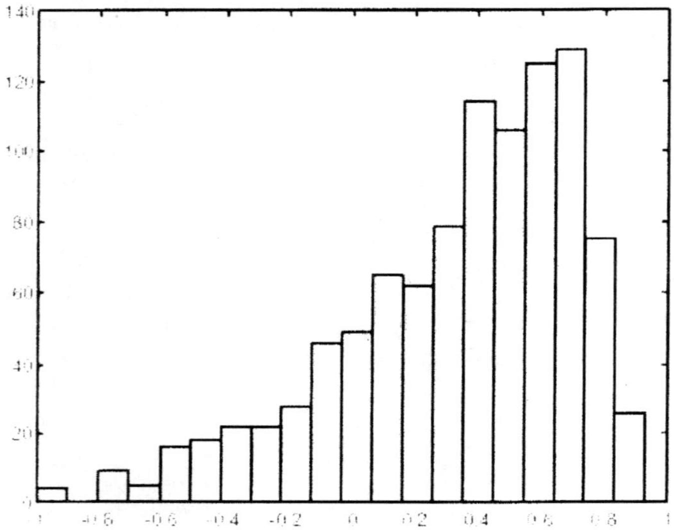

Figure 4.3b: Aggregate Model Estimates of ρ (3x6)

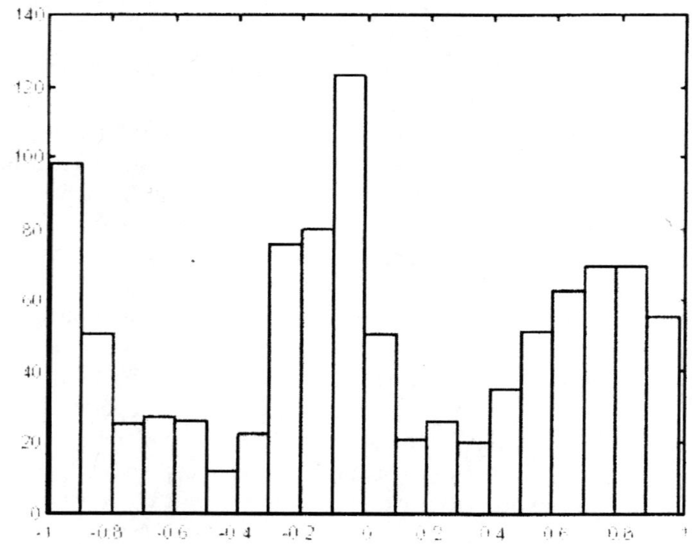

Figure 4.4a: Regions and Subregions (16x100)

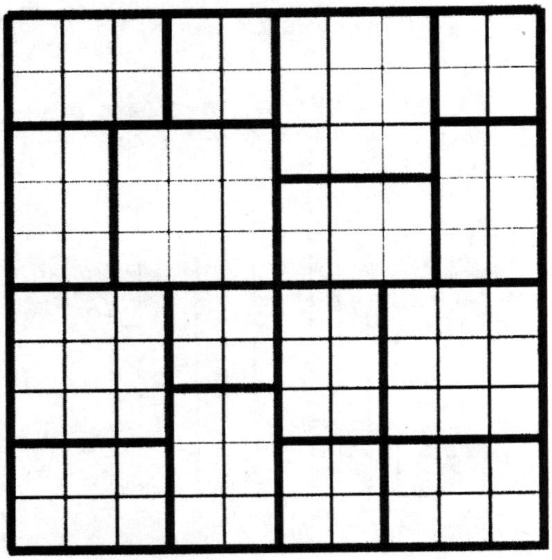

Figure 4.4b: Philadelphia Block Groups (43x312)

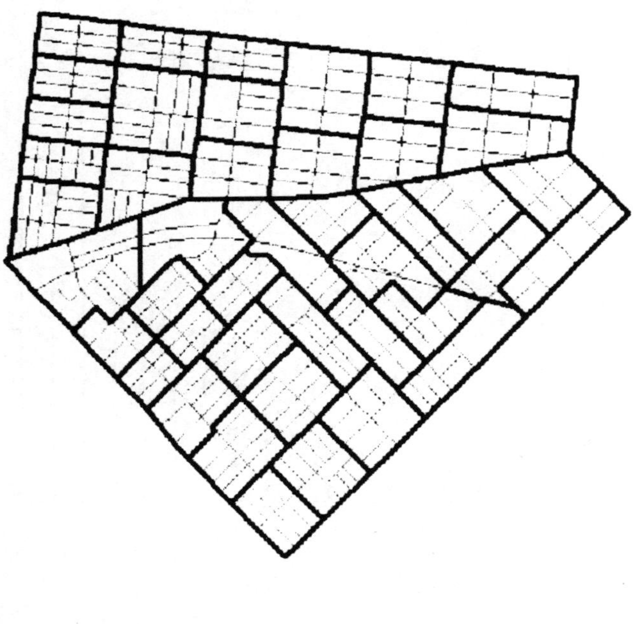

Figure 4.5a: Disaggregate Model Estimates of ρ (16x100)

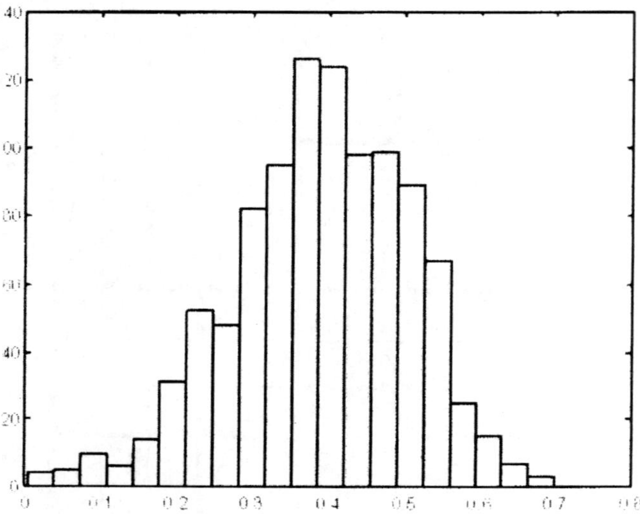

Figure 4.5b: Aggregate Model Estimates of ρ (16x100)

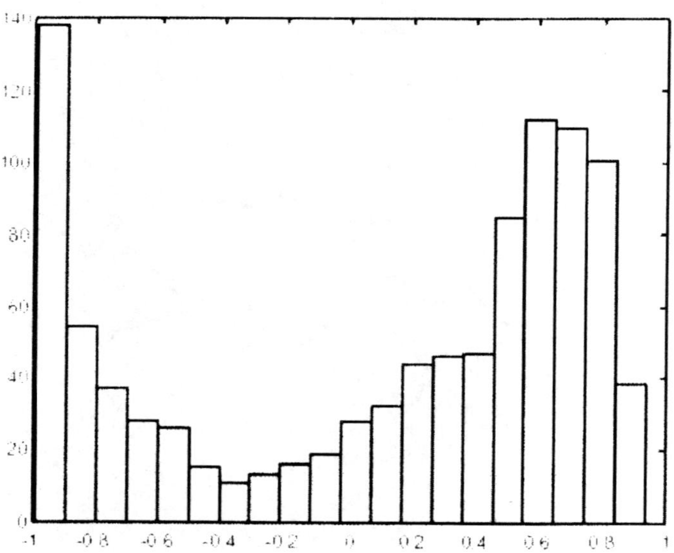

Figure 4.6a: Disaggregate Model Estimates of ρ (43x312)

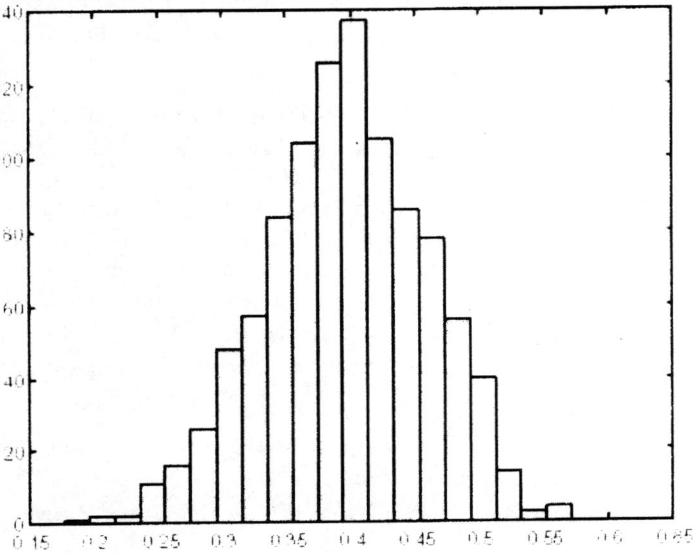

Figure 4.6b: Aggregate Model Estimates of ρ (43x312)

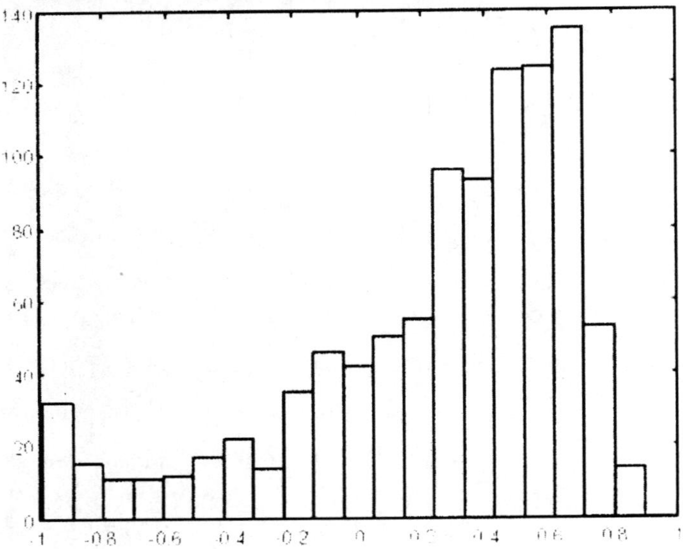

in the independence case can be applied. In particular, it follows from the results of Wald (1949) that for a random vector x distributed with density $f(x;\theta)$, if $\hat{\theta}_N$ denotes any choice of a global maximum of the associated likelihood function for N independent random samples, then the sequence $\hat{\theta}_N$ converges in probability (and in fact converges almost surely) to the true value of θ. Hence by applying this result to the sequence of N replicates with parameter $\theta = (\rho, \sigma^2)$, we could establish consistency of such estimators for this replicated case.

However, for our present purposes it is much more insightful to employ the 'extremum estimator' approach of Amemiya (1985). In particular, Amemiya shows (Theorem 4.1.1) that in our case if any positive monotone transformation of the concentrated likelihood function can be constructed which converges (uniformly in probability) to a nonstochastic function for which the true ρ value yields the unique global maximum, then the sequence of *maximal-root estimators*, ρ_N , will converge in probability to ρ. The advantage of this approach is that if the limiting nonstochastic objective function can be computed, then the qualitative nature of this convergence can be examined in some detail[7]. To do so, we begin by observing that if the true values of ρ and σ^2 are denoted respectively by ρ_0 and σ_0^2, and if $\mathbf{D_0} = (\mathbf{I_n} - \rho_0\mathbf{W})^{-1}$, then by (4.7) it follows that $x = \mathbf{AD_0}\varepsilon$. Hence for a single replication, the concentrated likelihood, $l_k(\cdot;x)$, in (4.11) is a random function of the form

$$
\begin{aligned}
l_k(.;\varepsilon) &= -\frac{1}{2}\ln(\varepsilon'\mathbf{D_0'}\mathbf{A'}\left(\mathbf{ADD'A'}\right)^{-1}\mathbf{A}\,\mathbf{D_0}\,\varepsilon) - \frac{1}{2k}\ln\left|\mathbf{ADD'A'}\right| \\
&= -\frac{1}{2}\ln\left[tr(\mathbf{D_0'}\mathbf{A'}\left(\mathbf{ADD'A'}\right)^{-1}\mathbf{A}\,\mathbf{D_0}\,\varepsilon\varepsilon')\right] - \frac{1}{2k}\ln\left|\mathbf{ADD'A'}\right|
\end{aligned} \tag{4.17}
$$

Next let us replicate this system N times, so that the $n\times n$ weight matrix $\mathbf{W}$ is replaced by the $Nn\times Nn$ block diagonal weight matrix $\langle\mathbf{W}\rangle_N$ with diagonal blocks all equal to $\mathbf{W}$ (indicating in particular that there are no spatial connections between replicates). The $k\times n$ aggregation matrix A also becomes an $Nk\times Nn$ block diagonal matrix $\langle\mathbf{A}\rangle_N$. It can then be easily verified that matrices $\mathbf{D_0'A'}\left(\mathbf{ADD'A'}\right)^{-1}\mathbf{A\,D_0}$ and $\mathbf{ADD'A}$ in (4.17) are both replaced by their block diagonal counterparts $\langle\mathbf{D_0'A'(ADD'A')^{-1}A\,D_0}\rangle_N$ and $\langle\mathbf{ADD'A}\rangle_N$. If for N independent samples ε_1, ..., ε_N from $N(0, \sigma_0^2\,\mathbf{I_n})$ we denote the $Nn\times 1$ stacked vector by $\varepsilon(N) = (\varepsilon_1', ..., \varepsilon_N')'$, then the resulting concentrated likelihood function takes the form

$$l_{Nk}(\cdot;\varepsilon_{(N)}) = -\frac{1}{2}\ln\left[tr\left(\left\langle \mathbf{D}_0'\mathbf{A}'(\mathbf{ADD'A'})^{-1}\mathbf{A}\,\mathbf{D}_0\right\rangle_N \varepsilon_{(N)}\varepsilon_{(N)}'\right)\right] - \frac{1}{2k}\ln\left|\left\langle \mathbf{ADD'A}\right\rangle_N\right|$$

$$= -\frac{1}{2}\ln\left[\sum_{i=1}^{N}\left(\mathbf{D}_0'\mathbf{A}'(\mathbf{ADD'A'})^{-1}\mathbf{A}\,\mathbf{D}_0\,\varepsilon_i\varepsilon_i'\right)\right] - \frac{1}{2Nk}N\ln\left|\mathbf{ADD'A}\right|$$

$$= -\frac{1}{2}\ln\left[\sum_{i=1}^{N}\left(\mathbf{D}_0'\mathbf{A}'(\mathbf{ADD'A'})^{-1}\mathbf{A}\,\mathbf{D}_0\left[\sum_{i=1}^{N}\varepsilon_i\varepsilon_i'\right]\right)\right] - \frac{1}{2k}\ln\left|\mathbf{ADD'A}\right| \qquad (4.18)$$

Hence if we define the new function, $z_N(\cdot,\ \varepsilon_{(N)})$ by

$$z_N(\rho,\varepsilon_{(N)}) = 2\cdot l_{Nk}(\rho,\varepsilon_{(N)}) + \frac{1}{2}\ln\left(\frac{N}{\sigma_0^2}\right)$$

$$= -\ln\left[\frac{1}{\sigma_0^2}tr(\mathbf{D}_0'\mathbf{A}'(\mathbf{ADD'A'})^{-1}\mathbf{A}\,\mathbf{D}_0\left[\frac{1}{N}\sum_{i=1}^{N}\varepsilon_i\varepsilon_i'\right])\right] - \frac{1}{k}\ln\left|\mathbf{ADD'A'}\right| \qquad (4.19)$$

then it follows from the first line of (4.19) that $z_N(\cdot,\ \varepsilon_{(N)})$ is a positive monotone transformation of $l_{Nk}(\cdot,\varepsilon_{(N)})$. Moreover, since it is well known that

$$\frac{1}{N}\sum_{i=1}^{N}\varepsilon_i\varepsilon_i' \xrightarrow[prob]{} \sigma_0^2\mathbf{I_n} \qquad (4.20)$$

it also follows that $z_N(\cdot,\varepsilon_{(N)})$ converges in probability to the nonstochastic function, $z(\cdot)$, defined by

$$z(\rho) = -ln\left[tr(\mathbf{D}_0'\mathbf{A}'(\mathbf{ADD'A'})^{-1}\mathbf{A}\,\mathbf{D}_0)\right] - \frac{1}{k}ln\left|\mathbf{ADD'A}\right| \qquad (4.21)$$

As is shown in Section 2 of the Appendix, $z(\cdot)$ is the desired limiting function.

It thus remains to examine the properties of this limiting function in specific cases. Here it is instructive to consider once again the simple example in Figure 4.1. In this case, the limiting function $z(\cdot)$ has the bimodal form shown in Figure 4.7a. More importantly, this example shows that while the global maximum is indeed at the true value ρ_0=.5, the secondary mode is *strongly negative*, ρ_{00}=.71, and has a *z*-value very close to the maximum value ($z(\rho_{00})$=1.0954 ≈ 1.0782=$z(\rho_0)$)). Hence it should be clear in this case that even for large numbers of replications, *N*, the maximal-root estimate, $\hat{\rho}_N$, may well be close to ρ_{00} rather than ρ_0. This is illustrated by the histogram of Figure 4.7b, which shows the estimation results for 1000 simulated samples of the replicated process with $N = 100$. As expected from the general consistency result above, at sample sizes this large ($k = 300$) the majority of maximal-root estimates $\hat{\rho}_N$ cluster around ρ_0. However, a substantial fraction (more that 12%) still cluster around the secondary mode ρ_{00}. Hence, even at these large sample sizes, the values $z(\rho_0)$ and $z(\rho_{00})$ are sufficiently close to ensure that the global maximum of the concentrated likelihood still has a significant chance of occurring near ρ_{00} rather than ρ_0[8].

It should also be noted that an additional consequence of this bimodal behaviour is the failure of standard significance tests based on asymptotic normality. So if one is unlucky enough to come up with strongly negative estimates of ρ in such situations where positive autocorrelation is expected, then standard significance tests will only serve to reinforce these negative findings[9].

Figure 4.7a: Limiting Likelihood Function (3x6)

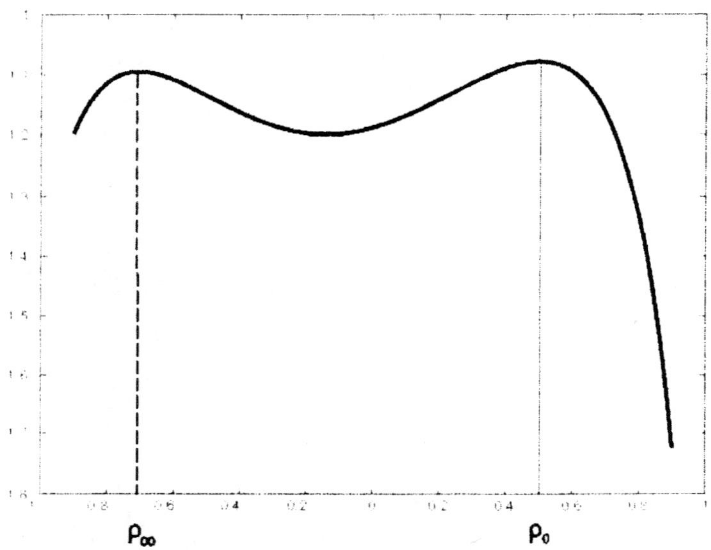

Figure 4.7b: Histogram for 100 Replications (3x6)

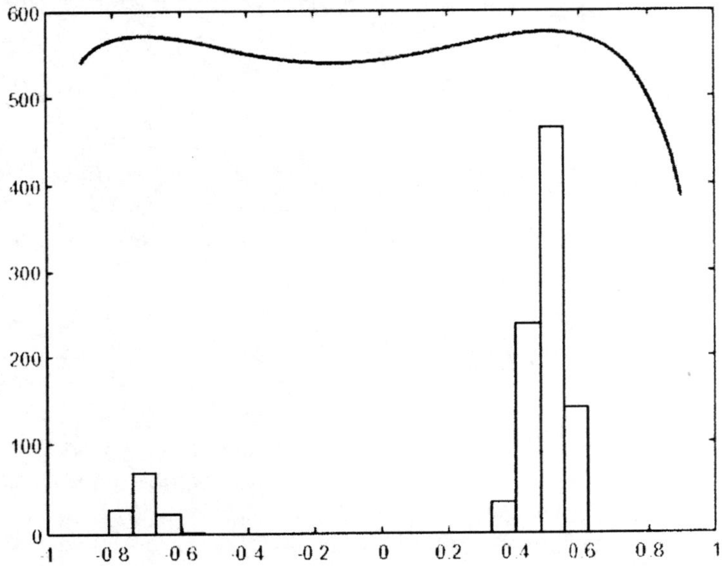

4.3 Analysis of Aggregation Bias

The negative results above raise the obvious questions: What is going on? What can we do about it? While there appear to be no definitive answers to these questions, some insight can be gained by studying the behaviour of the log-likelihood function in (4.9). Since our main focus is on ρ, it is of interest to consider the asymptotic behaviour of (4.9) with respect to ρ as the variance parameter, σ^2, be- comes large. To do so, observe that if for any aggregate data value, x, and distinct ρ-values, ρ_1 and ρ_2, we let $\mathbf{D}_i = (\mathbf{I}_n - \rho_i\mathbf{W})^{-1}$, i=1, 2, then

$$L_k(\rho_1, \sigma^2; x) - L_k(\rho_2, \sigma^2; x) = -\frac{1}{2}\ln\left|\mathbf{A}\,\mathbf{D}_1\mathbf{D}_1'\mathbf{A}'\right| + \frac{1}{2}\ln\left|\mathbf{A}\,\mathbf{D}_2\mathbf{D}_2'\mathbf{A}'\right|$$
$$-\frac{1}{2\sigma^2}x'\left[\left(\mathbf{A}\mathbf{D}_1\mathbf{D}_1'\mathbf{A}'\right)^{-1} - \left(\mathbf{A}\mathbf{D}_2\mathbf{D}_2'\mathbf{A}'\right)^{-1}\right]x$$

so that as σ^2 becomes large,

$$\lim_{\sigma^2\to\infty}\left[L_k(\rho_1, \sigma^2; x) - L_k(\rho_2, \sigma^2; x)\right] = -\frac{1}{2}\ln\left|\mathbf{A}\,\mathbf{D}_1\mathbf{D}_1'\mathbf{A}'\right| + \frac{1}{2}\ln\left|\mathbf{A}\,\mathbf{D}_2\mathbf{D}_2'\mathbf{A}'\right|$$

Hence we may conclude that the limiting form of L_k is a positive monotone function of the negative log determinant,

$$L(\rho) = -\ln|\mathbf{ADD'A'}| + \frac{1}{2}\ln|\mathbf{AD_2D_2'A'}| \tag{4.22}$$

which may be viewed as the *asymptotic log-likelihood of ρ under infinite dispersion.* Notice also that this asymptotic function is nonstochastic and hence can be analyzed independently of any *x*-data. For later use, we also note that the concentrated likelihood function in (4.11) can be written in terms of *L* as

$$l_k(\rho;x) = -\frac{1}{2}ln(x'(\mathbf{ADD'A'})^{-1}x) + \frac{1}{2k}L(\rho) \tag{4.23}$$

Given these observations, we next ask: how should this as asymptotic function behave? An examination of (4.1) suggests that as the dispersion of ε becomes arbitrarily large, any autocorrelation effects among the *y* value should eventually be overwhelmed. This conjecture is confirmed by analyzing *L* for the disaggregate model with $\mathbf{A} = \mathbf{I_n}$. In this case *L* reduces to

$$L(\rho) = -ln\,|\mathbf{DD'}| = -ln\,(|\mathbf{D}|^2) = const - ln\,|\mathbf{D}| \tag{4.24}$$

which is well known to be strictly concave[10]. Hence the unique maximum of *L* is given by the first-order condition in (4.14) (with $A = I_n$) as

$$0 = \frac{\partial L}{\partial \rho} = -2tr\left[(\mathbf{DD'})^{-1}\mathbf{DWDD'}\right] = -2tr(\mathbf{WD}) \tag{4.25}$$

But at ρ=0 we see that $\mathbf{D} = \mathbf{I_n} \Rightarrow tr(\mathbf{WD}) = tr(\mathbf{W}) = 0$[11], and hence may conclude that the unique solution to (4.25) is given by ρ=0. Thus for the disaggregate model we obtain the intuitively satisfying result that as dispersion of ε becomes arbitrarily large, the most likely value of ρ converges to zero.

However, for the aggregate model this is not the case. While a full analysis of this problem appears to be quite difficult, some insight can be gained by examining the derivative of *L* at $\rho = 0$ for the dissagregate model. Here again we see from (4.14) (with $\mathbf{D} = \mathbf{I_n}$) that

$$\left.\frac{\partial L}{\partial \rho}\right|_{\rho=0} = -2tr\left[\left(\mathbf{AA'}\right)^{-1}\mathbf{AWA'}\right] \tag{4.26}$$

But since **A** is nonnegative with orthogonal rows, it follows that **AA'** is a positive diagonal matrix, and hence that $(\mathbf{AA'})^{-1}$ is positive diagonal. This together with the nonnegativity of **W** implies that tr $[(\mathbf{AA})^{-1}\mathbf{AWA'}] \geq 0$, and hence that $(\partial L/\partial \rho)_{\rho=0} \leq 0$. Moreover (barring exceptional cases) this derivative is strictly negative. Hence there must always be at least a local maximum of L at negative values of ρ. While sharper results here are somewhat elusive, it can be shown (Section 3 of the Appendix) that if **W** is *symmetric* then L is strictly concave, and hence that the unique global maximizer of L is *always negative*. For *row normalizations* of symmetric weight matrices, it also appears that all maxima are achieved at negative ρ values[12].

What this means from a practical viewpoint is that at least for highly dispersed aggregate models, the asympotically most likely values of ρ tend to be *negative*. This leads naturally to the next question of why this should be true. Here again there seems to be no completely satisfactory answer. However, some insight can be gained by looking at the simplest possible case.

4.3.1 The Case of One Region and Two Subregions

At first glance the case of single region (k=1) divided into two subregions (n=2) would appear to be degenerate from a spatial viewpoint. With only one region in the aggregate model, there can be no *observable* spatial interaction. However, the model is still influenced by the *unobservable* interaction between subregions. Hence this case serves to emphasize the effects of this unobservable spatial component. One additional advantage here is that the regional covariance matrix reduces to simple variance, which is more readily interpretable.

To formalize this case, let the weight matrix **W** be given by

$$\mathbf{W} = \begin{pmatrix} 0 & 1 \\ 1 & 0 \end{pmatrix} \tag{4.27}$$

so that **B** and **D** have the respective forms

$$\mathbf{B} = \begin{pmatrix} 1 & -\rho \\ -\rho & 1 \end{pmatrix} \qquad \mathbf{D} = \frac{1}{1-\rho^2}\begin{pmatrix} 1 & \rho \\ \rho & 1 \end{pmatrix} \tag{4.28}$$

Next let the aggregation matrix be an unspecifed positive linear combination **A** $= a' = (a_1, a_2)$, so that for normally distributed $\varepsilon = (\varepsilon_1, \varepsilon_2)'$ in (4.2), the single aggregate regional variate x has the form

$$x = a'\mathbf{D}\varepsilon = \frac{1}{1-\rho^2}\left[(a_1 + \rho a_2)\varepsilon_1 + (a_1\rho + a_2)\varepsilon_2\right] \tag{4.29}$$

with corresponding variance given by

$$\begin{aligned} \mathrm{var}(x) &= \sigma^2 a'\mathbf{DD}a \\ &= \sigma^2 \frac{(a_1 + \rho a_2)^2 + (a_1\rho + a_2)^2}{\left(1-\rho^2\right)^2} \end{aligned} \tag{4.30}$$

Note that influence of ρ on the variance of x is summarized by the value, a'**D'D**a which may here be designated as the *relative variance* of x (since it defines variance up to a scalar multiple). With this terminology it is clear that the asymptotic log-likelihood $L(\rho)$ in (4.22) depends entirely on this relative variance:

$$\begin{aligned} L(\rho) &= -\ln\left[a'\mathbf{DD}'a\right] \\ &= const. - \ln\left[\frac{1}{\left(1-\rho^2\right)^2}\right] - \ln\left[(a_1 + \rho a_2)^2 + (a_1\rho + a_2)^2\right] \end{aligned} \tag{4.31}$$

Hence the value of ρ which is asymptotically most likely (as dispersion of ε becomes large) is precisely that value that *minimizes the relative variance of x*. More generally, if the determinant |**ADD'A'**| in (4.8) is designated as the *relative generalized variance of x*[13], then this same interpretation holds for (4.22) as well.

In this context, it should be clear from (4.30) that determining the minimizing value of ρ is complex even in this simple case. But here one can gain qualitative insight by observing from (4.24) that for the *disaggregate model*, the asymptotic log-likelihood function L becomes

$$L(\rho) = const. - \ln|\mathbf{D}|$$

$$= const. - \ln\left[\frac{1}{\left(1-\rho^2\right)^2}\right] \tag{4.32}$$

Hence this disaggregate likelihood is seen to be essentially the first term of (4.31), which corresponds to the denominator of relative variance in (4.30). It is thus reasonable in this case to interpret the denominator of relative variance (first term of L) as the 'disaggregate' effect of ρ on relative variance, arising from interaction between the individual subregions. The numerator of relative variance (second term of L) then represents the additional 'aggregate' effect of ρ arising from regional aggregation. Moreover, since the terms $(a_1+\rho a_2)^2$ and $(a_1\rho+a_2)^2$ in (4.30) are essentially the contributions to relative variance of the two aggregate variates $(a_1+\rho a_2)\varepsilon_1$ and $(a_1\rho+a_2)\varepsilon_2$ in (4.29), it is appropriate to designate this numerator as *aggregation variance*,

$$\upsilon(\rho) = (a_1+\rho a_2)^2 + (a_1\rho+a_2)^2 \tag{4.33}$$

By observing that the denominator of (4.30) achieves its maximum at $\rho=0$ (yielding minimum relative variance for the disaggregate case), it is clear that the key effect of aggregation on relative variance is in terms of $\upsilon(\rho)$. Moreover by differentiating $\upsilon(\rho)$, this aggregation variance is seen to be minimized at

$$\rho_\upsilon = \frac{2a_1a_2}{a_1^2+a_2^2} < 0 \tag{4.34}$$

This implies that the value of ρ minimizing relative variance in (4.30) must lie between the extremes, $\rho=\rho_\upsilon$ and $\rho=0$, and hence must be *negative*.

Thus in this simple case, the effect of aggregation on minimum relative variance is clear: the positivity of a_1 and a_2 imply that any positive correlation effect, ρ, in (4.33) must necessarily *increase* aggregation variance, and hence that relative variance can only be minimized at *negative* ρ values. This is of course completely analogous to standard 'variance reduction' techniques in sample design, where one reduces the variance of sums by creating negatively correlated samples. In the present case, the relevant 'sums' are simply the aggregated data. While this interpretation is less clear in the general case, where variance must be replaced by 'generalized variance' (as in endnote 13 above), it nonetheless appears that this type of variance reduction effect is at the root of the aggregation-bias problem in the present case. The maximum-likelihood values of ρ tend to be those which yield smaller relative (generalized) variances, and in the presence of aggregation, these values tend to be negative.

4.3.2 A Possible Bias-Correction Procedure

While there appears to be no quick fix for this type of aggregation bias, the above observations suggest at least one possible approach. For if one focuses on the asymptotic case when the dispersion of ε becomes large, then it is reasonable to require that (as in the bias-free disaggregate case) the aggregate maximum- likelihood estimate of ρ approach zero. As observed above, this is equivalent to requiring that relative generalized variance, |**ADD'A'**|, achieve its minimum at $\rho = 0$. There are many ways to implement such a modification. For example, by simply subtracting the derivative, $(\partial L/\partial\rho)_{\rho=0}$, of the asymptotic log-likelihood at zero (expression 4.26) from $L_k(\rho,\sigma^2;x)$, one produces a modified log-likelihood function

$$L_k^0(\rho,\sigma^2;x) = -\frac{k}{2}\ln(\sigma^2) - \frac{1}{2\sigma^2}x'\left(\mathbf{ADD'A'}\right)^{-1}x \\ -\frac{1}{2}\ln\left|\mathbf{ADD'A'}\right| + tr\left[\left(\mathbf{AA'}\right)^{-1}\mathbf{AWA'}\right] \tag{4.35}$$

for which the associated asymptotic log-likelihood must have zero derivative at $\rho = 0$.

However, after experimenting with numerous variations on this theme, the most successful approach found is simply to rescale ρ in the asymptotic log- likelihood, $L(\rho)$, in a manner which achieves the same effect. In particular, if the boundary values of ρ are denoted by $\rho_{min}=1/\lambda_{min}$ and $\rho_{max}=1/\lambda_{max}$, and if ρ^* denotes the point in (ρ_{min}, ρ_{max}) where L achieves its maximum, then the appropriate rescaling of ρ amounts to a piecewise linear transformation, $\rho \rightarrow \tilde{\rho}$ which moves ρ^*. to the origin while leaving the boundaries fixed, i.e.,

$$\tilde{\rho} = \begin{cases} \rho + \left(1 - \dfrac{\rho}{\rho_{\min}}\right)\cdot\rho^* & ,\rho_{\min} < \rho \le 0 \\ \rho + \left(1 - \dfrac{\rho}{\rho_{\max}}\right)\cdot\rho^* & ,0 < \rho < \rho_{\max} \end{cases} \tag{4.36}$$

If we denote the resulting transformation of L by $\tilde{L}(\rho) = L(\tilde{\rho})$, then the corresponding modification of the concentrated likelihood function in (4.24) now given by

$$\tilde{l}_k(\rho; x) = -\frac{1}{2}\ln(x'(\mathbf{ADD'A'})^{-1}x) + \frac{1}{2k}\tilde{L}(\rho) \qquad (4.37)$$

In the present case, the relevant values (ρ_{min} = .1.16, ρ^{*}. = ..226, ρ_{max} = 1) yield the transformed asymptotic log-likelihood, $\tilde{L}$, shown in Figure 4.8a, where the dotted curve represents the corrected values (achieving a maximum at $\rho = 0$).

The limiting likelihood function, $\tilde{\ell}$, under this rescaling is shown in Figure 4.8b. Notice first that while there are still two modes, the secondary mode in the negative range of ρ has now diminished markedly. As a consequence, the histogram of 1000 simulated estimates of ρ shown in Figure 4.9a no longer exhibits a second mode, yielding a dramatic improvement over Figure 4.7a. Figure 4.9b shows a similar result for the example in Figure 4.5 above.

However, one should hasten to add that this correction is meaningful only for 'small' samples, and is surely not consistent. In the present case, the primary mode is at $\rho_1 = .52 > .5 = \rho_0$, so that estimates will eventually exhibit a small upward bias. The is already seen in Figure 4.9a, where the sample mean is .513. Moreover, while this bias is small in the present case, this need not be true in general. In the present case, the table below shows the ρ_1 values calculated for a selection of nonnegative ρ_0 values. As ρ_0 approaches zero, bias clearly increases (and is even worse for negative ρ_0).

ρ_0	0	.1	.2	.3	.4	.5	.6	.7	.8
ρ_1	.278	.299	.331	.378	.440	.517	.605	.700	.800

Moreover, while the higher values look promising, the secondary mode in these cases is more severe (indicating simply that corrections of the likelihood function are relatively slight for large ρ_0). So if ρ_0 is close to zero in the present example (with σ=1), then this correction procedure will reject the null hypothesis much too often.

Hence a better correction procedure (assuming $\rho_0 \geq 0$) might involve some compromise between l_k and $\tilde{l}_k$ In particular, for any convex combination of these functions, $\alpha_k \tilde{l}_k + (1-\alpha_k) l_k$, with $lim_{k \to \infty} \alpha_k = 0$, it is clear that the maximum-root estimate of ρ must necessarily be consistent. The task here is to find a choice of α_k's which avoids the serious bimodality properties of l_k while at the same time minimizing the positive bias introduced by $\tilde{l}_k$. Such possibilities will be considered in subsequent work.

Figure 4.8a: Corrected Asymptotic Log-Likelihood (3x6)

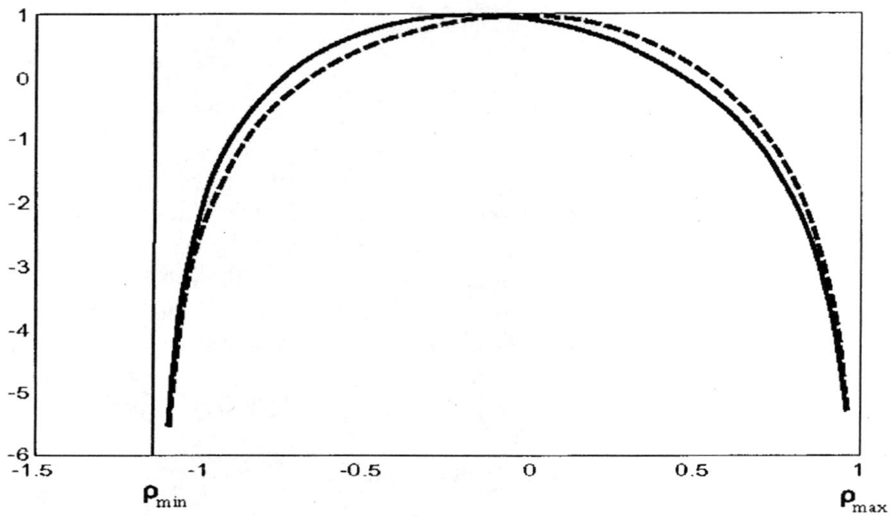

Figure 4.8b: Corrected Limit Likelihood (3x6)

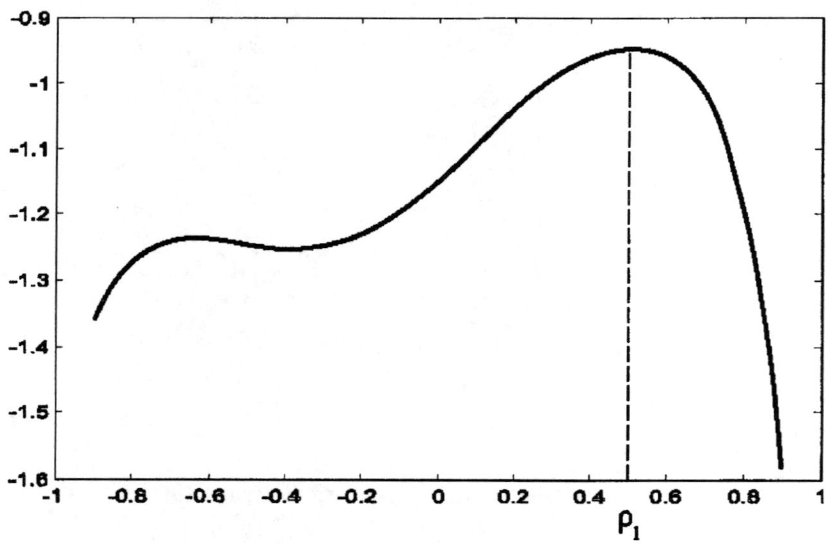

Figure 4.9a: Histogram of Corrected Estimates (3x6, N=100)

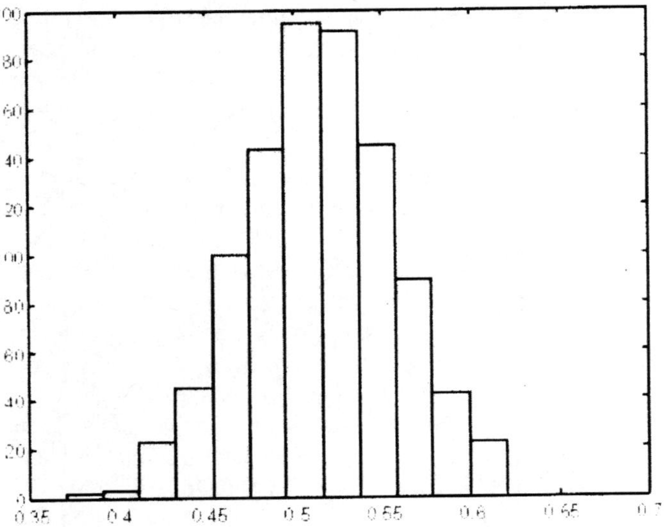

Figure 4.9b: Histogram of Corrected Estimates (16x100)

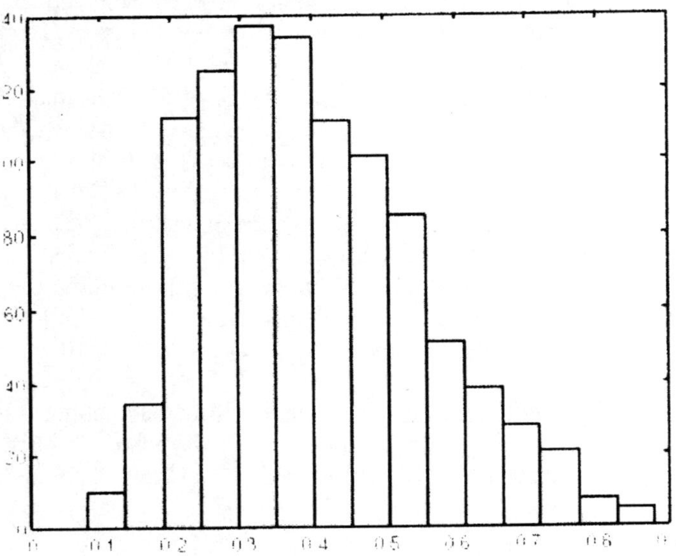

4.4 Conclusion

The central task of this chapter has been to illuminate the difficulties of estimating parameters in the presence of 'misalligned' data. Here we have focused on the simple case of a pure spatial autoregressive process in which the observable data is at a higher level of aggregation than the actual process itself, as represented by the weight-matrix data. In this setting, the resulting sampling distributions of maximum likelihood estimates for selected examples were shown to exhibit serious bias (even for large sample sizes), and a possible correction procedure was proposed. This procedure is admittedly *ad hoc*, and clearly represents only a first pass at the problem.

However, it can be argued that in the simple context of spatial autoregressive processes, there is perhaps no need for such correction procedures at all. Given model (4.1),(4.2),(4.6) for a specific weight matrix, **W**, and aggregation matrix, **A**, one can in principle simulate the sampling distribution of ρ estimates for a selected range of ρ_0 values and simply observe how they behave. For example, in the case illustrated by Figure 4.1 above, the sampling distribution (for 1000 simulations) under null hypothesis, $\rho_0=0$, is shown in Figure 4.10. Hence it is possible to test the null hypothesis of 'no spatial interaction' directly in terms of this sampling distribution. The fact that the sample mean is significantly negative (about -.10 in this case) is of no consequence. One can still test whether a given estimate is 'significantly big' with respect to this distribution, and hence test for the presence of spatial autocorrelation even though the estimates are themselves very biased[14].

But in the more important cases of multiple regression models with spatial autoregressive errors or spatial lags, the situation is far more complex. Here the estimate of ρ is mainly of interest as an intermediate step in estimating and testing the significance of the key β parameters. Hence the *value* of the ρ estimate is crucial for estimating the primary β parameters. An even more fundamental issue in these models is whether consistent estimation of parameters is even possible. In the case of spatial lag models for example, even if explanatory variables are observed at the same level of aggregation as the dependent variable, there exists no simple reduced form such as (4.7) above. Hence the possibility of consistent estimation in such cases is questionable. In these more complex models it may thus only be meaningful to introduce micro spatial lags in combination with other micro data that allow the possibility of consistent estimation (such as the 'auxiliary variable' techniques of Holt, Steel, Tranmer and Wrigley, 1996, and others)[15].

Alternatively, one might consider the general Bayesian approach to misaligned data recently proposed by Zhu, Gelfand, and Carlin (2000) in which micro spatial lags are treated simply as another type of data misalignment. Such possibilities will be explored in subsequent work.

Figure 4.10: Histogram for $\rho_0 = 0$ (100 replications of 3x6)

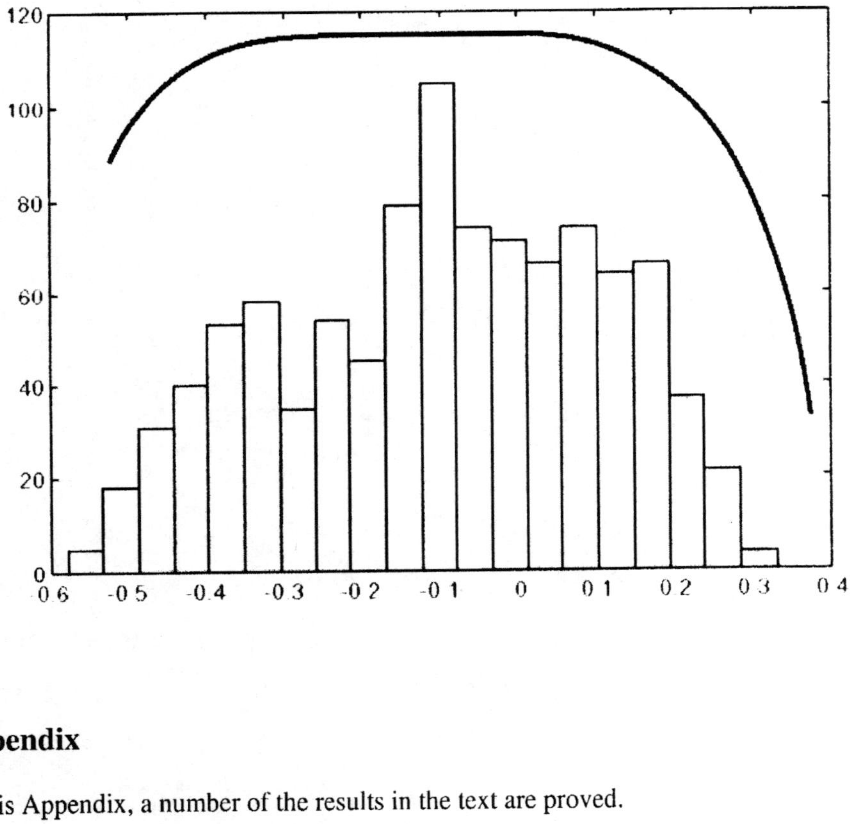

Appendix

In this Appendix, a number of the results in the text are proved.

A.1 Uniqueness of MLEs for the Disaggregate Case

In view of the central role plays by multiple roots of the likelihood function in aggregate case, it is appropriate to document uniqueness of roots for the standard disaggregate case (see also Ripley, 1988, section 2.2). To do so, it suffices to show that the second derivative of l_k with respect to ρ is always negative whenever the first derivative is zero. We begin by differentiating (4.16) to obtain

$$\frac{\partial^2}{\partial \rho^2} l_k = 2\left(\frac{y'\mathbf{B}'\mathbf{W}y}{y'\mathbf{B}'\mathbf{B}y}\right)^2 - \frac{y'\mathbf{W}'\mathbf{W}y}{y'\mathbf{B}'\mathbf{B}y} - \frac{1}{k} tr(\mathbf{WDWD}) = A_1 + A_2 \quad \text{(A.1)}$$

where

$$A_1 = \left(\frac{y'\mathbf{B}'\mathbf{W}y}{y'\mathbf{B}'\mathbf{B}y}\right)^2 - \frac{y'\mathbf{W}'\mathbf{W}y}{y'\mathbf{B}'\mathbf{B}y} \quad \text{(A.2)}$$

and

$$A_2 = \left(\frac{y'\mathbf{B}'\mathbf{W}y}{y'\mathbf{B}'\mathbf{B}y}\right)^2 - \frac{1}{k} tr(\mathbf{WDWD}) \quad \text{(A.3)}$$

Turning first to A_1, observe that if $x=\mathbf{B}y$ and $w=\mathbf{W}y$ then

$$A_1 = \frac{(x'w)^2}{\|x\|^4} - \frac{\|w\|^2}{\|x\|^2} = \frac{\|w\|^2}{\|x\|^2}\left(\frac{(x'w)^2}{\|x\|^2\|w\|^2} - 1\right) \leq 0 \quad \text{(A.4)}$$

by the Cauchy-Schwartz Inequality. Moreover, if $\partial l_k / \partial\rho = 0$, then by (4.16) it follows that

$$\frac{y'\mathbf{B}'\mathbf{W}y}{y'\mathbf{B}'\mathbf{B}y} = \frac{1}{k} tr(\mathbf{WD}) \quad \text{(A.5)}$$

and hence from (A.3) that

$$A_2 = \left(\frac{1}{k} tr(\mathbf{WD})\right)^2 - \frac{1}{k} tr(\mathbf{WDWD}) = \frac{1}{k^2}\left[tr(\mathbf{WD})^2 - k\,tr(\mathbf{WDWD})\right] \quad \text{(A.6)}$$

But since the spectrum, $\sigma(\mathbf{W}) = (\lambda_1, \ldots, \lambda_n)$, of $\mathbf{W}$ is real, and since the corresponding spectrum of $\mathbf{WD}$, say $\sigma(\mathbf{WD}) = (\omega_1, \ldots, \omega_n)$, is easily seen to be given by

$$\omega_i = \frac{\lambda_i}{1-\rho\lambda_i}, i = 1, \ldots, n$$

it follows that $\lambda(\mathbf{WD})$ is also real. Hence, observing by definition that $\lambda(\mathbf{WDWD}) = (\varpi_1^2,\ldots,\varpi_n^2)$, and recalling the trace of a matrix is the sum of its spectrum, it follows (from an application of Hölder's Inequality) that

$$A_2 = \frac{1}{k^2}\left[(\Sigma_i \varpi_i)^2 - k\Sigma_i \varpi_i^2\right] \le 0 \tag{A.7}$$

with strict inequality holding unless $\omega_1 = \cdots = \omega_n$. To see that the latter case is not possible, observe that since our condition, $\rho \in (1/\lambda_{min}, 1/\lambda_{max})$, implies that $1-\rho\lambda_i > 0$ for all $i = 1, .., n$, it follows that each ω_i has the same sign as λ_i. Finally since $\lambda_{max} > 0$ for every nonnegative (nonzero) matrix and since $w_{ii} \equiv 0$ implies that $0 = \text{tr}(\mathbf{W}) = \Sigma_i\lambda_i$, we must also have $\lambda_{min} < 0$, and may conclude that $\omega_{min} < 0 < \omega_{max}$.

A.2 Limiting Objective Function

To show verify that $z(\cdot)$ is the desired limit function, it must be shown that (i) the sequence of random functions $(z_N(\rho,\varepsilon_{(N)}))$ in (4.19) exhibit appropriate uniform probabilistic convergence to $z(\cdot)$, and that (ii) the unique global maximum of this function is at $\rho = \rho_0$. To establish (i), we begin by observing that while $z(\cdot)$ diverges at the boundaries of the parameter space $(1/\lambda_{min}, 1/\lambda_{max})$, it is easily seen to be uniformly continuous on every closed interval $\Gamma \subset (1/\lambda_{min}, 1/\lambda_{max})$. Hence it is enough to establish uniform probabilistic convergence on every closed interval $\Gamma \subset (1/\lambda_{min}, 1/\lambda_{max})$ containing ρ_0. To do so, let the random function $h(\cdot,\varepsilon)$ be defined on each Γ by

$$h(\rho,\varepsilon) = tr\left[\mathbf{D}_0'\mathbf{A}'(\mathbf{ADD'A'})^{-1}\mathbf{A}\,\mathbf{D}_0\,\varepsilon\varepsilon'\right] \tag{A.8}$$

and observe that

$$\begin{aligned} E[h(\rho,\varepsilon)] &= tr\left[\mathbf{D}_0'\mathbf{A}'(\mathbf{ADD'A'})^{-1}\mathbf{A}\,\mathbf{D}_0\,E(\varepsilon\varepsilon')\right] \\ &= tr\left[\mathbf{D}_0'\mathbf{A}'(\mathbf{ADD'A'})^{-1}\mathbf{A}\,\mathbf{D}_0(\sigma_0^2\mathbf{I_n})\right] \\ &= \sigma_0^2\, tr\left[\mathbf{D}_0'\mathbf{A}'(\mathbf{ADD'A'})^{-1}\mathbf{A}\,\mathbf{D}_0\right] \end{aligned} \tag{A.9}$$

Hence letting the nonstochastic mean-value function $h(\cdot)$ be defined by the right hand side, i.e., by

$$h(\rho) = \sigma_0^2 tr\left[\mathbf{D}_0' \mathbf{A}'(\mathbf{ADD'A'})^{-1}\mathbf{A}\,\mathbf{D}_0\right] \tag{A.10}$$

it follows easily that the difference function

$$g(\rho, \varepsilon) = h(\rho, \varepsilon) - h(\rho) \tag{A.11}$$

has zero mean for all $\pi \in \Gamma$, and is continuous in both ρ and ε. In addition if we write the matrix $\mathbf{D}_0' \mathbf{A}'(\mathbf{ADD'A'})^{-1}\mathbf{A}\,\mathbf{D}_0$ as $\mathbf{M}(\rho) = [m_{ij}(\rho)]$, then

$$h(\rho,\varepsilon) = tr[\mathbf{M}(\rho)\varepsilon\varepsilon'] = tr[\varepsilon'\mathbf{M}(\rho)\varepsilon] = \sum\nolimits_{i,j} \varepsilon_i m_{ij}(\rho)\varepsilon_j$$

$$\Rightarrow |h(\rho,\varepsilon)| \le \left|\sum\nolimits_{i,j} \varepsilon_i m_{ij}(\rho)\varepsilon_j\right| \le \sum\nolimits_{i,j} |\varepsilon_i\varepsilon_j||m_{ij}(\rho)|$$

$$\Rightarrow \sup_{\rho\in\Gamma}|h(\rho,\varepsilon)| \le \sum\nolimits_{i,j} |\varepsilon_i\varepsilon_j| \sup_{\rho\in\Gamma}|m_{ij}(\rho)| = \sum\nolimits_{i,j} |\varepsilon_i\varepsilon_j|\, m_{ij} \tag{A.12}$$

where $m_{ij} = sup_{\rho\in\Gamma}|m_{ij}(\rho)| < \infty$ for all i, j = 1, ..., n. Hence the finiteness of $E(|\varepsilon_i\varepsilon_j|)$ for all i, j = 1, ..., n implies that

$$E\left(\sup_{\rho\in\Gamma}|h(\rho,\varepsilon)|\right) = \Sigma_{i,j}\, m_{ij}\, E\left(|\varepsilon_i\varepsilon_j|\right) < \infty \tag{A.13}$$

and the function g is seen to satisfy all conditions of Theorem 4.2.1 in Amemiya (1985) for the *iid* sequence of random vectors (ε_i). Thus $\frac{1}{N}\sum_{i=1}^{N} g(\cdot,\varepsilon_i)$ converges uniformly in probability to zero on Γ, implying from (A.11) that $\frac{1}{N}\sum_{i=1}^{N} h(\cdot,\varepsilon_i)$ converges uniformly in probability to $h(\cdot)$ on Γ. Finally since (4.19) shows that $z_N(\cdot, \varepsilon_{(N)})$ can written as

$$z_N(\cdot,\varepsilon_{(N)}) = -ln\left[\frac{1}{\sigma_0^2}\left\{\frac{1}{N}\Sigma_{i=1}^{N} h(\cdot,\varepsilon_i)\right\}\right] - \frac{1}{k}\, ln|\mathbf{ADD'A'}| \tag{A.14}$$

and hence is uniformly continuous in both ρ and $\frac{1}{N}\sum_{i=1}^{N} h(\cdot,\varepsilon_i)$, it follows that uniform probabilistic convergence of $\frac{1}{N}\sum_{i=1}^{N} h(\cdot,\varepsilon_i)$ to $h(\cdot)$ implies uniform probabilistic convergence of $z_N(\cdot,\ \varepsilon_{(N)})$ to the function

$$z(\cdot) = -ln\left[\frac{1}{\sigma_0^2}h(\cdot)\right] - \frac{1}{k}ln|\mathbf{ADD'A'}| \tag{A.15}$$

which is precisely (4.21).

Establishing (*ii*) is somewhat more delicate (as should be clear from Figure 4.7a, which shows that other local maxima not only exist, but can also be very close in value to the global maximum). The following argument starts with the full log-likelihood function in (4.9) and makes use of the fundamental 'information inequality' which asserts that for any parameter pair (ρ,σ^2) distinct from the true values $(\rho_0,\ \sigma_0^2)$, if the functions $L_k(\rho,\sigma^2;\ \cdot)$ and $L_k(\rho_0,\ \sigma_0^2;\ \cdot)$ differ on a set of positive probability measure, then

$$E[L_k(\rho,\sigma^2;x)] < E[L_k(\rho_0,\sigma_0^2;x)] \tag{A.16}$$

where expectation is taken with respect to x under the true parameter values $(\rho_0,\ \sigma_0^2)$[16]. In the present case, it is enough to require that for all $(\rho,\sigma^2)\neq(\rho_0,\ \sigma_0^2)$, the covariance matrices, $\sigma^2\mathbf{ADD'A'}$ and $\sigma_0^2\mathbf{A\,D_0\,D_0'\,A'}$ be distinct[17]. To evaluate the left hand side of (A.16) we employ the arguments in (4.17) and (A.9) to obtain,

$$E[L_k(\rho,\sigma^2;x)] = E[L_k(\rho,\sigma^2;\mathbf{A\,D_0}\,\varepsilon)] =$$

$$= -\frac{k}{2}ln(\sigma^2) - \frac{1}{2}ln|\mathbf{ADD'A'}| - \frac{1}{2\sigma^2}tr[\mathbf{D_0'}\,\mathbf{A'}(\mathbf{ADD'A'})^{-1}\mathbf{A\,D_0}\,E(\varepsilon\varepsilon')]$$

$$= -\frac{k}{2}ln(\sigma^2) - \frac{1}{2}ln|\mathbf{ADD'A'}| - \frac{\sigma_0^2}{2\sigma^2}tr[\mathbf{D_0'}\,\mathbf{A'}(\mathbf{ADD'A'})^{-1}\mathbf{A\,D_0}] \tag{A.17}$$

where the constant term has been dropped for convenience. Next we maximize (A.17) with respect to σ^2 for each fixed value of ρ to obtain the unique value

$$\sigma_\rho^2 = \frac{\sigma_0^2}{k} tr\left[\mathbf{D}_0' \mathbf{A}'(\mathbf{ADD'A'})^{-1} \mathbf{A} \mathbf{D}_0\right] \tag{A.18}$$

Since (A.16) holds for all $(\rho,\sigma^2) \neq (\rho_0, \sigma_0^2)$, it follows in particular that for each $\rho \neq \rho_0$,

$$E\left[L_k\left(\rho, \sigma_\rho^2; \mathbf{A}\,\mathbf{D}_0\,\varepsilon\right)\right] < E\left[L_k\left(\rho_0, \sigma_0^2; \mathbf{A}\,\mathbf{D}_0\,\varepsilon\right)\right] \tag{A.19}$$

To evaluate the right hand side of (A.19) we next observe that $(\rho,\sigma^2) = (\rho_0, \sigma_0^2)$ implies $\mathbf{D}=\mathbf{D}_0$, so that

$$tr\left[\mathbf{D}_0' \mathbf{A}'\left(\mathbf{A}\,\mathbf{D}_0\mathbf{D}_0'\mathbf{A}'\right)^{-1} \mathbf{A}\,\mathbf{D}_0\right] = tr\left[\mathbf{A}\,\mathbf{D}_0\,\mathbf{D}_0' \mathbf{A}'\left(\mathbf{A}\,\mathbf{D}_0\mathbf{D}_0'\mathbf{A}'\right)^{-1}\right] = tr(\,\mathbf{I}_k\,) = k \tag{A.20}$$

Hence for these true values, we have

$$E\left[L_k\left(\rho_0, \sigma_0^2; \mathbf{A}\,\mathbf{D}_0\,\boldsymbol{\varepsilon}\right)\right] = -\frac{k}{2} ln\left(\sigma_0^2\right) - ln\left|\mathbf{A}\,\mathbf{D}_0\,\mathbf{D}_0'\,\mathbf{A}'\right| - \frac{k}{2} \tag{A.21}$$

and may use (A.17),(A.18),(A.20), and (A.21) to rewrite (A.19) as follows

$$-\frac{k}{2} ln\left(\frac{\sigma_0^2}{k} tr\left[\mathbf{D}_0' \mathbf{A}'(\mathbf{ADD'A'})^{-1} \mathbf{A}\,\mathbf{D}_0\right]\right) - \frac{1}{2} ln\left|\mathbf{ADD'A'}\right| - \frac{k}{2}$$

$$< -\frac{k}{2} ln(\,\sigma_0^2\,) - \frac{1}{2} ln\left|\mathbf{A}\,\mathbf{D}_0\,\mathbf{D}_0'\,\mathbf{A}'\right| - \frac{k}{2}$$

$$\Rightarrow ln(\,\sigma_0^2\,) - \frac{k}{2} ln\left(\frac{1}{k} tr\left[\mathbf{D}_0' \mathbf{A}'(\mathbf{ADD'A'})^{-1} \mathbf{A}\,\mathbf{D}_0\right]\right) - \frac{1}{2} ln\left|\mathbf{ADD'A'}\right|$$

$$< -\frac{k}{2} ln(\,\sigma_0^2\,) - \frac{1}{2} ln\left|\mathbf{A}\,\mathbf{D}_0\,\mathbf{D}_0'\,\mathbf{A}'\right|$$

$$\Rightarrow \frac{k}{2} ln(\,k\,) - \frac{k}{2} ln\left(\frac{1}{k} tr\left[\mathbf{D}_0' \mathbf{A}'(\mathbf{ADD'A'})^{-1} \mathbf{A}\,\mathbf{D}_0\right]\right) - \frac{1}{2} ln\left|\mathbf{ADD'A'}\right|$$

$$< -\frac{1}{2} ln\left|\mathbf{A}\,\mathbf{D}_0\,\mathbf{D}_0'\,\mathbf{A}'\right|$$

$$\Rightarrow -\frac{k}{2} ln\left(tr\left[\mathbf{D}_0' \mathbf{A}'(\mathbf{ADD'A'})^{-1} \mathbf{A}\,\mathbf{D}_0\right]\right) - \frac{1}{2} ln\left|\mathbf{ADD'A'}\right.$$

$$< -\frac{k}{2} ln(k) - \frac{1}{2} ln|\mathbf{A}\,\mathbf{D_0}\,\mathbf{D_0'}\,\mathbf{A'}|$$

$$\Rightarrow -\frac{1}{2} ln\left(tr\left[\mathbf{D_0'}\,\mathbf{A'}(\mathbf{ADD'A'})^{-1}\,\mathbf{A}\,\mathbf{D_0}\right]\right) - \frac{1}{2k} ln|\mathbf{ADD'A'}|$$

$$< -\frac{1}{2} ln\left(tr\left[\mathbf{D_0'}\,\mathbf{A'}\left(\mathbf{A}\,\mathbf{D_0}\mathbf{D_0'}\mathbf{A'}\right)^{-1}\mathbf{A}\,\mathbf{D_0}\right]\right) - \frac{1}{2k} ln|\mathbf{A}\,\mathbf{D_0}\,\mathbf{D_0'}\,\mathbf{A'}|$$

$$\Rightarrow z(\rho) < z(\rho_0) \qquad \text{(A.22)}$$

Hence $z(\rho)$ achieves its unique global maximum at $\rho = \rho_0$.

A.3 Concavity of L for Symmetric W

To establish concavity of **L** for the case of symmetric weight matrices, **W**, we begin by observing that if the orthogonal projection onto *span*(**D'A'**) is denoted by

$$\mathbf{P} = \mathbf{D'A'}(\mathbf{ADD'A'})^{-1}\mathbf{AD} \qquad \text{(A.23)}$$

then it follows at once from (4.14) and (4.22) together with the identity $tr(\mathbf{M_1M_2})=tr(\mathbf{M_2M_1})$ that

$$\frac{\partial L}{\partial \rho} = -2tr\left[\mathbf{D'A'}(\mathbf{ADD'A'})^{-1}\mathbf{ADWD}\right] = -2tr[\mathbf{PWD}] \qquad \text{(A.24)}$$

Moreover, by using (4.13) we see that

$$\frac{\partial}{\partial \rho}\mathbf{P} = \left(\frac{\partial}{\partial \rho}\mathbf{D'}\right)\mathbf{A'}(\mathbf{ADD'A'})^{-1}\mathbf{AD} + \mathbf{D'A'}\left[\frac{\partial}{\partial \rho}(\mathbf{ADD'A'})^{-1}\right]\mathbf{AD}$$

$$+ \mathbf{D'A'}(\mathbf{ADD'A'})^{-1}\mathbf{A}\left(\frac{\partial}{\partial \rho}\mathbf{D}\right) =$$

$$= \mathbf{D'W'P} - \mathbf{P}(\mathbf{WD} + \mathbf{D'W'})\mathbf{P} + \mathbf{PWD} =$$

$$= (\mathbf{I_n} - \mathbf{P})\mathbf{D'W'P} - \mathbf{PWD}(\mathbf{I_n} - \mathbf{P}) \qquad \text{(A.25)}$$

Hence the second derivative of L is given by

$$\frac{\partial^2 L}{\partial \rho^2} = -2tr\left[\left(\frac{\partial}{\partial\rho}\mathbf{P}\right)\mathbf{WD} + \mathbf{PW}\left(\frac{\partial}{\partial\rho}\mathbf{D}\right)\right] =$$

$$= -2tr[(\mathbf{I_n} - \mathbf{P})\mathbf{D'W'PWD} + \mathbf{PWD}(\mathbf{I_n} - \mathbf{P})\mathbf{WD} + \mathbf{PWDWD}] =$$

$$= -2tr\{[(\mathbf{I_n} - \mathbf{P})\mathbf{D'W'PWD}] + tr[\mathbf{PWD}(\mathbf{I_n} - \mathbf{P})\mathbf{WD}] + tr[\mathbf{WDPWD}]\} \quad \text{(A.26)}$$

To show that this expression is negative for symmetric **W**, we next observe from (4.3) and (4.4) that by definition, **W** = **W'** ⇒ **B** = **B'** ⇒ **D** = **D'**. Moreover, since

$$\mathbf{WB} = \mathbf{W} - \rho\mathbf{W}^2 = (\mathbf{I_n} - \rho\mathbf{W})\mathbf{W} = \mathbf{BW} \Rightarrow \mathbf{DW} = \mathbf{WD} \quad \text{(A.27)}$$

we see that (**WD**)' = **D'W'** = **DW** = **WD**, and hence that **WD** is also symmetric. Finally, since **P** and $\mathbf{I_n}$ - **P** are each orthogonal projections, they are symmetric, idempotent and positive semidefinite. Hence by letting $\mathbf{M_1} = \mathbf{WD(I_n\text{-}P)}$, the first term in the brackets of (A.26) is seen to be of the form

$$tr[(\mathbf{I_n} - \mathbf{P})\mathbf{D'W'PWD}] = tr\left[(\mathbf{I_n} - \mathbf{P})^2\mathbf{D'W'PWD}\right] =$$

$$= tr[(\mathbf{I_n} - \mathbf{P})\mathbf{D'W'PWD}(\mathbf{I_n} - \mathbf{P})] = tr\left[\mathbf{M_1'PM_1}\right] \quad \text{(A.28)}$$

But since the symmetric positive semidefiniteness of **P** implies that $\mathbf{M_1'PM_1}$ is also symmetric positive semidefinite, it then follows that all eigenvalues are non-negative, and hence that tr $[\mathbf{M_1'PM_1}] \geq 0$. Similarly, by setting $\mathbf{M_2} = \mathbf{WDP}$, the second term in the brackets of (A.26) is of the form

$$tr[\mathbf{PWD}(\mathbf{I_n} - \mathbf{P})\mathbf{WD}] = tr\left[\mathbf{P}^2\mathbf{WD}(\mathbf{I_n} - \mathbf{P})\mathbf{WD}\right] =$$

$$= tr[\mathbf{PWD}(\mathbf{I_n} - \mathbf{P})\mathbf{WDP}] = tr\left[\mathbf{M_2'}(\mathbf{I_n} - \mathbf{P})\mathbf{M_2}\right] \quad \text{(A.29)}$$

and hence is nonnegative by the same arguments. Finally, since the third term the brackets of (A.26) is also of this same form with $\mathbf{M_3} = \mathbf{WD}$, it follows that all terms are nonnegative.

It thus remains only to show that at least one term is positive. But for the third term, we see from the orthogonal projection properties of **P** (including the equality **ADP** = **AD**) together with the identity **WD** = **DW** and the nonsingularity of **D** that

$$
\begin{aligned}
tr\left[(\mathbf{WD})'\mathbf{PWD}\right] = 0 &\Rightarrow (\mathbf{WD})'\mathbf{PWD} = \mathbf{O} \\
&\Rightarrow \mathbf{PWD} = \mathbf{O} \Rightarrow \mathbf{ADPWD} = \mathbf{O} \\
&\Rightarrow \mathbf{ADWD} = \mathbf{O} \Rightarrow \mathbf{ADW} = \mathbf{O} \\
&\Rightarrow \mathbf{AWD} = \mathbf{O} \Rightarrow \mathbf{AW} = \mathbf{O}
\end{aligned} \tag{A.30}
$$

which is not possible given our definitions of **A** and **W**. Hence $tr\ [(\mathbf{WD})'\mathbf{PWD}] > 0$ and the result is established.

Endnotes

[1]For notational simplicity we have suppressed the dependency of both **B** and **D** on ρ.

[2] It should be noted that this interpretation is for convenience only, and is not meant to imply that a spatial autoregressive process in (4.1) should represent even a first approximation to the distribution of housing values.

[3]The symmetric normalization, $diag\,(\mathbf{W}'u)^{-1/2}[\mathbf{W}]diag\,(\mathbf{W}'u)^{-1/2}$ (with unit vector, $u'=(1, .., 1)$) was employed in these examples. However, the standard row normalization, $diag\,(\mathbf{W}'u)^{-1/2}[\mathbf{W}]$, of **W** yields essentially the same results.

[4]The respective *y*-vectors for Figures 4.2a and 4.2b are (.0043,.0.318, 1.095,.1.874, 0.428, 0.896) and (.0.399, 0.690, 0.816, 0.712, 1.290, 0.669).

[5]Such conditions must for example include the requirement that the average number of subregions per region converge (i.e., that k_n/n have a positive limit in (0,1)), and that the sequence of aggregation matrices ($\mathbf{A_n}$) as well as the weight matrices ($\mathbf{W_n}$) exhibit appropriate 'growth, covergence and continuity' properties.

[6]It is worth noting in particular that the general result of Sweeting (1980) shows only that there is a consistent root of the likelihood equations. Hence in cases of multiple local maxima (such as those illustrated below) such general arguments offer no help in picking a consistent root. (See endnote 7 below for further discussion of this point)

[7]An additional advantage of this approach is that it permits consistent estimation in the case of multiple local maxima. For as long as the global maximum is unique (generally a reasonable assumption), one can be assured that a global search of the parameter space will yield the consistent estimator.

[8]Simulations of larger replication numbers show that by $N = 1000$ ($k = 3000$) the second mode has finally disappeared, so that classical consistency and asymptotic normality properties are in full force. However, it should be emphasized that this case was chosen mainly for its simplicity. A local search for 'worst cases' in the neighbourhood of this example produced a case in which $|z(\rho_0)-z(\rho_{00})|$ was so small that even for $N = 1000$ the fraction of 1000 samples in the ρ_{00}-cluster exceeded 40%.

[9]This was verified in the present case by calculating the asymptotic covariance matrix and constructing (Wald) significance tests for the simulated estimates. Not surprisingly, for the given sample size of k = 300 all negative estimates near ρ_{00} were shown to be highly significant. These results are not reported here.

[10]In section (A.3) of the Appendix it is shown that $\partial^2 L / \partial \rho^2 = -k \sum_i \varpi_i^2 < 0$, where the ω_i's are the (real) eigenvalues of **WD**.

[11]Recall that $w_{ii} \equiv 0$ by assumption.

[12]Extensive studies of numerical examples for such matrices show that L need not be concave, but that the derivative $\partial L / \partial \rho$ is always negative for $\rho > 0$. It thus seems reasonable to conjecture in this case that all maxima continue to occur at negative values of ρ.

[13]This terminology follows the generalized variance interpretation of covariance matrix determinants first introduced by Wilks (1932).

[14]It is also of interest to note in the present case that the behaviour of these sampling distributions is well predicted by the corresponding asymptotic form of the concentrated likelihood function (4.21), as shown in Figure 4.7. This is also seen in Figure 4.10 where the associated asymptotic function is plotted above the histogram. Notice in particular, that at ρ_0=0, the secondary mode has now merged with the primary mode, to produce a single 'flat' mode extending significantly into the negative range of ρ. Again this form is roughly mirrored by the sampling distribution below. Hence one can (very quickly) plot these asymptotic functions for a selected range of ρ_0 values, and see how the estimates behave.

[15]Some initial results along these lines using Best Linear Unbiased Prediction methods have been obtained by James LeSage (personal communication).

[16]It is worth noting that one of the earliest applications of this basic inequality was by Wald (1949) in his proof of consistency of maximal roots for the independent sampling case. Hence its relevance here is not surprising.

[17]This requires that the (highly overdetermined) system of $k(k+1)/2$ distinct polynomial equations in the two unknowns (ρ, σ^2))implied by the matrix equality $\sigma^2 \mathbf{ADD'A'} = \sigma_0^2 \mathbf{A D_0 D_0' A'}$) have no solutions in $(1/\lambda_{min}, 1/\lambda_{max}) \times R_+$ other than (ρ_0, σ_0^2).

5 Econometric Estimation of a Spatial Dynamic Model in Continuous Space and Continuous Time: an empirical demonstration

Kieran P. Donaghy
Maria Plotnikova
University of Illinois at Urbana-Champaign

Abstract

This chapter follows in a series of others that have presented and demonstrated a general approach to 1) solving spatial dynamic models in continuous space and continuous time that characterize the behaviour of intertemporally and interspatially optimizing agents, and 2) estimating from discrete data the parameters of such models. The approach involves the use of a projection method to solve systems of partial differential equations subject to boundary conditions and a quasi-Newton (or a hybrid optimization) method to update quasi-FIML estimates of the systems' parameters. In a previous paper, simulated data were employed to demonstrate solution, estimation, and stability analysis of a macroeconomic model of optimal space-time development introduced by Isard and Liossatos. The present chapter reports on research in which the Isard and Liossatos model is estimated with published data on the economies of ten spatially contiguous western European countries. In addition to other analyses of the estimated residuals, we conduct a cursory analysis of spatio-temporal autocorrelation. We also discuss difficulties that have come to light in applying the approach to solution and estimation to larger problems and suggest refinements of the algorithm that increase computational tractability and extend the scope of models to which the approach may be applied.

5.1 Introduction

Much, if not most, of the extant spatial econometrics literature concerns identification of the conditions of spatial dependence and spatial heterogeneity in cross-sectional data and control for their influence so that the business of applied econometrics (specifying, estimating, testing and simulating behavioural models) might be conducted as usual. That spatial econometricians have made so much progress in understanding and coping with these conditions over the last two decades is surely a remarkable achievement.

(See, e.g., Anselin, 1988, and Anselin and Florax, 1995.) While the literature on the econometrics of spatial dynamic models is thinner, perhaps because of the need to identify and control for the influence of spatio-temporal dependence in spatial time series and the less well developed state of theoretical explanations of spatial dynamic processes, it *is* developed nonetheless. (See Bennett, 1979.) Little has been published, however, on the econometrics of spatial dynamic models that characterize the behaviour of intertemporally and interspatially optimizing agents in continuous-time and continuous-space. It is true that the set of such models is only sparsely populated by the contributions of, say, Beckmann (1957), Isard and Liossatos (1979), Puu (1982), Lesse (1983), Puu (1987), and more recently Fujita, Krugman, and Venables (1998). But the challenges to researchers in this area are also great, because, even if one can identify the nature of spatial-temporal dependence in spatial time series, solving models comprising partial-differential equations (usually subject to boundary-point conditions in both space and time) is difficult enough without estimating them in a manner such that spatio-temporal dependence and spatial heterogeneity are taken into account[1].

In view of the degree of difficulty in working with such models, it is reasonable to ask: 'Why not work with spatial dynamic models formulated in discrete space and discrete time?' Apart from the fact that in most of the mature spatial sciences (e.g., physics, geomorphology, or meteorology) modeling in continuous space and continuous time is viewed as the more general approach, virtually all spatial dynamic models in the literature of regional science have continuous spatial and temporal formulation. There are good reasons for this. Very few interesting spatial processes are manifested only at lattice points. And from the temporal perspective, continuous time formulations are analytically more tractable and provide more realistic depictions of ongoing aggregate behaviour than do discrete-time formulations. They also permit proper handling of stock and flow phenomena and adjustment or transitional dynamics. Several considerations from the perspective of applied work that commend their use are the fact that there is no natural time unit for periodic observation of aggregate economic variables and we may well wish to model behaviour in markets that adjust in a shorter length of time than the period over which they are observed (e.g., a month or a quarter). Estimates of the mean time lags of continuous-time models are independent of the observation interval, and once the parameters of such models have been estimated with observations of a given interval, they can be used to forecast the levels of variables at any observation interval. (On all these points see Gandolfo, 1981.)

Such considerations suggest that spatial dynamic models ought to be estimated and analyzed by methodological approaches that take into account both the nature of the theoretical model and the manner in which aggregate spatial time series have been generated. Continuous time econometric methods have been available for mixed-order, mixed-sample linear models for over thirty years and for non-linear models, nearly twenty (see discussions in Wymer, 1972 and Wymer, 1993). Extending existing continuous-time methods to continuous space, Donaghy (2001) presents and demonstrates a general approach to 1) solving spatial dynamic models that characterize the behaviour of intertemporally and interspatially optimizing agents, and 2) estimating from discrete data the parameters of such models. The approach involves the use of a projection method to solve systems of partial differential equations (PDEs) subject to boundary conditions and a quasi-Newton (or a hybrid optimization) method to update quasi-FIML estimates of the systems' parameters. In Donaghy (2001) simulated data are employed to demonstrate solution, estimation, and stability analysis of a macroeconomic model of optimal space-time development introduced by Isard and Liossatos (1979).

The present chapter reports on research in which the Isard and Liossatos (1979) model is estimated with published data on the economies of ten spatially contiguous western European countries and the estimated model is analyzed and evaluated. Apart from the demonstration of feasibility of the general approach with empirical data, this work may be of interest for at least two other reasons. Recent work by Parker (2000) and others has focused on physical explanations for divergence in long-run economic growth performance and remarked upon the spatial correlation of performance measures of economies that share the same latitude and political system. In common with this line of inquiry, the model of Isard and Liossatos here investigated explicitly relates productivity of aggregate economic output to distance from a given location that is deemed to be most propitious for production. On a somewhat related theme, Baumont et al. (2000) have identified patterns of spatial correlation in economic indicators for European regions and shown the importance of taking spatial effects into account in econometric studies of convergence of measures of economic performance. Donaghy (2001) conjectured that the approximations introduced to facilitate solution of the Isard and Liossatos model might, by explicitly accounting for not only spatial but spatio-temporal correlation between system variables, control for the influence of such correlation on the estimates of parameters and their standard errors. In addition to other analyses of the estimated residuals, we conduct a cursory analysis of spatio-temporal autocorrelation.

In applying the approach set out in Donaghy (2001) to a larger number of locations, certain potential difficulties in implementation of the approach have come to light. A second contribution of the chapter is to provide a discussion of these difficulties and to suggest refinements of the algorithm for solving and estimating systems of PDEs in space and time that increases computational tractability and extends the scope of models to which the approach may be applied.

In the second section we briefly review the model of Isard and Liossatos and the general approach to its solution and estimation demonstrated in Donaghy (2001). We then discuss in the third section potential problems in implementing this approach and refinements of it, and in the fourth, present and discuss empirical findings and a brief analysis of spatio-temporal autocorrelation in the residuals of the estimated model. We conclude the chapter with suggestions of how the research reported might be extended.

5.2 Isard and Liossatos's Macro Model of Space-Time Development, Solution and Estimation

The two equations of motion of Isard and Liossatos's (1979) macro model of space-time development are

$$\dot{K}(x,t) = F(K(x,t),x) - C(x,t) - \overset{x}{U}(x,t) - \sigma U(x,t) \tag{5.1}$$

$$l\dot{U}(x,t) = -\xi \overset{x}{K}(x,t) - vU(x,t) \tag{5.2}$$

Equation (5.1) is a generalization of the primary equation of motion in an aspatial economy. It gives the *time-rate of change* in the capital stock at some location x at time t, $\dot{K}(x,t)$ as the difference between aggregate output, $F(k(x,t),x)$, and aggregate consumption, $C(x,t)$, the *space-rate of change* in the flow of goods through location x at time t, $\overset{x}{U}(x,t)$, and goods consumed in transportation, $\sigma U(x,t)$, where σ denotes the transport rate. Equation (5.2) characterizes the time rate of change in the flow of goods through a location, x, where l is a coefficient of inertia, the first term on the right-hand side represents the gradient of development and the second term represents a dissipation of the force $-\xi \overset{x}{K}(x,t)$, a negative function of the development gradient. This second equation of motion is associated with the movement of goods generated by differences in productivity of capital among markets. (See Chapter 6 of Isard and Liossatos, 1979 for full discussion.)

Isard and Liossatos frame the optimal space-time development problem as one in which a representative agent or social planner chooses trajectories of consumption flows, capital stocks, and flows of goods so as to maximize a utility function subject to the two equations of motion, initial values of the state variables, and terminal conditions. They write the objective functional corresponding to this problem as:

$$\max_{\{C,K,U\}} \left[\int_0^{t_1} \int_0^B f(C(x,t)) dx dt + \int_0^B \Phi(K(x,t_1), U(x,t_1)) dx \right] \tag{5.3}$$

and construct the Lagrangian function, from which the first-order necessary conditions for an optimal solution can be generated, as

$$L = f(C(x,t)) + p(x,t)\left[F(K(x,t),x) - C(x,t) - \overset{x}{U}(x,t) - \sigma U(x,t) - \dot{K}(x,t) \right] +$$

$$\omega(x,t)\left[-\xi \overset{x}{K}(x,t) - vU(x,t) - l\dot{U}(x,t) \right]$$

(5.4)

in which $p(x,t)$ and $\omega(x,t)$ are costate variables associated with (5.1) and (5.2) respectively. Suppressing arguments x and t, the first-order necessary conditions for an optimal solution are, in addition to equations (5.1) and (5.2),

$$
\begin{aligned}
&f_C = p \\
&p\,F_K(K,x) + \overset{x}{\xi}\,\omega = -\dot{p} \qquad (5.5)\\
&\overset{x}{p} - \sigma p - v\omega = -l\dot{\omega}
\end{aligned}
$$

and the terminal conditions

$$
\begin{aligned}
p(x,t_1) &= p_1(x) = \frac{\partial \Phi}{\partial K}(K(x,t_1),U(x,t_1)) \\
l\omega(x,t_1) &= l\omega_1(x) = \frac{\partial \Phi}{\partial U}(K(x,t_1),U(x,t_1))
\end{aligned}
\qquad (5.6)
$$

The terminal conditions in (5.6) give the boundary values that the two costate variables must satisfy at $t1$. K and U must satisfy the initial-value conditions

$$
\begin{aligned}
K(x,t_0) &= K_0(x) \\
U(x,t_0) &= U_0(x)
\end{aligned}
\qquad (5.7)
$$

Isard and Liossatos (1979) provide full interpretations of these conditions, which Donaghy (forthcoming) reviews.

A numerical-solution approach that allows us to avail ourselves of the power of contemporary computational capacity and explore a broader range of solution space than closed form approaches permit is that employed in projection methods, which were developed partly to aid solution of problems posed in terms of PDEs (see Judd, 1998). Whereas Donaghy (2001) presents and motivates the general algorithm followed in projection methods, we proceed to a review of its implementation in the case of Isard and Liossatos's model.

Note first that a solution to the set of PDEs, (5.1), (5.2), and (5.5), subject to boundary conditions (5.6) and (5.7), is a set of functions that solves functional equations. In a projection-method approach we are not looking for functions that solve the original problem exactly but rather functions that provide serviceable approximations. Hence in the first step we choose as a candidate solution basis a set of finite polynomial approximations of K, U, p, and ω of the forms

$$\hat{K}(x,t) = K_0(x) + \sum_{j=1}^{n} a_j(t)\,x^j \qquad (5.8)$$

$$\hat{U}(x,t) = U_0(x) + \sum_{j=1}^{n} b_j(t)x^j \tag{5.9}$$

$$\hat{p}(x,t) = p_1(x) + \sum_{j=1}^{n} c_j(t)x^j \tag{5.10}$$

$$\hat{\omega}(x,t) = \omega_1(x) + \sum_{j=1}^{n} d_j(t)x^j \tag{5.11}$$

Boundary conditions (5.6) and (5.7) and function definitions (5.8)-(5.11) imply the further conditions on the time-varying coefficients (TVCs) of the approximations,

$$\sum_{j=1}^{n} a_j(t_0)x^j = 0 \tag{5.12}$$

$$\sum_{j=1}^{n} b_j(t_0)x^j = 0 \tag{5.13}$$

$$\sum_{j=1}^{n} c_j(t_1)x^j = 0 \tag{5.14}$$

$$\sum_{j=1}^{n} d_j(t_1)x^j = 0 \tag{5.15}$$

The second step is to choose a degree of approximation for the polynomials. This choice will be dictated in part by considerations of feasibility and adequacy of the solution's accuracy. In the case of spatial dynamic models data availability will also play a role[2].

In the third step we rewrite our first-order conditions in residual-function form, take appropriate partial derivatives of our polynomial approximations and make substitutions into the residual functions. Noting and making use of Isard and Liossatos's assumption of an iso-elastic (or constant relative-risk aversion) instantaneous utility function,

$$f(C(x,t)) = \frac{C(x,t)^{1-\alpha}}{1-\alpha}, \tag{5.16}$$

and a space-varying production function,

$$F(K(x,t),x) = \frac{\gamma K(x,t)^{\beta}}{[1+\eta(x)]}, \tag{5.17}$$

and choosing for $\eta(x)$ the simplest of forms, ηx, then for some location x the resulting expressions for the residual functions would be as follows.

$$R^1(x,t) = \sum_{j=1}^{n} \dot{a}_j(t)\,x^j - \frac{\gamma\hat{K}(x,t)^{\beta}}{[1+\eta x]} + \hat{p}(x,t)^{1/\alpha} + U_{0x}(x) + \sum_{j=1}^{n} j\,b_j(t)\,x^{j-1} + \sigma\hat{U}(x,t) = 0$$

$$R^2(x,t) = \sum_{j=1}^{n} \dot{b}_j(t)\,x^j + (\xi\xi/I\left[K_{0x}(x) + \sum_{j=1}^{n} j\,a_j(t)\,x^{j-1}\right] + (\upsilon\upsilon/I\hat{U}(x,t) = 0$$

$$R^3(x,t) = \sum_{j=1}^{n} \dot{c}_j(t)\,x^j + \frac{\hat{p}(x,t)\beta)\hat{K}(x,t)^{\beta-1}}{[1+\eta x]} + \xi\left[\omega_{1x}(x) + \sum_{j=1}^{n} j\,d_j(t)\,x^{j-1}\right] = 0$$

$$R^4(x,t) = l\sum_{j=1}^{n} \dot{d}_j(t)\,x^j + p_{1x}(x) + \sum_{j=1}^{n} j\,c_j(t)\,x^{j-1} - \sigma\hat{p}(x,t) - \upsilon\hat{\omega}(x,t) = 0$$

(5.18)

We now have 4x*n* unknown functions in *a(t), b(t), c(t),* and *d(t),* which can be solved for by imposing in the fourth step 4x*n* projection conditions of the form

$$< R(x,t); \psi_l(x) >= 0, l = 1,\ldots,4n \qquad (5.19)$$

in which ψ_l denotes a test function against which the residual function *R(x,t)* is projected[3].

The test function can be the transpose of the vector of values of the evaluated residual function itself, in which case the criterion of what counts as a good enough approximation to a solution of the problem's original functional equation is a 'least squares' criterion, or any number of things. (See Judd, 1998.) When a projection method for solving a set of PDEs subject to boundary point conditions is incorporated in a multiple-step solution-estimation algorithm, it seems logical to use differences between observed estimated values of true state variables in what is termed a 'collocation' approach, in which solutions of the residual functions are forced to zero at specific points.

The following algorithm for solving and estimating systems of PDEs in continuous-space and continuous-time has been proposed in Donaghy (2001 and forthcoming).

1. Specify an initial set of parameter values.
2. Employing the projection method sketched above, solve the PDEs subject to all conditions given by theory.
3. Compute the residuals for the variables on which observations exist. (There will be no observations on the co-state variables of the original problem or the time-varying coefficients of the transformed system of ordinary differential equations (ODEs).)

4. Employ an optimization procedure to update the parameter estimates by choosing new values to minimize the natural log of the determinant of the variance-covariance matrix[4].
5. Check whether convergence criteria and tests of optimality of the likelihood function are satisfied. If so, stop. If not return to step two.

5.3 Potential Difficulties in Implementation and Refinements

As demonstrated in Donaghy (2001), the algorithm sketched above works quite well with small and well-behaved problems. To anticipate the discussion of the next section, two difficulties in its implementation come to light, however, as the size of the problem (number of locations considered) increases and the spatial economies modelled are less uniform in structure.

1. It is difficult to initiate iterations of the algorithm without some prior knowledge of either initial values of the TVCs or the values of the structural parameters. (It is best to have some knowledge of both.) Problems are most apparent in step two where a shooting algorithm is employed to solve for the values of the TVCs.
2. As the number of locations increases, and increasingly higher powers of distance measures, x_l^j, appear in the polynomial approximations, the values of the TVCs can become quite large. This presents two problems. As the scales of the original variables and the TVCs diverge, the system of ODEs in (5.18) and (5.19) equations can quickly become 'stiff,' that is, possess variables whose solutions are of different orders of magnitude, and unsolvable by most numerical methods. Also, the partial derivatives of the approximations taken w.r.t. space depart in behaviour from their analytical counterparts, dominating the structural equations in which they appear[5].

Regarding the first difficulty, note that the TVCs of approximations of observed state variables can be solved for (almost) exactly in terms of the original time series. For example, differentiating approximation (5.1) completely w.r.t. time yields

$$\dot{K}(x,t) = \sum_{j=1}^{n} \dot{a}_j(t) x^j \tag{5.20}$$

Since is known, the set of linear ODEs

$$\sum_{j=1}^{n} \dot{a}_j(t) x^j = z_l(t), \text{ where } z(t) = \dot{K}_l(x,t), \text{ for } l = 1, ..., m \tag{5.21}$$

can be solved for the levels of $a_j(t)$, given the initial values, $K_l(x,t_0)$. The data so derived can then be used in the estimation of the parameters. The computational complexity of the solution/estimation problem will be reduced, since the derived time series data on the TVCs for the approximations of the state variables will already satisfy boundary conditions (5.5) and (5.6)[6]. Taking note of the first of the first-order necessary conditions in (5.5), the co-state variable p can be eliminated entirely from the model, since $p(x,t) = C(x,t)^{-\alpha}$ and the second two first-order necessary conditions (the costate equations) can be rewritten (suppressing arguments x and t) as

$$C F_{\kappa} + \xi \overset{x}{\omega} C^{1+\alpha} = -\alpha \dot{C} \tag{5.22}$$

$$-\alpha C^{-\alpha-1} \overset{x}{C} - \sigma C^{-\alpha} - \upsilon\omega = -l\dot{\omega}$$

In which case, $C(x,t)$ would be approximated by

$$\hat{C}(x,t) = C_0(x) + \sum_{j=1}^{n} c_j'(t) x^j \tag{5.10'}$$

for which

$$\sum_{j=1}^{n} c_j'(t_0) x^j = 0 \tag{5.14'}$$

and the second two residual functions in (5.18) would be replaced by

$$R^3(x,t) = \alpha \sum_{j=1}^{n} \dot{c}_j'(t) x^j + \hat{C}(x,t) \frac{\beta\gamma \hat{K}(x,t)^{\beta-1}}{[1+\eta x]} + \xi \left[\omega 1_x(x) + \sum_{j=1}^{n} j d_j(t) x^{j-1} \right] \hat{C}(x,t)^{1+\alpha} = 0$$

$$R^4(x,t) = l \sum_{j=1}^{n} \dot{d}_j(t) x^j - \alpha \hat{C}(x,t)^{-\alpha-1} \left[C_{0x}(x) + \sum_{j=1}^{n} j c_j'(t) x^{j-1} \right] - \sigma \hat{C}(x,t)^{-\alpha} - \upsilon\hat{\omega}(x,t) = 0$$

Observations on aggregate consumption, $C(x,t)$, and plausible initial values of estimates of the most important parameters (obtained by estimating the state equations separately) can then be used to generate feasible initial values of the TVCs for the approximations of $C(x,t)$ and $\omega(x,t)$ to initiate the shooting algorithm in step 2 of the solution/estimation process.

Regarding the second difficulty noted above, we conjecture that it arises because evaluations of the partial derivatives of the approximations at points x_l are closer to ratios of *differences* instead of *differentials* (i.e., in the case of $\overset{x}{K}$ evaluating the expression $K_{0x}(x)+\sum_{j=1}^{n} j\,a_j(t)\,x^{j-1}$ yields a value closer to $\Delta K/\Delta x$ than to $\partial K/\partial x$). This might be addressed either by parameterizing the derivative expression, e.g.,

$$\overset{x}{K} = K_{0x} + \delta_K \sum_{j=1}^{n} j\,a_j(t)\,x^{j-1} \tag{5.23}$$

or by adopting an alternative approximation with partial derivatives that are better behaved. (In the work discussed below, we take the first tack).

5.4 An Empirical Application

Employing the approach set out above and Wymer's (1997) software for estimating differential equation systems subject to boundary-point conditions, we estimated the transformed model of Isard and Liossatos with real aggregate time series (1975-1999) on net investment, consumption and net trade for the ten countries of Denmark, France, Germany, Greece, Italy, the Netherlands, Portugal, Spain, Sweden, and the United Kingdom[7]. Estimates of the root-means square errors for static (or one-period) in-sample forecasts of the model's 'observed' variables are given in Table 5.1 in proportionate terms. In view of the model's level of abstraction, its fit is reasonably good for Germany, Greece, Portugal, Spain and the United Kingdom; it is quite poor for the other countries.

Table 5.1 Proportionate Root-Mean-Square Errors of In-Sample, One-Period Forecasts

Country	x*	K	U
Denmark	1.11367	0.12802	0.42952
France	0.97667	0.12870	0.45900
Germany	1.05000	0.04194	0.00727
Greece	0.75933	0.00970	0.00413
Italy	0.83800	0.16100	0.57933
Netherlands	1.04767	0.11823	0.14234
Portugal	0.77400	0.11081	0.02186
Spain	0.80833	0.04984	0.02984
Sweden	1.18600	0.30167	2.26334
United Kingdom	1.03200	0.10471	0.03023

*x denotes the degrees latitude of the capital city of the country, relative to 50 degrees north.

We are not aware of any statistical tests for temporal autocorrelation in the residuals of *non-linear* simultaneous equation systems. But for heuristic purposes and to gain indications of possible model misspecification, we have computed portmanteau (or Q) test statistics for autocorrelations in the residuals of the levels of the variables of the model for up to an order of 3.

Table 5.2 Portmanteau (or Q) Test Statistics for Autocorrelations in Residuals of Variables*

	Order of Autocorrelation		
Variable	1	2	3
K1	3.31	3.53	3.59
K2	3.43	3.52	3.80
K3	11.87	15.49	16.02
K4	5.80	6.04	6.30
K5	7.50	9.68	9.81
K6	7.05	7.45	7.51
K7	12.24	15.88	15.98
K8	13.66	18.24	18.39
K9	11.03	13.90	13.95
K10	9.87	11.19	11.40
U1	0.41	0.55	0.64
U2	0.42	0.67	1.64
U3	2.63	2.65	3.15
U4	0.05	1.11	1.43
U5	0.11	0.60	0.61
U6	0.62	0.75	0.87
U7	1.11	1.12	2.48
U8	0.28	0.30	0.35
U9	1.33	1.33	1.42
U10	1.09	1.37	1.37

* The asymptotic distribution of the Q(k) statistic is Chi-square with k degrees of freedom. Approximate critical values of Chi-square distribution for 1, 2, and 3 degrees of freedom are at the 1 per cent level: 6.6, 9.2,11.3, and at the 5 per cent level: 3.8, 6.0, 7.8.

Although this test may be inappropriate for the type of model here considered, the results seem to suggest that there are systematic patterns in the errors of the investment equations of Germany, Portugal, Spain, Sweden and the United Kingdom (i.e., that the model is perhaps overly general in its specification). Owing to the dearth of data points and the non-linear simultaneous-equation nature of the model, we have not computed any of the standard test statistics for spatial correlation[8].

A statistic that *would* seem appropriate to compute is an autocorrelation measure for space-time contiguity systems (Bennett, 1979), which is obtained by the following formula for a lag of order τ in time and s in one dimension of space:

$$RUU(\tau\tau s) = \frac{\sum_{t=1+\tau}^{T} \sum_{i=1}^{N} (U_{ti}-\bar{U})(L^{s}U_{t-\tau i}-\bar{U})}{\left\{\left[\sum_{t=1+\tau}^{T} \sum_{i=1}^{N} \left(U_{ti}-\bar{U}\right)^{2}\right]\left[\sum_{t=1+\tau}^{T} \sum_{i=1}^{N} \left(L^{s}U_{ti}-\bar{U}\right)^{2}\right]\right\}^{1/2}} \tag{5.24}$$

Table 5.3: Quasi-FIML Estimates of Parameters and their Asymptotic Standard Errors Parameter Estimate

Parameter	Parameter Estimate	Estimate of A.S.E.
α	-0.22517	0.00175
β	0.77168	0.00457
η	0.09485	0.00046
γ	0.15608	0.00028
σ	0.13545	0.00175
ξ	0.00001*	----
l	1.37519	0.00132
υ	-0.00570	0.00001
δ_U	0.00001	0.95E-7
δ_K	1.00000*	----
δ_ω	1.16619	0.00284
δ_p	0.22710	0.00161
K_{0x1}	-0.01171	0.00002
K_{0x2}	-0.00519	0.00005
K_{0x3}	-0.00816	0.00005
K_{0x4}	-0.00780	0.00003
K_{0x5}	-0.00796	0.39E-5
K_{0x6}	-0.00812	0.00004
K_{0x7}	-0.00764	0.00001
K_{0x8}	-0.00821	0.00003
K_{0x9}	-0.01162	0.00002
K_{0x10}	-0.01364	0.00007
U_{0x1}	-0.00566	0.12E-5
U_{0x2}	-0.00546	0.49E-5
U_{0x3}	-0.00535	0.21E-5
U_{0x4}	-0.00346	0.29E-5
U_{0x5}	-0.00635	0.42E-5
U_{0x6}	-0.00509	0.18E-5
U_{0x7}	-0.00780	0.00003
U_{0x8}	-0.00598	0.00002
U_{0x9}	-0.00573	0.00002
U_{0x10}	-0.00489	0.00001

* denotes value imposed in estimation

Table 5.3 (cont'd): Quasi-FIML Estimates of Parameters and their Asymptotic Standard Errors

Parameter	Parameter Estimate	Estimate of A.S.E.
C_{1x1}	-0.00421	0.15E-5
C_{1x2}	0.00018	0.19E-5
C_{1x3}	-0.00332	0.00002
C_{1x4}	0.00026	0.00001
C_{1x5}	-0.00428	0.00001
C_{1x6}	0.00010	0.33E-5
C_{1x7}	-0.00390	0.00003
C_{1x8}	0.00010	0.34E-5
C_{1x9}	-0.00413	0.55E-6
C_{1x10}	-0.00114	0.00002
ω_{1x1}	0.01184	0.00020
ω_{1x2}	-0.00041	0.00008
ω_{1x3}	-0.00219	0.59E-5
ω_{1x4}	-0.01427	0.00040
ω_{1x5}	-0.00484	0.00021
ω_{1x6}	0.00917	0.00019
ω_{1x7}	-0.00509	0.00001
ω_{1x8}	0.00032	0.69E-5
ω_{1x9}	-0.00219	0.00025
ω_{1x10}	-0.01143	0.00021

Where

$$L^s U_{ti} = \sum_{j=1}^{k} w_{tij} U_{tj}$$

in which w_{tij} is the *i,j*th element of a spatial weights matrix with neighbours of order (or distance) s at time t. In (5.24) U_{ti} is the error in the estimated level of the true structural variable for location *i* at time *t* and $\overline{U}$ is the mean such errors over space and time. For τ=1,2 and *s*=1,2 and constant first- and second-order contiguity spatial weight matrices, estimates of spatiotemporal system autocorrelation for estimated residuals of the investment equations are $R_{UU}(1,1)=0.07130$, $R_{UU}(1,2)=0.02332$, $R_{UU}(2,1)=0.111452$, and $R_{UU}(2,2)=0.05956$, which, on the basis of *t* tests for any conventional level of statistical significance, would not lead one to reject null hypotheses of no spatio-temporal autocorrelation of the orders considered. While this matter needs to be explored at greater length, this finding suggests that the approximations introduced to facilitate solution of the Isard and Liossatos model might, by explicitly accounting for not only spatial but spatio-temporal correlation between system variables, control for the influence of such correlation in the estimation of parameters and their standard errors. Estimates of the model's parameters and their standard errors are presented in Table 5.3.

The estimates of the structural parameters of the theoretical (pre-transformation) model seem plausible and are super-efficient, as one would expect of continuous-time estimators (Phillips, 1991 and Wymer, 1993). The value of the parameter ξ was set to 0.00001, the value of a preliminary estimate from which it did not later differ significantly. The signs of the estimates of the partial derivatives of K and U with respect to space at t_0 all seem to be correct. (See Isard and Liossatos, 1979.)

We are unaware of any statistical tests for cointegration of the variables of non-linear simultaneous equation systems. Although they are perhaps inappropriate, we have nonetheless conducted single-equation augmented Engle-Granger (AEG) residual-based tests of cointegration because of the increasing sensitivity in applied econometrics to the charge that the institutional structures captured by structural-equation econometric models are 'fragile' artifacts of the data, do not represent 'long-run equilibrium relationships,' and hence cannot, when used in simulations, support robust theoretical inferences. The AEG tests we conduct are based on a three-period-lag auto-regression of the residuals of the 'constant, trend, and squared trend' type, i.e.,[9]

$$\Delta e_t = a + b_1 t + b_2 t^2 + c_0 e_{t-1} + c_1 \Delta e_{t-1} + c_2 \Delta e_{t-2} + c_3 \Delta e_{t-3} \tag{5.25}$$

These tests are not invalidated by non-normality or heteroscedasticity in the residuals and account for the presence of serial correlation of order three or lower by virtue of their lag structure. The results of the AEG tests, tabulated below, suggest that at the 95% level of confidence, the hypothesis of noncointegration (that the residuals have a unit root, or that c_0 is not significantly different from zero) can be rejected for only the net trade equations for Greece and the Netherlands, and at the 1% level *cannot* be rejected for any of the equations. This finding would seem to suggest that the model does *not* adequately characterize long-run equilibrium relationships and that inferences drawn from experiments conducted with the model would not be valid.

5.5 Conclusion

In this chapter we have presented an empirical demonstration of a general approach to solving and estimating spatial dynamic models in continuous space and continuous time as applied to the most basic of the Isard and Liossatos models of optimal space-time development. The demonstration has shown that the general approach is feasible. It has also shown that the model does not capture particularly well space-time development at the aggregate level in the economies of ten contiguous western European countries over a twenty-five-year period and there are indications on the basis of analyses of residuals that the model may be misspecified. This should not be surprising given the level of abstraction of the model's specification and the complexity of growth dynamics over the sample period considered[10]. Richer, competing specifications of this and other types of spatial dynamic models can be explored at the national, regional, and lower levels of spatial aggregation by the approach proposed, and that, after all is the point of empirical work. The demonstration has also brought to light several potential difficulties with the implementation of the proposed estimation/solution approach, and several refinements have been discussed. Although relations of spatial and spatio-temporal dependence between estimated residuals of the model obtained by other approaches to its solution and estimation are not known, it does appear that the approximations introduced to facilitate the solution of the model might, by explicitly accounting for spatial and spatio-temporal dependence between system variables, control for the influence of such dependence in the residuals, hence in the estimation of parameters and their standard errors. Further exploration of these matters is needed.

Table 5.4. Augmented Engel-Granger Test of Noncointegration

Variable	Estimate of c_0	Phillips' Non- parametric z*†
K_1	-0.874	-14.0
K_2	-0.947	-15.2
K_3	-0.403	-6.46
K_4	-1.070	-17.1
K_5	-0.903	-14.5
K_6	-0.959	-15.3
K_7	-0.640	-10.2
K_8	-0.665	-10.6
K_9	-1.040	-16.7
K_{10}	-0.517	-8.28
U_1	-1.530	-24.4
U_2	-1.630	-26.1
U_3	-1.100	-17.7
U_4	-1.940	-31.1
U_5	-1.470	-23.5
U_6	-1.860	-29.8
U_7	-1.430	-23.0
U_8	-1.310	-21.0
U_9	-1.340	-21.4
U_{10}	-1.100	-1.75

†Davidson and McKinnon (1993) report that the asymptotic critical value of this statistic is -28.10 at a 5% significance level, and is -36.6 at a 1% significance level.

Endnotes

[1]To capture adequately the nature of the problem, we might need a more generous conception of spatial econometrics that, like Frisch's (1933) original definition of econometrics, encompasses statistics, economic theory, and mathematics. In his oft-cited editorial in the inaugural issue of *Econometrica,* Frisch observed that '...each of these three view points, that of statistics, economic theory, and mathematics, is a necessary, but not by itself sufficient, condition for a real understanding of the quantitative relations in modern economic life. It is the *unification* of all three that is powerful. And it is the unification that constitutes econometrics' (Frisch, 1933, p.2).
[2]Judd (1998) has found n=10 to work well for applications of projection methods to macroeconomic growth models.
[3]In any given application, the nature of the test function determines the nature of the projection method and gives the method its name. Judd (1998) notes that projection methods are also called 'weighted residual methods' since the test functions weight the residual functions.
[4]A genetic algorithm might be used initially to search parameter space, followed by a Newton algorithm which employs formal criteria of optimality.
[5]Note that the Weierstrass theorem, which sanctions the introduction of polynomial approximations for variables in the model, doesn't vouch for the fidelity of derivatives of a polynomial approximation at a point, only the convergence of the value of the approximation to that of the true continuous function in a closed interval as the number of terms increases.
[6]Let $\dot{a}(t)$ denote an $n \times 1$ vector and let **X** denote the $m \times n$ matrix $\left[x_j^j\right]$, where m is the number of locations l and n is the order (number of variable terms) in the polynomial approximations. When $m=n$, $\dot{a}(t) = X^{-1}\dot{z}(t)$, and when $m>n$, $\dot{a}(t) = (X'X)^{-1}X'\dot{z}(t)$. When $m>n$, $\dot{a}(t) = (X'X)^{-1}X'\dot{z}(t)$ the consistency of parameter estimates obtained with data derived for a_j(t) may be questionable, however.
[7]Data were obtained from various publications of the OECD and International Monetary Fund.
[8]Luc Anselin advised us against doing so because of the paucity of data points in space.
[9]The rule of thumb for the number of lags discussed in Davidson and McKinnon (1993) is that there should be at *least* $n^{1/3}$ lags, which for a sample size of 25 would be 3.
[10]The specification considered is the most simple of those developed in Isard and Liossatos (1979). The authors develop the model in considerable detail in that volume. See also Donaghy (forthcoming) for a discussion of how Isard and Liossatos's research program could be extended with more recent developments in modelling economic growth.

6 Using Matrix Exponentials to Estimate Spatial Probit/Tobit Models

James P. LeSage
University of Toledo

R. Kelley Pace
Louisiana State University

Abstract

Spatial data samples that contain binary dependent variables (probit) or sample truncation (tobit) are quite common. Tobit models arise because censoring is often used for confidentiality purposes resulting in sample truncation. Probit models arise because spatial problems often involve the absence or presence of some phenomena of interest at various points in space. Given the pending release of Census 2000 information, there is a need for computationally efficient estimation methods for spatial tobit and probit models. We set forth a Bayesian approach based on the matrix exponential spatial specification (MESS) introduced in Pace and LeSage (2000) that can produce estimates for both these models. In the case of tobit estimation, our method produces posterior inferences regarding the usual model parameters as well as imputations for the censored observations. In the case of probit models the method can produce a generalization of traditional logit/probit models. In addition, the methods described employ a flexible spatial weight structure rather than follow the usual practice that assumes the spatial weight matrix represents fixed sample information. Our flexible specification for the spatial weights introduces hyperparameters to control the number of neighbouring entities as well as decay of influence over space, and our approach allows posterior inference for these parameters. We illustrate the tobit model using a sample of housing prices from Westchester County, New York where high-priced homes were censored, and the probit model using a sample of presidential election results for US counties.

6.1 Introduction

Spatial data samples that contain binary dependent variables (probit) or sample truncation (tobit) are quite common. Tobit models arise because censoring is often used for confidentiality purposes resulting in sample truncation. In addition, some problems inherently involve censoring. For example, spatial analysis of home mortgage applications might record positive magnitudes representing loan amounts for approved loans and zero amounts for loans that were denied. Probit models arise because spatial problems often involve the absence or presence of some phenomena of interest at various points in space.

We show how a Bayesian extension of the matrix exponential spatial specification (MESS) introduced in Pace and LeSage (2000) can be used to produce estimates for spatial tobit and probit models. Despite the prevalence of binary and truncated dependent variables in spatial samples, there have been few suggestions for estimation of spatial tobit and probit models. McMillen (1992, 1995) introduced an EM approach to estimating spatial probit models and LeSage (2000) introduced a Bayesian approach to estimating both spatial tobit and probit models for the basic spatial econometric models introduced in Anselin (1988).

The approach taken here is similar to that used in LeSage (2000) to extend traditional Bayesian spatial autoregressive models to the case of tobit and probit specifications. There are several advantages to a Bayesian approach in this setting. First, the Bayesian approach introduced here relies on the same prior/likelihood-based inferential paradigm as LeSage and Pace (2000) despite the fact that the dependent variable is either truncated or binary in nature. This is in contrast to the EM approach introduced by McMillen (1992) that requires adopting an alternative inferential paradigm.

Another advantage is the computational speed arising from the matrix exponential specification introduced in Pace and LeSage (2000) relative to more traditional specifications. They show that matrix exponentials can be used to spatially transform the dependent and/or independent variables to eliminate the computationally troublesome log-determinant term from the log likelihood. LeSage (2000) shows that this same term arises in the conditional distribution for the spatial dependence parameter in Bayesian variants of traditional spatial autoregressive models slowing evaluation of this conditional distribution during Markov Chain Monte Carlo (MCMC) estimation. MCMC estimation of the Bayesian MESS model is around six times faster than MCMC estimation of more traditional spatial autoregressive models, with the speed improvements partially derived from elimination of the log-determinant term associated with the matrix exponential spatial specification.

Yet another advantage was pointed out by Albert and Chib (1993) who note that assumptions made regarding the distributional form of the disturbances determine whether a logit or probit specification is used. They note that a logit model corresponds roughly to an assumption that the disturbances take the form of an independent $\chi^2(r)$ distribution with $r = 7$ or 8. In contrast, the probit model is associated with assumed normal disturbances that correspond to a $\chi^2(\infty)$ distribution. Implementation of the Bayesian MESS probit model introduced here treats the degrees of freedom parameter r as a hyperparameter in the model, allowing estimates corresponding to both logit and probit models as well as others based on a range of small to large r values.

Another advantage of the Bayesian MESS tobit/probit models is that we can employ a more flexible spatial weight structure than conventional spatial autoregressive models. Model parameters are included that allow inferences regarding the number of important neighbouring entities as well as decay of influence from neighbouring units over space. This is in contrast to traditional specifications that treat the spatial weight matrix as fixed, resulting in parameter inferences for the spatial regression relationship that are conditional on the specific form of weight matrix used in the model. In many spatial applications the spatial structure itself may be a subject of interest. As an example, spatial hedonic pricing relationships for homes might intuitively include influences from neighbouring homes that were sold recently. Both the magnitude and spatial extent of influence from neighbouring homes might be of interest for real estate appraisal or tax assessment purposes. Appraisal practices typically include an examination of 'comparable housing units' where one dimension of comparability is the location. Defining the relevant spatial region in which appraisers should seek comparable units would be facilitated by estimates that describe the magnitude and extent of spatial influence from neighbouring homes sold. Other examples where inferences about spatial structure might be important would include: cases where spatial influence arises from externalities or spillovers, models that examine the impact of environmental hazards, and models that assess multiplier impacts from economic development projects.

A final advantage to the Bayesian MESS tobit/probit models is that prior information regarding the regression parameters, spatial weight matrix parameters, the spatial dependence parameter and disturbance variances can be introduced. Spatial modeling typically involves large samples, so prior information regarding regression parameters is unlikely to exert much influence on the regression estimates. However, priors placed on the spatial weight structure used in the model or the spatial dependence parameter could exert an important influence, since these dimensions of spatial models remain relatively important even in large samples.

We illustrate the spatial tobit method using an application based on generated data constructed using median home prices from a sample of census block groups for Westchester county, New York. Using a generated data sample allows us to examine the accuracy of inferences regarding the imputed values for censored observations, a subject of interest in censored dependent variable applications.

To illustrate the spatial probit model we use a sample of 1996 presidential voting decisions in each of 3110 US counties in the 48 contiguous states. The dependent variable was set to 1 for counties where Clinton won the majority of votes and 0 for those where Dole won the majority.

In Section 5.2 we introduce binary choice models in the context of utility maximization behaviour. Section 5.3 summarizes relevant aspects of a Bayesian spatial model based on the matrix exponential spatial specification introduced in LeSage and Pace (2000). The relatively simple extension to the case of tobit and probit models along with estimation via Markov Chain Monte Carlo methods is also discussed. In Section 5.4 we illustrate the spatial tobit and probit methods with applications.

5.2 Binary Choice Models

Markov Chain Monte Carlo (MCMC) estimation of the Bayesian probit regression model introduced by Chib (1992) and Albert and Chib (1993) relies on a latent variable during estimation. This section motivates this latent variable in the context of the traditional utility maximization derivation of probit models. For simplicity we develop these ideas in the context of a regression model and avoid use of a spatial weight matrix. For our purposes, we might envision that the individuals faced with choice decisions are located at different points in space. Note there should be little problem converting a regression model $y = \mathbf{X}\beta + \varepsilon$ to a spatial autoregressive model $y = \rho\mathbf{W}y + \mathbf{X}\beta + \varepsilon$ in this setting. This translation that introduces spatial dependence in the binary choices represented by y simply indicates that individuals located at nearby points in space should exhibit similar choices. (Of course, the presence of spatial dependence can be subjected to the usual test that $\rho = 0$ for any particular application.)

Albert and Chib (1993) assume that we observe choices for a set of individuals, $i = 1, \ldots. n$ between two (mutually exclusive and collectively exhaustive) alternatives, which we label '0', and '1'. The observed choice for each individual i is treated as the realization of a random choice variable, Y_i, where $Y_i = 1$ if individual i chooses alternative 1 and $Y_i = 0$ otherwise. If we now let the joint vector of all utility differences for individuals be denoted by $z' = (z_1, \ldots, z_n)$, and similarly, let $\mathbf{X}' = (x_1, \ldots, x_n)$, and $\varepsilon' = (\varepsilon_1, \ldots, \varepsilon_n)$, the regression model in (6.1) arises.

$$z = \mathbf{X}\beta + \varepsilon \tag{6.1}$$

Finally, if $\delta(A)$ denotes the indicator function for each event A (in the appropriate underlying probability space), so that $\delta(A) = 1$ for all outcomes in which A occurs and $\delta(A) = 0$ otherwise, then by definition

$$Pr(Y_i = 1|z_i) = \delta(z_i > 0) \tag{6.2}$$
$$Pr(Y_i = 0|z_i) = \delta(z_i \leq 0) \tag{6.3}$$

If the outcome value of Y_i that we observe is denoted by y_i {0, 1}, then (following Albert and Chib, 1993) these relations may be combined as follows:

$$Pr(Y_i = y_i) = \delta(y_i = 1)\,\delta(z_i > 0) + \delta(y_i = 0)\,\delta(z_i \le 0) \tag{6.4}$$

Hence, letting $Y = (Y_1, \ldots, Y_n)$, and assuming that the choices, $Y_1, \ldots, Y_n$ are conditionally independent given $z_1, \ldots, z_n$, it follows that for each possible observed set of choice outcomes, $y \in \{0,1\}^n$,

$$Pr(Y = y|z) = \prod_{i=1}^{n} \{\delta(y_i = 1)\delta(z_i > 0) + \delta(y_i = 0)\delta(z_i \le 0)\} \tag{6.5}$$

We will see later that this view of the probit model (and a similar approach in the case of the tobit model) will allow us to introduce the latent variable vector z during MCMC estimation. This greatly simplifies estimation of these models since the conditional distribution of z takes the form of a truncated normal distribution that is easy to sample from.

6.3 The Bayesian MESS Model

The Bayesian MESS model is based on the specification shown in (6.6). Note that we rely on y^T to represent the censored dependent variable in the tobit model and y^P to indicate the binary dependent variable of the probit model.

$$\begin{aligned}
&Sy = \mathbf{X}\beta + \varepsilon\\
&S = e^{\alpha \mathbf{W}}\\
&y^T = L,\ Z \ge L,\ y^T = y,\ Z < L\\
&y^P = 1,\ Z > 0,\ y^P = 0,\ Z \le 0\\
&W = \sum_{i=1}^{m} \left(\rho^i N_i / \sum_{i=1}^{m} \rho^i \right)\\
&\varepsilon \sim N(0, \sigma^2 \mathbf{V})\quad \mathbf{V} = diag(v_1, \ldots, v_n)
\end{aligned} \tag{6.6}$$

Initially, we focus our discussion on the model equation: $\mathbf{S}y = \mathbf{X}\beta + \varepsilon$, where y is the usual nx1 vector of observations on the dependent variable which we take to be censored (y^T) or binary (y^P). $\mathbf{X}$ represents the nxk matrix of explanatory variables, $\mathbf{S}$ is a positive definite nxn matrix that we describe in the sequel, and the n-element vector ε is assumed distributed $N(0, \sigma^2\mathbf{V})$, where $\mathbf{V} = diag(v_1, \ldots, v_n)'$ represent a set of variance scalars to be estimated (also described later).

Pace and LeSage (2000) use the matrix exponential defined in (6.7) to model **S** as:

$$\mathbf{S} = e^{\alpha W} = \sum_{i=0}^{\infty} \frac{\alpha^i \mathbf{W}^i}{i!} \quad \mathbf{W} = \frac{\sum_{i=1}^{m} \rho^i \mathbf{N_i}}{\sum_{i=1}^{m} \rho^i} \tag{6.7}$$

where **W** represents an *nxn* non-negative matrix with zeros on the diagonal and α represents a scalar real parameter. The terms $\mathbf{N_i}$ represent individual nearest neighbour matrices containing the ith nearest neighbour constructed in the following fashion. The main diagonal of N_i contains zeros and each row contains zeros for non-neighbouring observations and a value of unity for the observation containing the *i*th nearest neighbour. For example, the matrix N_3 contains information about the third nearest neighbours to each of the n observations in our sample. The spatial weight matrix **W** contains information about all neighbours ranging from 1 to *m*, where m represents the maximum number of neighbours considered. In typical spatial applications *m* might range between 4 and 30, with m = 30 reflecting the largest number of neighbours considered.

In (6.7), ρ^i weights the relative effect of the *i*th individual neighbour matrix relative to the weight received by the nearest neighbour. To see the role of the spatial decay parameter ρ, consider that a value of $\rho = 0.87$ implies a decay profile where the nearest neighbour receives a weight of $\rho / \sum_{i=1}^{m} \rho^i = 0.2295$ and the 6th nearest neighbour is assigned a weight of $\rho^6 / \sum_{i=1}^{m} \rho^i = 0.1144$, indicating that it exerts less than 1/2 the influence of the nearest neighbour. We might think of this value of ρ as indicating a 'half-life' of six neighbours. On the other hand, a value of $\rho = 0.95$ has a half-life between 14 and 15 neighbours. The flexibility arising from this type of weight specification adds to the burden of estimation requiring that we produce estimates for the parameters ρ and m. Together these parameters determine the nature and extent of the spatial weight structure based on the sample data.

This specification for **W** produces a weight matrix that works in conjunction with the matrix exponential transformation as follows.

$$\mathrm{S} = e^{\alpha W} = \frac{\alpha^1 \mathrm{W}^1}{1!} + \frac{\alpha^2 \mathrm{W}^2}{2!} + \frac{\alpha^3 \mathrm{W}^3}{3!} + \ldots \tag{6.8}$$

where by construction **W** represents an *nxn* non-negative matrix with zeros on the diagonal and α represents a scalar parameter to be estimated. Our specification of **W** results in a situation where $\mathbf{W}^2$ reflects neighbours to the nearest neighbours and similar relations hold for higher-order powers of **W**. Thus the matrix exponential **S**, associated with matrix **W**, can be interpreted as assigning rapidly declining weights for observations involving higher-order neighbouring relationships. We rely on an approximation based on a series expansion that is truncated after a sufficient number of terms to produce convergence in the Taylor's series expansion in (6.8) to values close to **S** based on the infinite series. Convergence depends on $\mathbf{W}^i$ not growing too rapidly and a value for α that is not too large. Pace and LeSage (2000) point out that in this case, the inverse factorial weights will rapidly converge to zero. For the row-stochastic weight matrix specification used here the maximum eigenvalue is 1, implying that powers of this weight matrix would also have a maximum eigenvalue of 1 which facilitates rapid convergence.

Estimates of α that are not significantly different from zero indicate a lack of spatial dependence in the model relationship whereas negative values indicate positive spatial autocorrelation, with larger (negative) magnitudes reflecting more dependence. This parameter plays the role of ρ in the more traditional spatial autoregressive formulation: $y = \rho\mathbf{W}y + \mathbf{X}\beta + \varepsilon$

Extending this model to the case of a tobit model where the dependent variable is truncated due to sample censoring or a probit model where the dependent variable is a binary outcome simply requires that we rely on the latent variable view of Chib (1992) for the tobit model and that of Albert and Chib (1993) set forth in Section 2 for the probit model.

6.3.1 Prior Distributions for the Parameters

Here we turn attention to the priors and assumptions set forth in (6.9). Note that in our exposition we indicate censoring at the right using L in (6.6) as the limiting value, but without loss of generality we might also consider cases where censoring occurs at the left or at both the left and right. Another point to note is that for the probit model $\sigma^2 = 1$, so it does not appear as a parameter to be estimated. For notational simplicity we ignore this point in (6.6).

$$
\begin{aligned}
&\beta \sim N(c,T) \\
&r / v_i \sim \chi^2(r), \forall i \\
&1/\sigma^2 \sim \Gamma(d,v) \\
&\alpha \sim U[-\infty,0], \text{ or } N(\alpha,B) \\
&\rho \sim U[0,1], \\
&m \sim U^D[1,m_{max}], \qquad (6.9)
\end{aligned}
$$

We rely on a normal-gamma prior for the parameters β, σ in (6.9) where the prior mean is c and variance $\mathbf{T}$ for the normal prior on β. Prior parameters d, υ are used in the gamma prior on σ (for the tobit model). In the case of very large samples, the normal-gamma priors for β, σ should exert relatively little influence on the estimates for these parameters. Because of this, one might typically rely on a relatively diffuse implementation of this prior in large samples that can be implemented by setting $\mathbf{T}$ to sufficiently large values.

In contrast, we need to be concerned about the prior assigned to the parameter α that reflects spatial dependence in the model, since it should exert an impact on the estimation outcomes even in large samples because of the important role played by the spatial structure in the model. In the absence of prior information one can assign a relatively diffuse prior by relying on a uniform prior over an interval that encompasses zero to allow for the case of no spatial dependence. Alternatively, if prior information suggests the existence of spatial dependence, one might rely on an informative prior based on a normal distribution centered on a with prior variance B as indicated in (6.9). We rely on a uniform $U[-\infty, 0]$ prior for this parameter in our applied illustrations.

In cases where interest centers on the nature and extent of the spatial structure, one would want to rely on a relatively non-informative prior for the parameters ρ and m that control the spatial weight structure. As in the case of the parameter α, prior information assigned for these parameters can exert an influence on the estimation outcomes even in large samples because of the importance of the spatial structure in the model. A relatively diffuse prior was used here in our applied illustrations in an effort to let the sample data determine the character of spatial influence. We rely on a uniform prior distribution for ρ over the open interval [0, 1] and a discrete uniform distribution for m, the number of nearest neighbours based on the open interval $[0, m_{max}]$, where m_{max} denotes a maximum number of neighbours to be considered in the spatial weight structure. We use U^D to represent the discrete uniform distribution that imposes an integer restriction on values taken by m.

In cases where practitioners have prior knowledge regarding the number of neighbouring observations that are important, or the extent to which spatial influence decays over neighbouring units, informative priors could be developed and used. An example would be a case where we are modeling appraised home prices determined by an appraisal methodology based on three comparables. Here we should be able to place a strong prior on the three nearest neighbours that sold recently and we would assign a prior value of $\rho = 1$ if the appraisal methodology involved a straight average of three comparables.

The relative variance terms $(v_1, v_2, \ldots, v_n)$, are parameters to be estimated, allowing the model to accommodate heteroscedasticity. The prior distribution for the vi terms takes the form of an independent $\chi^2(r)/r$ distribution. Recall that the χ^2 distribution is a single parameter distribution, where we have represented this parameter as r. This allows us to estimate the additional n parameters v_i in the model by adding the single parameter r to our estimation procedure. This type of prior has been used by Geweke (1993) in modeling heteroscedasticity and outliers in the context of linear regression. The specifics regarding the prior assigned to the v_i terms can be motivated by considering that the mean equals unity and the variance of the prior is $2/r$. This implies that as r becomes very large, the terms vi will all approach unity, resulting in $\mathbf{V} = \mathbf{I_n}$, the traditional assumption of constant variance across space. In our tobit application, the role of $\mathbf{V} \neq \mathbf{I_n}$ is to accommodate outliers and observations containing large variances by downweighting these observations. Even in large samples, this prior may exert an impact on the estimation outcome.

This assumed prior for these variance scalars takes on a different interpretation in the context of the probit model. As pointed out by Albert and Chib (1993), a value for r around 7 or 8 will produce estimates approximating a logit model, while large values of r will result in probit estimates. Greene (1997) indicates that the impact of heteroscedasticity in non-spatial probit on the estimates is a relatively unexplored area. The specification employed here allows for disturbances that exhibit 'fatter tails' than those arising from the probit specification where the disturbances are assumed normal. Green (1997) indicates that choice of logit versus probit specifications in applied practice is an unresolved issue, and our approach allows the practitioner to control this dimension of the model using the hyperparameter r.

Given these distributional assumptions, it follows that the prior densities for β, σ^2, α, ρ, m, v_i are given up to constants of proportionality by (6.10), (where we rely on a uniform prior for α).

$$
\begin{aligned}
&\pi(\beta) \propto exp\left[-\frac{1}{2}(\beta-c)'\mathbf{T}^{-1}(\beta-c)\right] \\
&\pi(\sigma^2) \propto \left(\sigma^2\right)^{-(d+1)} exp\left[-\frac{v}{\sigma^2}\right] \\
&\pi(\rho) \propto 1 \\
&\pi(\alpha) \propto 1 \\
&\pi(m) \propto 1 \\
&\pi(v_i) \propto v_i^{-(\frac{r}{2}+1)} exp\left[\frac{r}{2v_i}\right]
\end{aligned}
\qquad (6.10)
$$

6.3.2 Estimating the Model

Given the prior densities from section 6.3.1 along with the likelihood function and assumed prior independence of the parameters, we can establish the posterior joint density for the parameters, $p(\beta, \sigma^2, \rho, \alpha, m, z, V)$. Note that we are treating the latent dependent variable values z for binary or censored observations as parameters of the model that will be estimated. We estimate the posterior distribution for the parameters of our model using a methodology known as Markov Chain Monte Carlo (MCMC). This involves determining the sequence of conditional distributions for the parameters β, σ, z, V in the model and relying on a Metropolis-Hastings algorithm to produce samples for the remaining parameters α, ρ, m. A discussion of this methodology, as well as the conditional distributions for the parameters in the model not involving the latent variables z, are presented in the appendix. We focus here on the latent observations conditional on all other parameters in the model. This distribution is used to produce a random draw for each latent value of z_i.

First note that the model $\mathbf{S}z = \mathbf{X}\beta + \varepsilon$ can be expressed in terms of the vector z as:

$$
\begin{aligned}
z &= \mathbf{S}^{-1}\mathbf{X}\beta + \mathbf{S}^{-1}\varepsilon \\
E[uu'] &= \sigma^2[\mathbf{S'VS}]^{-1} \\
u &= \mathbf{S}^{-1}\varepsilon
\end{aligned}
\tag{6.11}
$$

Which makes it clear that z_i depends linearly on all ε_i, through $\mathbf{S}^{-1}$. If we let δ_{ij} represent the i,jth element of $\mathbf{S}^{-1}$, we can write (6.11) as:

$$
z_i = \sum_j \delta_{ij}\mathbf{X}_j\beta + \sum_j \delta_{ij}\varepsilon_j = \mathbf{X}_j^*\beta + \phi_i \tag{6.12}
$$

where ϕ_i follows a normal distribution when ε_i is normally distributed, and Albert and Chib (1993) show that the binary z_i values follow a truncated normal distribution. The variance of ϕ_i is $\sigma^2\sum_j \delta_{ij}^2 v_j^2$, where we note that in the case of the probit model, $\sigma^2 = 1$.

There are computational advantages associated with the matrix exponential transformation here. First, note that $\mathbf{S} = e^{\alpha\mathbf{W}}$ implies that $\mathbf{S}^{-1} = e^{-\alpha\mathbf{W}}$. Second, we can avoid explicitly forming the matrix $\mathbf{S}$ or $\mathbf{S}^{-1}$, by forming matrix-vector products such as $\mathbf{S}y$, or $\mathbf{S}^{-1}\mathbf{X}$, and one can carry out inverse transformations using the matrix-vector product $\mathbf{S}^{-1}(\mathbf{S}y)$.

One computationally efficient way to sample the latent vector z is to sample the vector **S**z, which has a truncated normal distribution with mean of $\mathbf{X}\beta$ and a variance of $\sigma^2\mathbf{V}$ in the case of the tobit model and a variance of **V** for the probit model (where $\sigma^2 = 1$). After obtaining the vector **S**z by sampling from a truncated normal distribution, we can pre-multiply the vector **S**z by $\mathbf{S}^{-1}$ to arrive at the vector z that we desire. Another approach would be to sample z directly using matrix-vector products to compute the mean and variance implied by (6.11) that involve $\mathbf{S}^{-1}\mathbf{X}\beta$ and $\mathbf{S}^{-1}\varepsilon$. In either case, the conditional means and variances required to carry out the truncated normal draw for the latent vector z are simple to compute, despite the linear dependence of each z_i on all other observations.

To illustrate the matrix-vector products in detail, we define the nxq matrix **B** comprised of powers of **W** times a vector b in (6.13).

$$\mathbf{B} = [b\ \mathbf{B}b\ \mathbf{W}^2 b \ldots \mathbf{W}^{q-1} b] \tag{6.13}$$

Note that, this simple form of sparse matrix-vector multiplication can be implemented without explicit use of sparse matrix multiplication algorithms. Given a list of m neighbours for observation i (i.e., $(i, j_1), (i, j_2), \ldots, (i, j_m)$) and assuming equal weights for the neighbours, the product **W**b for observation i is merely $(b_{j1} + b_{j2} + \ldots + b_{jm})/m$. That is, the sparse matrix-vector products needed to compute **B** in (6.13) require only indexing and addition, two of the fastest operations possible on digital computers.

Bearing in mind this computational approach to sampling for the latent vector z by transforming sampled vectors **S**z, we can formally follow Albert and Chib (1993) and make the crucial observation that based on the view of z as a utility-difference vector, the conditional posterior distribution is:

$$Sz|* = \begin{cases} N(X\beta(V) \text{ truncated to the left at } 0 \text{ if } y_i^P = 1 \\ N(X\beta(V) \text{ truncated to the right at } 0 \text{ if } y_i^P = 0 \end{cases} \tag{6.14}$$

Where * denotes all other parameters in the model on which we condition.

A similar approach can be taken with the tobit model where the latent variable z only applies to observations that are censored, leading to a similar regression interpretation:

$$Sz|* = \begin{cases} N(X\beta(\sigma^2 V) \text{ right-truncated at } L & \text{if } y_i^T \geq L \\ S\, y_i^T & \text{if } y_i^T < L \end{cases} \tag{6.15}$$

MCMC sampling proceeds as follows. Starting with arbitrary initial values for the parameters that we denote $\beta^0, \sigma^0, \alpha^0, \rho^0, m^0, z^0, \mathbf{V}^0$, we sample sequentially from the following set of conditional distributions for the parameters in our model.

1. $p(\beta^0 | \sigma^0, \alpha^0, \rho^0, m^0, z^0, \mathbf{V}^0)$, which is a multinormal distribution with mean and variance given in (A2) of the appendix. We label this up-dated value for the parameter vector β, β^1.
2. $p(\sigma | \beta^1, \sigma^0, \alpha^0, \rho^0, m^0, z^0, \mathbf{V}^0)$, which is chi-squared distributed $n + 2d$ degrees of freedom as shown in (A3) of the appendix. Note that we rely on the updated value of the parameter vector $\beta = \beta^1$ when evaluating this conditional density. We label the updated parameter $\sigma = \sigma^1$ and note that we will continue to employ the updated values of previously sampled parameters when evaluating the next conditional densities in the sequence. (In the case of the probit model this parameter is unity by definition.)
3. $p(Sz | \beta^1, \sigma^1, \alpha^0, \rho^0, \mathbf{V}^0)$, which we sample: using a truncated normal distribution implied by (6.15) for the tobit model and in the case of the probit model we use the left and right-truncated normal distributions suggested by (6.14). We then transform the sampled $\mathbf{S}z$ vectors using $\mathbf{S}^{-1}(\mathbf{S}z)$ as discussed in the text.
4. $p(\alpha | \beta^1, \sigma^1, \rho^0, m^0, z^1, \mathbf{V}^0)$, which we sample using a Metropolis step with a normal proposal density, along with rejection sampling to constrain ρ to the interval $(-\infty, 0)$. We can rely on the likelihood to evaluate the candidate value of α except in the case of a normal $N(a,B)$ prior on α. Here where we rely on: $-e'e/(2\sigma^2) - 0.5\{(\alpha - a)^2/B\sigma^2\}$, with $e = \mathbf{S}z - \mathbf{X}\beta$.
5. $p(\rho | \beta^1, \sigma^1, \alpha^1, m^0, z^1, \mathbf{V}^0)$, which we sample using a Metropolis step based on a uniform distribution that constrains ρ to the interval (0,1). Here again, we rely on the likelihood to evaluate the candidate value of ρ. As in the case of the parameter α it would be easy to implement a normal or some alternative prior distributional form for this parameter.
6. $p(m | \beta^1, \sigma^1, \alpha^1, \rho^1, z^1, \mathbf{V}^0)$, which we sample using a Metropolis step based on a discrete uniform distribution that constrains m to be an integer from the interval $(1, m_{max})$. As in the case of α and β, we rely on the likelihood to evaluate the candidate value of m.
7. $p(\mathbf{V} | \beta^1, \sigma^1, \alpha^0, \rho^0, m^0, z^1)$ which can be obtained from the chi-squared distribution presented in (A4) of the appendix. Note that this draw can be accomplished as a vector, providing greater speed.

Summarizing, we will rely on Metropolis sampling for the parameters α, ρ and m within a sequence of Gibbs sampling steps to obtain β, σ, $\mathbf{V}$ and z, a procedure that is often labeled 'Metropolis within Gibbs sampling' (Carlin and Louis, 1998, p. 181).

6.4 Applied Examples

In section 6.4.1 we illustrate the Bayesian MESS tobit model using a generated data set that allows us to examine the impact of sample censoring on alternative estimation procedures. In the spatial statistical literature some authors simply ignore sample truncation and proceed with standard spatial models. Yet another possibility is to ignore spatial dependence in the sample data and rely on a non-spatial tobit estimation procedure. Using our generated data sample, we explore the impact of these alternative approaches to estimation when the sample data are censored.

Section 6.4.2 provides an illustration of the Bayesian MESS probit model using county-wide results from the 1996 presidential election. In this illustration we see that the predictive performance of a spatial model is vastly superior to a non-spatial probit model.

6.4.1 A Generated Data Example

To illustrate the Bayesian MESS tobit model we generated a data set based on 839 census block-groups in Westchester County, New York. The latitude-longitude centroids for these census block-groups were used to generate a spatially dependent set of y values based on a spatial weight matrix, $W = \sum_{i=1}^{m} \left(\rho^i N_i / \sum_{i=1}^{m} \rho^i \right)$, where the number of nearest neighbours m was set to 10, the spatial decay parameter ρ was set to 0.9. Because the sample data was generated, we can compare the imputed dependent variable values to the actual values, a subject that is often of interest in tobit models.

The explanatory values **X** were based on an intercept term plus five normally distributed random variates with zero means. The variance of the **X** variables were set to produce two levels of signal-to-noise ratios. One based on $\sigma_x^2 = 5$, $\sigma_\varepsilon^2 = 1$, producing a signal-to-noise ratio of 5 and another where the signal-to-noise ratio was 2, generated by setting $\sigma_x^2 = 2$. The signal-to-noise ratio should have an impact on the ability of the tobit model to accurately impute the latent y values associated with censored values of the dependent variable. We report results for the lower signal-to-noise ratio that produced a fit measured by R^2 around 0.7 for the uncensored data. All parameters β associated with the five variables in **X** were set to unity.

A value of α = -1.2 was used that corresponds to a traditional spatial autocorrelation parameter in the spatial autoregressive model: $y = \rho \mathbf{W} y + \mathbf{X}\beta + \varepsilon$ around 0.76 suggesting strong spatial dependence. The generated model conformed to: $\mathbf{S}y = \mathbf{X}\beta + \varepsilon$, so our experiments are based on the correct model specification[1]. Since the data generating process involved homoscedastic disturbances, the vector V of variance scalars was set to unity in these experiments.

We use this experiment to compare estimates from a host of models to assess different issues that arise in practice.

1. A non-spatial tobit model, that we label TOBIT in reporting results. This aspatial tobit model will allow us to assess the impact of bias created by spatial dependence that results from using a non-spatial estimation procedure that takes censoring into account.
2. A MESS model estimated using maximum likelihood methods that relies on the full data sample where no values are censored, that we label MLMESS.
3. A Bayesian MESS model estimated using the full data sample where no values are censored, that we label BMESS. This model should provide an accuracy comparison of estimates based on the complete sample and those from the Bayesian tobit model produced using the censored sample.
4. A Bayesian MESS model inappropriately applied to the censored data sample that we label IMESS. This model is included to assess the impact of ignoring censoring, but taking spatial dependence into account.
5. The Bayesian MESS tobit model that we advocate in Section 6.3 based on the censored data sample. Ideally, estimates from this model that we label TMESS would be identical to those from the BMESS model fitted to the non-censored sample data.

Results from an experiment where 70 percent of the sample observations were censored are shown in Table 6.1 and Table 6.2 shows results for a case where 30 percent of the observations were censored. True values for the parameters used to generate the sample data are presented in parenthesis in the first column next to the parameter labels.

The non-spatial tobit model estimates in Tables 6.1 and 6.2 reflect biases in the β and σ^2 estimates, presumably arising from ignoring spatial dependence in the sample data.

Examining the impact of censoring on the spatial regression methods parameters β, we see a dramatic impact from simply ignoring the presence of sample truncation and estimating the model using the IMESS model. A similar impact arises with maximum likelihood estimation (not shown). A second point is that the BMESS and MLMESS estimates based on the full sample are very close, as we would expect with relatively diffuse or conjugate priors for all parameters. Note that MLMESS estimates were based on the actual number of neighbours and the true value for the spatial decay parameter ρ, whereas the BMESS, IMESS and TMESS models produced estimates for these parameters.

Turning attention to the TMESS estimates for the parameters β, we see that the model does a good job of providing estimates close to those based on the full sample in Table 6.1 where 70 percent of the sample was censored and Table 6.2 with 30 percent of the sample censored. Figure 6.1 illustrates the pattern of degradation in the estimates for

β graphically presenting estimates from an experiment where censoring of the generated sample data ranged from 10 percent to 90 percent of the sample in 10 percent increments. The figure presents only one of the β slope estimates which is representative of the other slope estimates. From the figure we see that the TMESS estimates remain close to BMESS based on the full sample even in cases where a high percentage of the sample is censored. The adverse impact of censoring is evident in the IMESS model that ignores censoring as well as the tobit model that ignores the spatial dependence in the data.

Table 6.1: A comparison of estimates based on 70% sample censoring

Parameters	TOBIT censored Parameters	MLMESS full sample	BMESS full sample	IMESS censored sample	TMESS censored sample
$\beta_1(1)$	2.4919	0.9871	0.9935	-0.7014	0.8992
$\beta_2(1)$	0.9916	0.9897	0.9977	0.2796	0.9137
$\beta_3(1)$	1.1057	1.0096	1.0090	0.3393	0.9589
$\beta_4(1)$	1.2058	0.9820	0.9897	0.3057	0.9586
$\beta_5(1)$	1.0579	0.9902	0.9911	0.2953	0.9446
$\alpha(-1.2)$		-1.1945	-1.2028	-0.9069	-1.1055
$\rho(0.9)$			0.9050	0.8797	0.9003
$m(10)$			10.0000	10.9820	10.0650
$\sigma^2(1)$	9.3107	1.1375	1.0982	2.3668	1.4774
#obs	839	839	839	839	839
#censored	588	0	0	588	588

The impact of ignoring censoring is also apparent in the estimates for the spatial structure represented by α, ρ and m in the tables. The IMESS estimate for α exhibited a distinct upward bias and the posterior distribution for this parameter exhibit a bi-modal character. Figure 6.2 shows the posterior distribution for α from an experiment where the signal-to-noise ratio was 5 and 90 percent of the sample was censored.

Similar inaccuracy arises in the posterior distribution for the spatial decay parameter ρ in the IMESS estimates. Figure 6.3 shows the posterior distributions for ρ from the IMESS model (labeled 'mess' in the figure) and TMESS model (labeled 'tobit') associated with the 70 percent censoring experiment whose results are presented in Table 6.2.

A similar adverse impact arises in the IMESS model for the posterior distribution on the parameter representing the number of neighbours shown for both IMESS (labeled 'mess') and TMESS (labeled 'tobit') models in Figure 6.4 for the case of 70 percent censoring.

Table 6.2: A comparison of estimates based on 30% sample censoring

Parameters	TOBIT censored Parameters	MLMESS full sample	BMESS full sample	IMESS censored sample	TMESS censored sample
$\beta_1(1)$	2.5404	0.9871	0.9944	0.6585	0.9978
$\beta_2(1)$	1.0362	0.9897	0.9987	0.7015	0.9704
$\beta_3(1)$	1.1174	1.0096	1.0088	0.7612	1.0049
$\beta_4(1)$	1.1694	0.9820	0.9892	0.7571	0.9771
$\beta_5(1)$	1.0788	0.9902	0.9912	0.7027	0.9899
$\alpha(-1.2)$		-1.1945	-1.2058	-1.0553	-1.1776
$\rho(0.9)$			0.9049	0.8901	0.8976
$m(10)$			10.0000	10.0000	10.0510
$\sigma^2(1)$	9.0142	1.1375	1.0945	2.6247	1.5296
#obs	839	839	839	839	839
#censored	253	0	0	253	253

Figure 6.1: The impact of censoring on β estimates

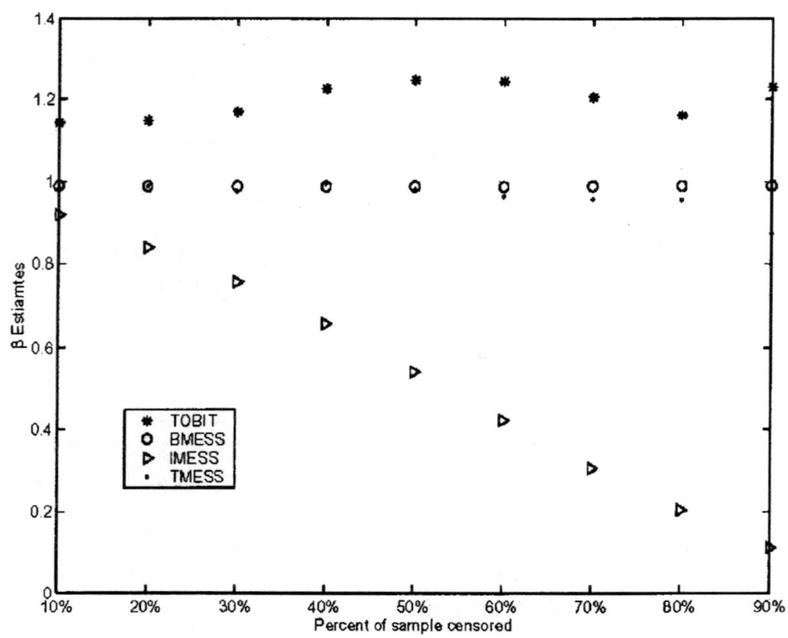

Figure 6.2: Posterior density for α from two models

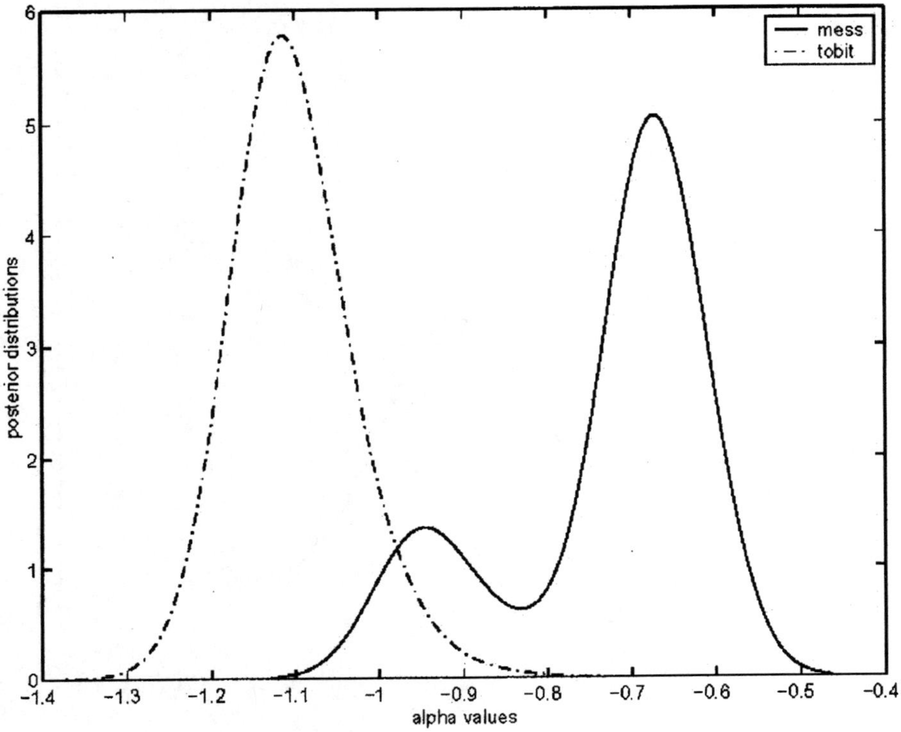

Intuitively, we might expect that censoring in a spatial sample may reflect the pattern of spatial dependence, making it more difficult to estimate the spatial structure when we ignore censoring as in the IMESS model.

Turning to the imputed values for y which can be compared to the actual values in this generated data example, we find that these estimates compare favourably to the actual values even in the case of 60 percent sampling censoring, shown in Figure 6.5. The imputed tobit values (labeled 'tobit imputed' in the figure) are based on the mean of the draws for censored y values from the TMESS model. For comparison, predicted values from the BMESS model (labeled 'mess predicted' in the figure) are also presented. In the face of excessive sample censoring such as 90 percent shown in Figure 6.6, the accuracy of imputed values declines, but is nevertheless far superior to imputation based on the IMESS model that ignores censoring.

Figure 6.3: Posterior density for ρ from two models

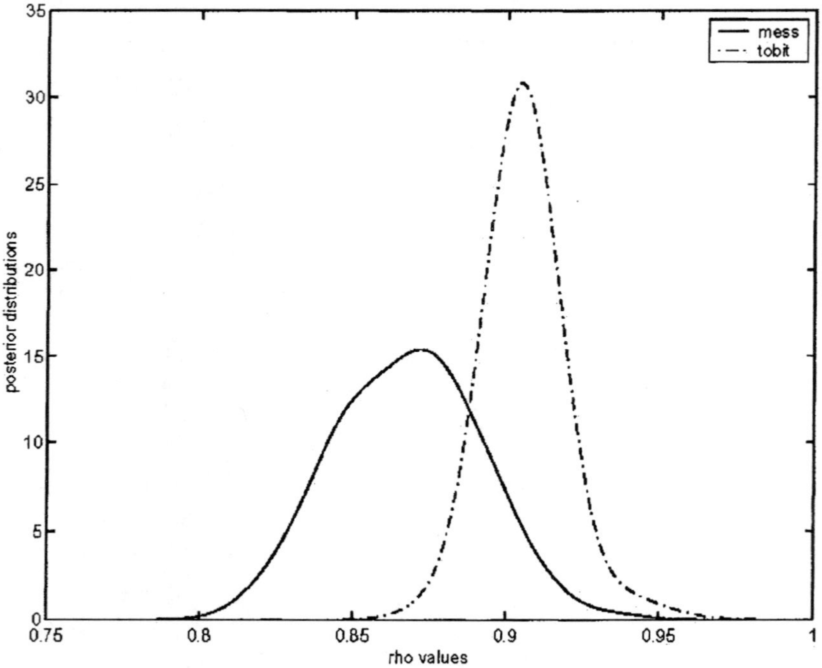

Summarizing the results from these experiments, we demonstrated that ignoring sample censoring in cases where the data exhibit spatial dependence will adversely impact estimates for regression parameters as well as the parameter measuring the extent of spatial dependence. Intuitively, sample censoring is likely to follow a spatial pattern that will adversely impact estimates of the regression relationship as well as the magnitude of spatial dependence. Bayesian MCMC methods provide a simple way to extend our model based on the matrix exponential spatial specification to appropriately take into account sample truncation and provide accurate estimates that avoid the bias inherent in estimates that simply ignore this feature of the sample data.

6.4.2 An Application to Binary Data

To illustrate the probit model in an applied setting we used data on the 1996 presidential voting decisions in each of 3110 US counties in the 48 contiguous states. The dependent variable was set to 1 for counties where Clinton won the majority of votes and 0 for those where Dole won the majority(2). As explanatory variables we used a constant term,

Figure 6.4: Posterior density for *m* (# neighbours) from two models

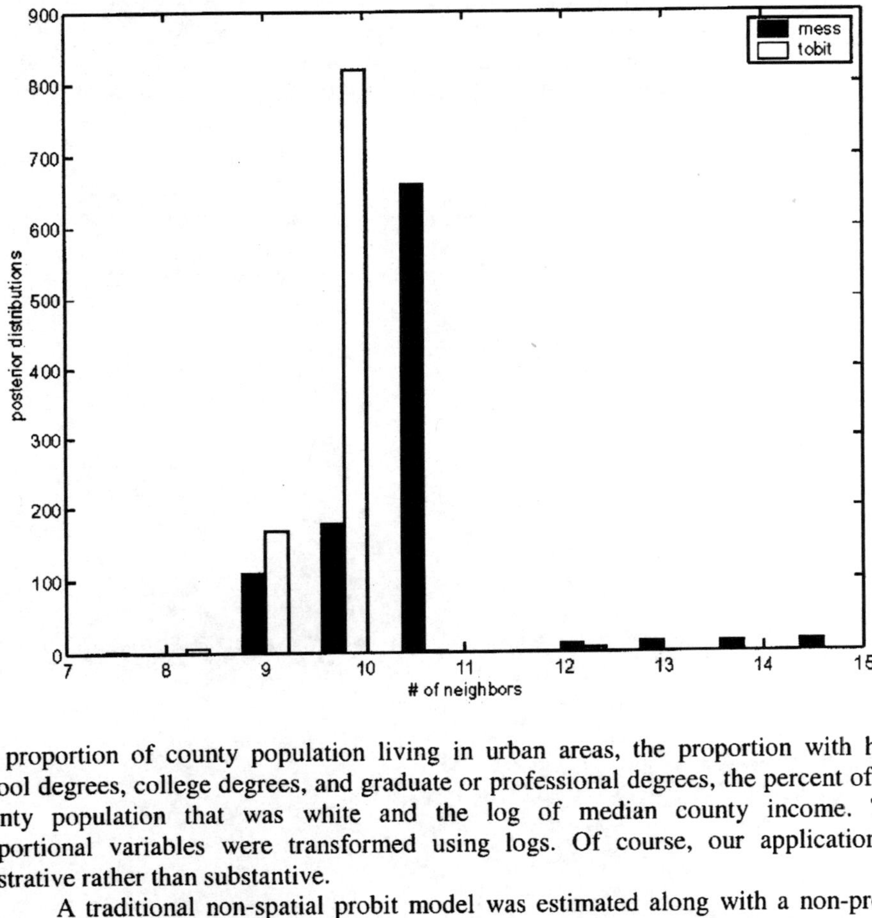

the proportion of county population living in urban areas, the proportion with high school degrees, college degrees, and graduate or professional degrees, the percent of the county population that was white and the log of median county income. The proportional variables were transformed using logs. Of course, our application is illustrative rather than substantive.

A traditional non-spatial probit model was estimated along with a non-probit maximum likelihood MESS model and the Bayesian probit MESS model. Maximum likelihood MESS estimates were produced by setting the number of neighbours equal to 14 and the decay parameter $\rho = 0.99$, based on the Bayesian posterior means for these parameters. Figure 6.7 shows the predictions from the traditional probit model and the non-probit maximum likelihood MESS model, which are in stark contrast to the accuracy of the probit MESS model predictions shown in Figure 6.8. In the figures counties were sorted into those where a majority voted for Dole ($y = 0$) and those that voted for Clinton ($y = 1$). It is clear that taking into account the spatial dependence as well as the binary character of the sample data results in predictions that are vastly superior.

Estimates from these three models are presented in Table 6.3, where marginal significant levels are in parentheses for the maximum likelihood estimates and Bayesian *p-levels* are in parentheses following the MCMC estimates. A marginal significance level of 0.05 indicates that the coefficient estimate was significantly different from zero at the 95% level of significance, and 0.000 was used to signify significance at the 99.9% level or higher.

Figure 6.5: Imputed *y* values for 60 percent censored observations

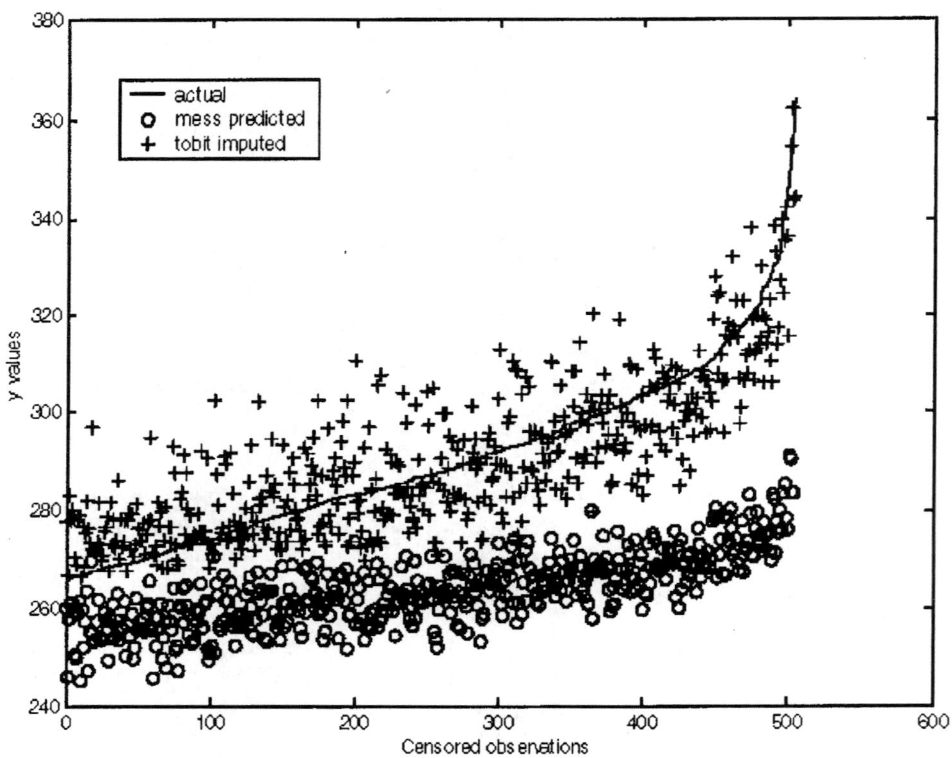

Figure 6.6: Imputed *y* values for 90 percent censored observations

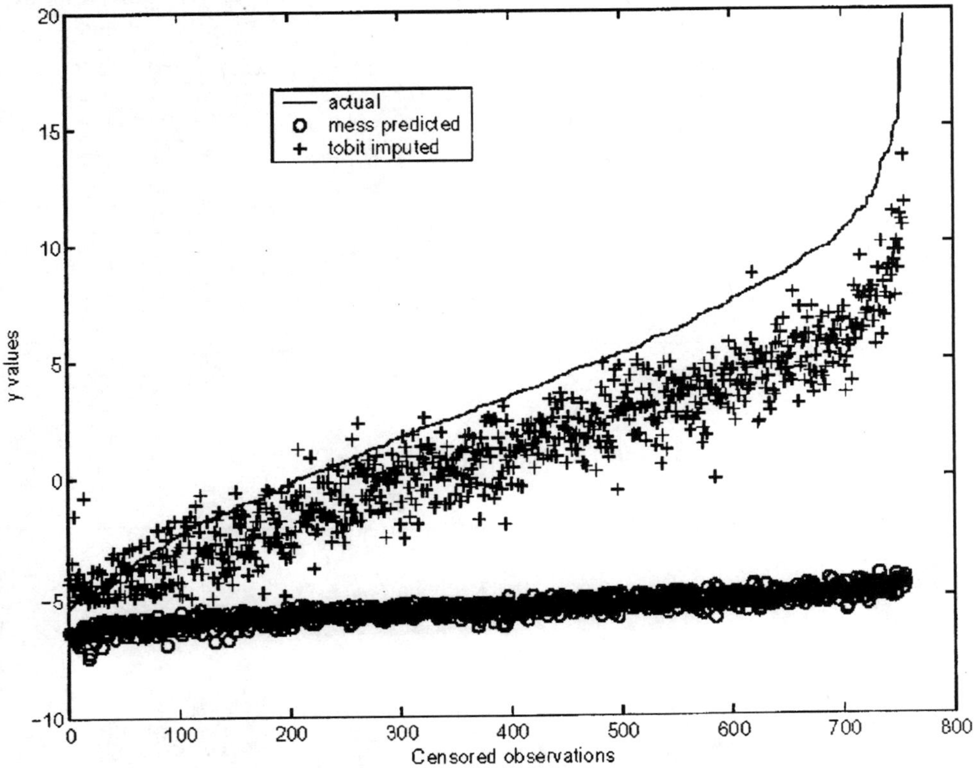

The traditional non-spatial probit model appears to overestimate the influence of education on the voting choices. Given that our educational explanatory variables represent percentage of county-level populations, we can interpret the coefficients on high school, college and graduate degrees as indicating that a 1 percent increase in county-level population with graduate degrees would result in a 14.7 percent higher probability that the county voted for Clinton, using the traditional non-spatial probit model. In contrast, the spatial probit model produced a coefficient estimate of 5.88. A similar situation arose for all other variables where the non-spatial probit model produced larger estimated impacts than the spatial probit model. Intuitively, ignoring spatial dependence leads the non-spatial probit model to attribute variation in the dependent variable to explanatory variables rather than assign this variation to spatial dependence. In contrast to this tendency of the non-spatial probit to overestimate the influence of explanatory variables, the non-tobit MESS model systematically underestimates the influence of explanatory variables.

Table 6.3: A comparison of Probit estimates

Variable	Maximum Likelihood Probit	Maximum Likelihood MESS	Bayesian MCMC MESS Probit
Constant	2.130 (0.071)	1.074 (0.035)	1.652 (0.063)
% urban	0.015 (0.905)	-0.026 (0.562)	-0.062 (0.305)
% high school	5.687 (0.000)	1.357 (0.000)	2.500 (0.000)
% college degrees	-4.694 (0.003)	-0.923 (0.115)	-1.756 (0.115)
% graduate degrees	14.701 (0.000)	3.123 (0.000)	5.888 (0.006)
% white	-2.502 (0.000)	-0.685 (0.000)	-1.300 (0.000)
Median income	-0.192 (0.135)	-0.071 (0.187)	-0.142 (0.117)
α		-0.720 (0.000)	-1.490 (0.000)
ρ			0.993 (0.000)
M			14.000 (0.000)

Figure 6.7: Predicted 0,1 values for non-spatial probit and MESS models

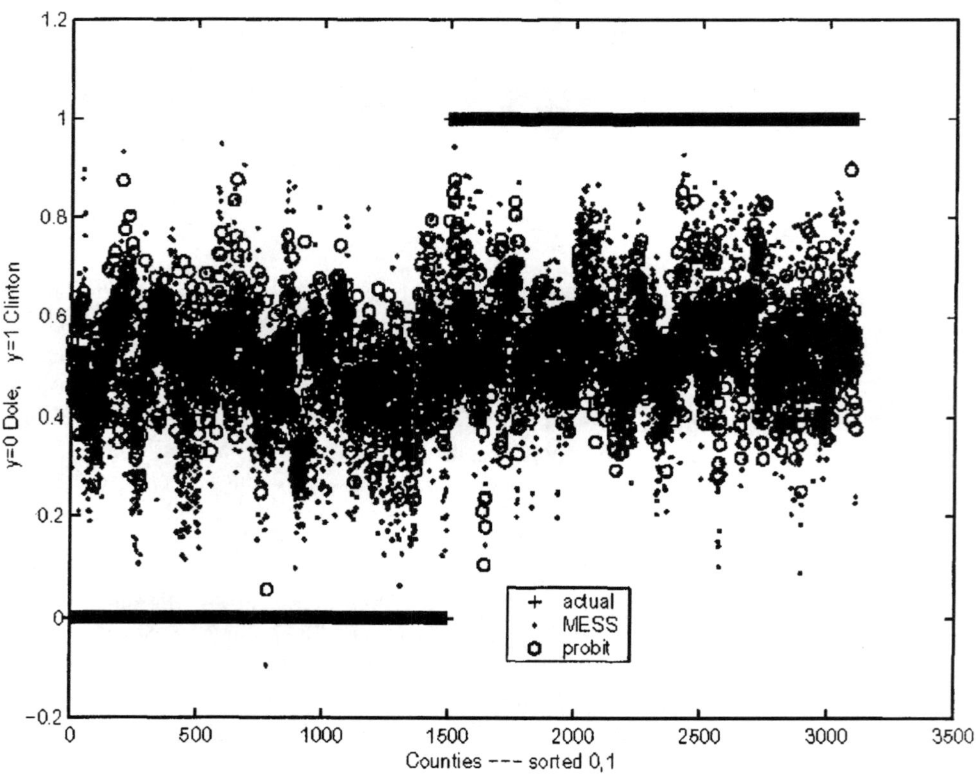

Figure 6.8: Predicted 0,1 values for spatial probit MESS models

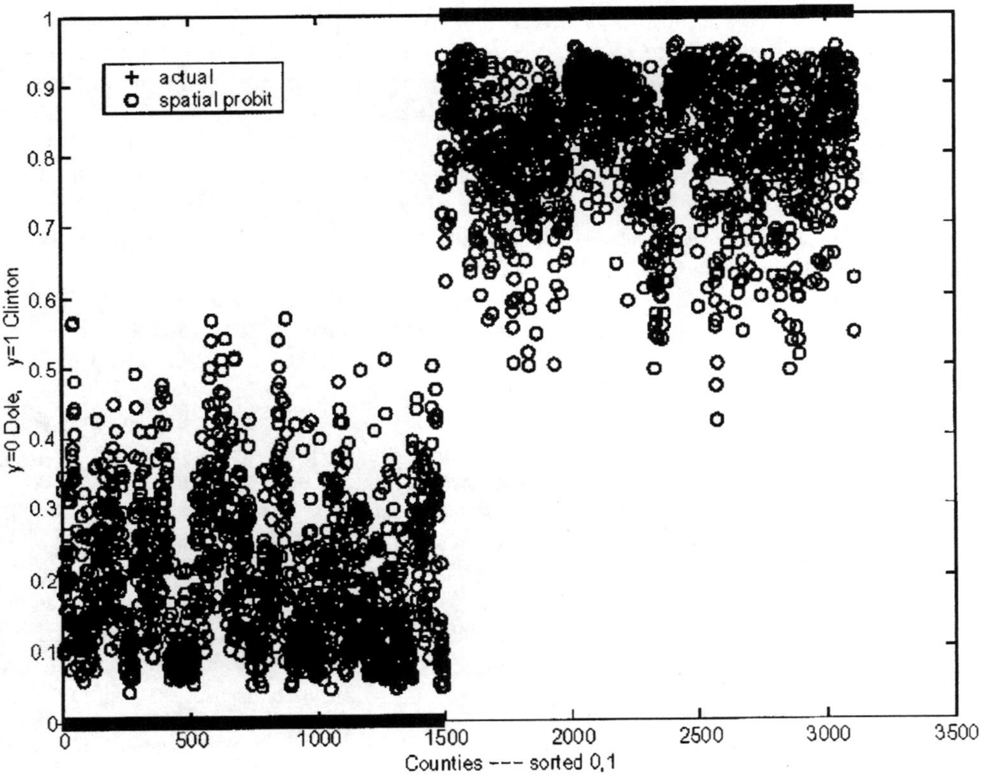

The estimates for α point to significant spatial dependence in the sample data with the non-probit MESS model producing an estimate around -0.7 and the probit MESS around -1.5 suggesting stronger dependence when one takes the binary character of the dependent variable into account.

6.5 Conclusion

Estimation of spatial tobit and probit models has been a relatively neglected area in the spatial econometrics literature. LeSage (2000) argues that Bayesian methods in conjunction with MCMC estimation represent a unified methodology for spatial models that involve continuous or dichotomous dependent variables as well as models that contain outliers or non-constant variance.

This work demonstrates the relative ease with which the new matrix exponential spatial specification introduced in Pace and LeSage (2000) could be extended to deal with dichotomous dependent variables as well as non-constant variances using this methodology.

The ability to work with truncated spatial data samples should allow applied work in situations that often arise due to confidentiality concerns. Experiments carried out here suggest that estimates comparable to those based on the full uncensored data sample can be achieved even in the face of 60 percent sample censoring.

Spatial probit models arise from marketing surveys that contain a spatial dimension or other situations where interest centers on the presence or absence of a quantity of interest at various points in space. An illustration here demonstrated a vast improvement in predictive power from taking into account spatial dependence in the binary sample of presidential election results from 1996.

The methods introduced here hold computational speed advantages over those introduced in LeSage (2000), and they allow a more flexible spatial weight structure to be used than those employed in conventional spatial autoregressive models. Algorithms to implement these methods programmed in the MATLAB matrix language are available at www.spatial-econometrics.com as part of the spatial econometrics toolbox available at that website.

Acknowledgements

The first author would like to thank Tony E. Smith for making clear the relationships between utility theory and spatial probit models as well as numerous other insights regarding Bayesian spatial probit estimation used in this study

Appendix

Given the prior densities from section 6.3.1, the Bayesian identity

$$p(\beta, \sigma^2, \mathbf{V}, \rho, \alpha, m)\, p(y) = p(y|\beta, \sigma^2, \mathbf{V}, \rho, \alpha, m)\, \pi(\beta, \sigma^2, \mathbf{V}, \rho, \alpha, m) \tag{A.1}$$

together with the assumed prior independence of the parameters allows us to establish the posterior joint density for the parameters, $p(\beta, \sigma^2, \mathbf{V}, \rho, \alpha, m)$. This posterior has no analytical solution, a problem that has often plagued Bayesian methods in the past. We can however derive a numerical posterior distribution for the parameters in our model using a methodology known as Markov Chain Monte Carlo (MCMC).

MCMC is based on the idea that rather than compute the posterior density of our parameters based on the expression in (A.1), we would be just as happy to have a large random sample from the posterior of our parameters, which we designate $p(\theta|D)$, using θ to denote the parameters and D the sample data. If the sample from $p(\theta|D)$ were large enough, we could approximate the form of the probability density using kernel density estimators or histograms, eliminating the need to know the precise analytical form of the density.

We rely on Metropolis-Hastings to compute the posterior distributions for the parameters α, ρ and m in the MESS model. A normal distribution is used as the proposal density for α and rejection sampling can be used to constrain α to the range (-1, 0) which imposes positive spatial autocorrelation. This prior restriction is often used in spatial modeling because negative spatial correlation is difficult to motivate. A uniform proposal distribution for ρ over the interval (0,1) was used along with a discrete uniform for m over the interval $(1, m_{max})$.

The parameters β, $\mathbf{V}$ and σ in the MESS model can be estimated using draws from the conditional distributions of these parameters, a process known as Gibbs sampling. Assume a parameter vector $\theta = (\theta_1, \theta_2)$, a prior $p(\theta)$, and likelihood $l(\theta|y, \mathbf{X})$, that produces a posterior distribution $p(\theta|D) = c\ p(\theta)l(\theta|y, \mathbf{X})$, with c a normalizing constant. The case we encounter in the MESS model is one where the posterior distribution over all parameters is difficult to work with. However, if we partition our parameters into two sets θ_1, θ_2 and had initial estimates for θ_1, we could estimate θ_2 conditional on θ_1 using $p(\theta_2|D, \hat{\theta}_1)$. As a concrete example, in our case an estimate of β conditional on σ, V, α, ρ, m is very easy to derive and compute. Denote the estimate, $\hat{\theta}_2$ derived by using the posterior mean or mode of $p(\theta_2|D, \hat{\theta}_1)$, and consider that we are now able to construct a new estimate of θ_1 based on the conditional distribution $p(\theta_1|D, \hat{\theta}_2)$. Note that for our problem it is also easy to compute estimates of σ and $\mathbf{V}$ conditional on the other parameters in the MESS model. This new estimate for θ_1 can be used to construct another value for θ_2, and so on.

Summarizing, we will rely on Metropolis sampling for the parameters α, ρ and m within a sequence of Gibbs sampling steps to obtain β, σ and **V**, a procedure that is often labeled 'Metropolis within Gibbs sampling' (Gelman, Carlin, Stern and Rubin, 1995).

To implement our Metropolis within Gibbs sampling approach to estimation we need the conditional distributions for β, σ and **V** which are presented here.

For the case of the parameter vector β conditional on the other parameters in the model, α, σ, **V**, ρ, m we find that:

$$p(\beta|\alpha,\sigma,\mathbf{V},\rho,m) \sim N(\bar{b},\sigma^2\bar{\mathbf{B}})$$
$$\bar{b} = (\mathbf{X}'\mathbf{V}^{-1}\mathbf{X}+\mathbf{T}^{-1})^{-1}(\mathbf{X}'\mathbf{V}^{-1}\mathbf{S}y+\sigma^2\mathbf{T}^{-1}c)$$
$$\bar{\mathbf{B}} = (\mathbf{X}'\mathbf{V}^{-1}\mathbf{X}+\mathbf{T}^{-1})^{-1} \qquad \text{(A.2)}$$

Note that given the parameters **V**, α, ρ, σ and m, the vector **S**y and $\mathbf{X'V}^{-1}\mathbf{X}$ can be treated as known, making this conditional distribution easy to compute and sample from. This is often the case in MCMC estimation, which makes the method attractive.

The conditional distribution of σ is shown in (18), (see Gelman, Carlin, Stern and Rubin, 1995).

$$p(\sigma^2|\alpha,\sigma,\mathbf{V},\rho,m) \propto (\sigma^2)^{-\left(\frac{n}{2}+d+1\right)} exp\left[-e'\mathbf{V}^{-1}e+\frac{2\upsilon}{2\sigma^2}\right] \qquad \text{(A.3)}$$

where e=**S**y-**X**β, which is proportional to an inverse gamma distribution with parameters $(n/2) + d$ and $e'\mathbf{V}^{-1}e + 2v$.

Geweke (1993) shows that the conditional distribution of **V** given the other parameters is proportional to a chi-square density with $r+1$ degrees of freedom. Specifically, we can express the conditional posterior of each v_i as:

$$\frac{e_i^2 + r}{v_i}\Bigg|(\beta,\alpha,\sigma^2,v_{-i},\rho,m) \sim \chi^2(r+1) \qquad \text{(A.4)}$$

As noted above, the conditional distributions for α, ρ and m take unknown distributional forms that require Metropolis-Hastings sampling.

Endnotes

[1]LeSage and Pace (2000) provide evidence that the matrix exponential specification produces estimates for β and σ as well as α estimates of spatial autocorrelation consistent with values of ρ in cases where the generating process is a more traditional spatial autoregressive form: $y = \rho \mathbf{W} y + \mathbf{X}\beta + \varepsilon$. We do not pursue this point here.

[2]The third party candidacy of Perot was ignored and only votes for Clinton and Dole were used to make this classification of 0,1 values.

Part III

Spatial Dependence and Spatial Errors

7 Properties of Tests for Spatial Error Components: A Further Analysis

Harry H. Kelejian
Yevgeny Yuzefovich
University of Maryland

Abstract

Recent Monte Carlo studies suggest that the power of a test for spatial error correlation introduced by Kelejian and Robinson (1992) is low compared to Lagrangian multiplier tests or to the test based on Moran's *I* statistic. The purpose of this chapter is to point out shortcomings in the designs of these Monte Carlo studies. We also give Monte Carlo results which are not subject to these shortcomings. Our results suggest that, for the case considered in Anselin and Moreno, the power of the Kelejian-Robinson test is much higher than that reported in Anselin and Moreno, but not as great as their Lagrangian multiplier test. This is important because our results imply a robustness of the Lagrangian multiplier test that was not recognized in Anselin and Moreno. An important by-product of this chapter is that we suggest a number of new error term processes which intertwine spatial correlation with heteroskedasticity that relates to exogenous variables. These processes should be of interest to researchers in the field.

7.1 Introduction

Monte Carlo methods are useful for studying and comparing the small sample properties of parameter estimators, and tests of hypotheses. Using such methods, Anselin and Florax (1995) and Anselin and Moreno (2000) compare, among other things, the power of a test for spatial correlation which was suggested by Kelejian and Robinson (1992) to that of Lagrangian multiplier tests, and the test based on Moran's *I* statistic, henceforth the *MI* test[1]. In both of these studies, the test suggested by Kelejian and Robinson (1992) was found to have lower power than the other tests considered. Indeed, in Anselin and Moreno (2000) a revised and more powerful version of the Kelejian-Robinson test had only 56% of the power of their Lagrangian multiplier test for sample sizes between 49 and 121.

In this chapter we point out shortcomings in the designs of these Monte Carlo studies. First, and most importantly, the Kelejian-Robinson test was based on the premise that if the error term is spatially correlated, that spatial correlation is intertwined with heteroskedasticity which relates to the exogenous variables of the model[2].The Monte Carlo models in Anselin and Florax (1995), and in Anselin and Moreno (2000) did not consider spatial correlation which involved such a heteroskedastic component. In such scenarios one would indeed expect the power of a test which presupposes heteroskedasticity to be low!

Our second criticism is more subtle, and undoubtedly relates to most Monte Carlo studies. Specifically, as is usually the case, Anselin and Florax (1995) and Anselin and Moreno (2000) conduct a wide variety of experiments in which the model parameters are varied in order to understand the relationship between their focus of attention, namely the power of tests for spatial correlation, and the model parameters. However, this was done without regard for the induced changes in the R^2 statistic. For example, by a technique described below, we analytically determine that the average of the R^2 statistics corresponding to one parameter selection in the Anselin and Moreno (2000) paper must have been roughly .71, but .87 for another parameter selection[3]. The power of their considered tests relates to this R^2 statistic since the test statistics involved are based on estimated parameters whose estimation precision increases as the model noise factors decrease; this, in turn, would correspond to an increase in the R^2 statistic. Thus, e.g., if the power of a test is high corresponding to a given selection of model parameters this could be due not to the relative magnitudes of the parameters but to the corresponding increase in the R^2 statistic. One way to determine the sensitivity of the Anselin-Moreno (2000) results to their corresponding R^2 statistics would be in terms of response functions, see e.g. Kelejian and Prucha (1999). Typically, the accurate estimation of such functions requires a large set of Monte Carlo experiments, which is beyond the scope and intent of this chapter.

As perhaps suggested above, the purpose of this chapter is limited. We describe the nature of the shortcomings in previous studies, and we give some Monte Carlo results which are not subject to these shortcomings. An important by-product of our analysis is that we introduce several disturbance processes which involve both spatial correlation and heteroskedasticity which relates to exogenous variables. We describe why the heteroskedasticity in some of these processes is 'muted' while in others it is not. This distinction should be recognized in further studies in this field[4]. Our Monte Carlo results are given only to highlight the issues involved and are not meant to be exhaustive. In doing this we adapt the general framework described in Anselin and Moreno (2000). We focus our attention on the Kelejian-Robinson test and the Lagrangian multiplier test described in Anselin and Moreno (2000)[5].

Our results are both expected and unexpected. Our expected result is that the power of the Kelejian-Robinson test is much higher than that reported in Anselin and Moreno (2000). As an illustration, for sample sizes between 49 and 121 we estimate the power of the Kelejian-Robinson test to be roughly 64% higher than that reported in Anselin and Moreno (2000). The unexpected result is that the Lagrangian multiplier test described in Anselin and Moreno (2000), which presupposes the absence of heteroskedasticity, remains very powerful even in the face of strong heteroskedasticity. For example, the Kelejian-Robinson test explicitly accounts for both spatial correlation as well as heteroskedasticty. Nevertheless, the power of the Kelejian-Robinson test in detecting spatial correlation, averaged over our experiments which incorporate both spatial correlation as well as heteroskedasticity, is only 92% of that of the Lagrangian multiplier test! This implies a certain robustness of that Lagrangian multiplier test which was not previously noted. There are, however, qualifications. Specifically, the Kelejian-Robinson test does not assume detailed knowledge of the weighting matrix; the Lagrangian multiplier test does. On the other hand, the Kelejian- Robinson test assumes a certain pattern of the heteroskedasticty which may be involved; the Lagrangian multiplier test assumes its absence. All of this suggests that the evaluation of power comparisons between these tests is not entirely straightforward. We return to these issues below.

7.2 The Model

We consider the following Monte Carlo model

$$y_i = b_0 + b_1 x_i + \varepsilon_i,\ i = 1, \ldots, N \tag{7.1}$$

where y_i is the i-th observation on the dependent variable, x_i is the i-th observation on the exogenous independent variable, b_0 and b_1 are parameters, and ε_i is the disturbance term. The matrix of values corresponding to (7.1) is

$$y = \mathbf{X}\beta + \varepsilon \tag{7.2}$$

where $y' = (y_1, \ldots, y_N)$, the i-th row of the $N \times 2$ matrix $\mathbf{X}$ is $(1, x_i)$, $\beta' = (b_0, b_1)$, and $\varepsilon' = (\varepsilon_1, \ldots, \varepsilon_N)$.

In our Monte Carlo experiments we consider three values of the sample size, namely N= 49, 81, and 121. Concerning the regressor, x_i, we generate 121 values as independent selections from the uniform distribution (0, 10). The values of the regressor in samples sizes N= 49, 81, 121 are taken as the first N values of the 121 values. We consider 5000 iterations of each Monte Carlo experiment. The values of the regressor are held fixed in the iterations. In all of the experiments, $b_0 = b_1 = 1$.

In our Monte Carlo experiments we consider the spatial error component heteroskedastic disturbance process: SECH

$$\varepsilon = \mathbf{D}^{\alpha}\mathbf{M}u + \psi, \ \alpha = 1, 2 \tag{7.3}$$

where $\mathbf{D}^{\alpha}$ is the diagonal matrix

$$\mathbf{D}^{\alpha} = \begin{bmatrix} x_1^{\alpha} & 0 & 0 & \cdots & 0 \\ 0 & x_2^{\alpha} & 0 & \cdots & 0 \\ 0 & 0 & x_3^{\alpha} & \cdots & 0 \\ \cdots & \cdots & \cdots & \cdots & \cdots \\ 0 & 0 & 0 & \cdots & x_N^{\alpha} \end{bmatrix} \tag{7.4}$$

M is a rook weighting matrix, u and ψ are $N \times 1$ random vectors which are, respectively, $N(0, \sigma^2\mathbf{I}_N)$ and $N(0, \sigma_\psi^2 \mathbf{I}_N)$, and u and ψ are independent[6].

The process defined by (7.3) is a generalization of the error component process introduced in Kelejian and Robinson (1995). The generalization is that the variances and covariances of the elements of the error term ε have a heteroskedastic component which relates to the independent variable. To see this, note first that $E(\varepsilon) = 0$ and, given the independence of u and ε,

$$\begin{aligned} E(\varepsilon\varepsilon') &= \mathbf{\Omega}_\varepsilon \\ &= \sigma_u^2 \mathbf{D}^{\alpha}\mathbf{M}\mathbf{M}'\mathbf{D}^{\alpha} + \sigma_\psi^2 \mathbf{I}_N \end{aligned} \tag{7.5}$$

The i,j-th element of $\mathbf{\Omega}_\varepsilon$, namely the covariance of ε_i and ε_j, is ω_{ij} where

$$\begin{aligned} \varpi_{ij} &= \sigma_u^2 x_i^{\alpha} x_j^{\alpha} m_{i.} m'_{j.} + \sigma_\psi^2 \delta_{ij} \\ \delta_{ii} &= 1 \\ \delta_{ij} &= 0, if \quad i \neq j \end{aligned} \tag{7.6}$$

where $m_{i.}$ is the i-th row of **M**. Clearly, unless $\sigma_u^2 = 0$ the elements of ε defined in (7.3) will be both spatially correlated as well as heteroskedastic. Clearly, the heteroskedastic component will be more severe, ceteris paribus, if σ_u^2 is large relative to σ_ψ^2. Also, this component will be more 'severe' when $\alpha = 2$ than when $\alpha = 1$. In Section 7.5 we introduce three other disturbance processes which contain both spatial correlation and heteroskedasticity. In two of those processes the heteroskedasticity is 'muted' in the sense that each non-zero covariance involves a weighted sum of all N values of the regressor; for the process in (7.3), as well as for the third one in Section 7.5 the heteroskedasticty is not muted in this sense.

Recall that $b_0 = b_1 = 1$, and that **M** is taken as a rook in all of our Monte Carlo experiments. Also, the values of x_i are independently and randomly generated from a uniform (0, 10) in the 5000 iterations. Thus, at this point we can describe a Monte Carlo experiment in terms of the set of parameters $(N, \alpha, \sigma_u^2, \sigma_\psi^2)$. Since we have already specified our experimental values of N and α, we now specify σ_u^2 and σ_ψ^2. The experimental values of these two parameters are taken to make the average R^2 statistic corresponding to the least squares estimation of (1), over the 5000 iterations for each experiment, approximately equal to .75. These values are described in Table 7.1. That table also describes the experimental values of σ_ψ^2 corresponding to H_0: $\sigma_u^2 = 0$ under the heading in the table $\lambda = \sigma_u^2/\sigma_\psi^2 = 0$. Again, the values of σ_ψ^2 are such that the average R^2 statistic over the 5000 iterations for each experiment is roughly 0.75.

The Determination of σ_u^2 u and σ_ψ^2

Let $\bar{y} = N^{-1}\sum_{i=1}^{N} y_i$, and $\hat{y}_i = \hat{b}_0 + \hat{b}_1 x_i$, $i=1,2,\ldots, N$, where $\hat{b}_0$ and $\hat{b}_1$ are the least squares estimators of b_0 and b_1 based on (7.1). Then the R^2 statistics[7]

$$R^2 = \frac{\sum_{i=1}^{N}(\hat{y}_i - \bar{y})^2}{\sum_{i=1}^{N}(y_i - \bar{y})^2} = \frac{\hat{b}_1^2 \sum_{i=1}^{N}(x_i - \bar{x})^2 / N}{\sum_{i=1}^{N}(y_i - \bar{y})^2 / N} \tag{7.7}$$

In our model $b_1 = 1$ and, unconditionally, x_i is *i.i.d.* from the uniform (0, 10). Among other things this implies $E(y_i) = E(\bar{y})$, $i = 1, \ldots, N$. Given this, the population value of R^2, say R_*^2, is[8]

$$R_*^2 = \frac{\sigma_x^2}{\sum_{i=1}^{N} E[y_i - E(y_i)]^2 / N} \tag{7.8}$$

Given our specifications

$$R_*^2 = \frac{\sigma_x^2}{\sigma_x^2 + \sum_{i=1}^{N}\left[\sigma_u^2 E(x_i^{2\alpha}) m_{i.} m_{i.}^{'}\right]/N + \sigma_\psi^2} = \frac{\sigma_x^2}{\sigma_x^2 + \sigma_u^2 E(x_i^{2\alpha})\sum_{i=1}^{N}\left[m_{i.} m_{i.}^{'}/N\right] + \sigma_\psi^2} \tag{7.9}$$

Our assumptions imply that $\sigma_x^2 = 25/3$ and $E(x_i^{2\alpha}) = 10^{2\alpha}/(2\alpha+1)$. The value of $\sum_{i=1}^{N} m_{i.} m_{i.}^{'}/N$ is easily calculated from the rook specification for each sample size. Therefore pairs of values of σ_u^2 and σ_ψ^2 corresponding to a given value of R_*^2, in our case 0.75, can be determined from (7.9). The pairs of values in Table 7.1 are consistent with (7.9). In the Anselin and Moreno (2000) 'rook' models $\alpha = 0$ and the regression contains two regressors which are independent of each other, of the error terms, and are each *i.i.d.* as a uniform (0, 10). Therefore inferences concerning their R^2 statistics can be obtained from (7.9) by multiplying σ_x^2 everywhere by 2, and setting $\alpha = 0$.

7.3 The Considered Tests

The Kelejian-Robinson Tests

We consider two versions of the test described in Kelejian and Robinson (1992). Kelejian and Robinson (1992) assumed that if the researcher suspects spatial correlation that researcher assumes that the non-zero covariances are of the form

$$E(\varepsilon_i \varepsilon_j) = z_{ij}\,\gamma \tag{7.10}$$

where z_{ij} is a vector whose values are functions of the exogenous variables relating to the i-th and j-th cross-sectional units. In the framework of our Monte Carlo model the Kelejian-Robinson formulation would be $z_{ij} = (1, x_i\, x_j)$. In the context of the SECH model in (5), one would consider all values of $i < j$ such that $m_{i.} m_{j.}^{'} \neq 0$[9]. Note that this selection only requires that the researcher know whether or not the i-th and j-th units have neighbours in common; it does not require a detailed knowledge of the weighting matrix **M**. It should also be clear that a specification such as that in (7.7) involves both spatial correlation and heteroskedasticity.

Table 7.1: Experimental values of σ_u^2 and σ_ψ^2

	$\alpha=1$				$\alpha=2$		
N=49	λ^*	σ_u^2	σ_ψ^2	*N*=49	λ	σ_u^2	σ_ψ^2
	0.0	0.0000	2.7778		0.0	0.0000	2.7778
	0.5	0.2287	0.4573		0.5	0.0045	0.0091
	1.0	0.2492	0.2492		1.0	0.0046	0.0046
	4.0	0.2672	0.0668		4.0	0.0046	0.0011
	8.0	0.2704	0.0338		8.0	0.0046	0.0006
N=81				*N*=81			
	0.0	0.0000	2.7778		0.0	0.0000	2.7778
	0.5	0.2373	0.4746		0.5	0.0048	0.0095
	1.0	0.2595	0.2595		1.0	0.0048	0.0048
	4.0	0.2790	0.0698		4.0	0.0048	0.0012
	8.0	0.2826	0.0353		8.0	0.0048	0.0006
N=121				*N*=121			
	0.0	0.0000	2.7778		0.0	0.0000	2.7778
	0.5	0.2429	0.4858		0.5	0.0049	0.0098
	1.0	0.2662	0.2662		1.0	0.0049	0.0049
	4.0	0.2868	0.0717		4.0	0.0049	0.0012
	8.0	0.2906	0.0363		8.0	0.0049	0.0006

(*) $\lambda = \sigma_u^2 / \sigma_\psi^2$. Due to rounding and omitted digits, the figures in the table are not exact.

Let ε_i the *i*-th least-squares residual from the regression in (7.1). Then the Kelejian-Robinson test corresponding to the selection $z_{ij} = (1, x_i x_j)$ is in terms of the least squares estimation of the model

$$\varepsilon_i \varepsilon_j = c_0 + c_1 x_i x_j + \eta_{ij}, \quad i < j, \; m_{i.} m_{j.}' \neq 0 \tag{7.11}$$

where η_{ij} is an error term. The null hypothesis is $H_0 : c_0 = c_1 = 0$. Let $\hat{c}' = (\hat{c}_0, \hat{c}_1)$ be the least squares estimator of $c' = (c_0, c_1)$ based on (7.11). Then the most direct form of the Kelejian-Robinson test is to

$$\text{Reject } H_0 \text{ if}: \hat{c}' \hat{V}C^{-1} \hat{c} > \chi_2^2(.95) = 5.991 \tag{7.12}$$

where $\hat{V}C$ is the estimated variance-covariance matrix of $\hat{c}$ obtained in the usual way. We henceforth refer to this test as $KR(m_{i.} \cup m_{j.}')$.

The Anselin and Moreno (2000) suggestion for the implementation of the Kelejian-Robinson test based on the estimation of a model such as (7.11) would correspond to the sample selection *i<j*, and ($m_{i.}m_{j.}^{'} \neq 0$ and/or $m_{ij} \neq 0$). We henceforth refer to their version of the Kelejian-Robinson test as $KR\left(m_{ij} \cup m_{i.}m_{j.}^{'}\right)$.

The Lagrangian multiplier test considered in Anselin and Moreno

The Lagrangian multiplier test statistic, which was derived in Anselin (2000) and considered in Anselin and Moreno (2000) is[10]

$$LM = \frac{\left[\frac{\varepsilon^{'}\mathbf{MM}'\varepsilon}{\sigma^2} - T_1\right]^2}{2\left[T_2 - \frac{T_1^2}{N}\right]} \tag{7.13}$$

where $\varepsilon^{'} = (\varepsilon_1, \ldots, \varepsilon_N)$, $\sigma^2 = \varepsilon^{'}\varepsilon / N$, T_1 = *tr*(**MM'**), and T_2 = *tr*(**MM'M'**), and *tr*(.) denotes the trace of a matrix. This Lagrangian multiplier test was developed in Anselin (2000) for the spatial error component model in the absence of heteroskedasticity, e.g., for the case in which **D** = **I** in (7.3). At the 5% level the test is to

$$\text{Reject } H_0 \text{ if: } LM > \chi_1^2(.95) = 3.841 \tag{7.14}$$

7.4 Monte Carlo Results

Results relating to the size of the tests are given in Table 7.2. At the theoretical 5% level, with 5000 Monte Carlo iterations, table entries which are outside the interval (.044, .056) are significantly different than the theoretical level. The $KR\left(m_{i.} \cup m_{j.}^{'}\right)$ test is the only test whose entries are all within this interval for all three sample sizes considered. The rejection frequencies for the *LM* test are within this interval for the larger samples *N* = 81, 121 while those of the variation on the Kelejian-Robinson test suggested in Anselin and Moreno, namely $KR\left(m_{ij} \cup m_{i.}m_{j.}^{'}\right)$, are outside the interval for all three values of the sample size considered. However, even for this test the rejection frequencies increase toward the theoretical level as the sample size increases, as one would expect on the basis of large sample theory. For the sample sizes *N* = 81 and *N* =

121 the rejection frequencies under H_0 of the Kelejian-Robinson $KR(m_{i.} \cup m'_{j.})$ test and those of the *LM* test are virtually identical!

Rejection frequencies under H_1 are given in Table 7.3. For all three tests the rejection frequencies increase as the sample size increases, as one would expect from large sample theory. In all cases considered, the Anselin and Moreno variation of the Kelejian-Robinson test, $KR(m_{ij} \cup m_{i.}m'_{j.})$, has the lowest power while *LM* has the highest power. However the difference in power between $KR(m_{i.} \cup m'_{j.})$ and *LM* is somewhat minor, especially for the larger sample size cases in which $N = 121$. For all of these considered cases, the power of $KR(m_{i.} \cup m'_{j.})$ is at least .978 while that of *LM* is at least .996.

Table 7.2: Rejection frequencies under H_0

N	$KR(m_{i.}m'_{j.})$	$KR(m_{ij} \cup m_{i.}m'_{j.})$	LM
49	0.046	0.033	0.038
81	0.045	0.036	0.046
121	0.046	0.040	0.045

Averaged over all of the cases reported in Table 7.3, the power of the $KR(m_{i.} \cup m'_{j.})$ test is 96% of that of the *LM* test. This result should be interpreted with two considerations in mind. First, the $KR(m_{i.} \cup m'_{j.})$ test is based on less information than that of the *LM* test since it does not require complete knowledge of the weighting matrix, which the *LM* test does. Therefore, in a sense, its power would be expected to be somewhat less. Secondly, the *LM* test is based on the assumption that the model does not contain heteroskedasticity which is related to the exogenous variables. Despite this, its rejection frequencies are very high even for our considered cases in which such heteroskedasticty is 'strong', as parameterized by $\alpha = 2$ in the table. These results suggest a certain robustness of the *LM* test which was not noted in the literature.

Table 7.3: Rejection frequencies under H_1

$\alpha = 1$					
	N	λ^*	$KR(m_{i.}m_{j.}')$	$KR(m_{ij} \cup m_{i.}m_{j.}')$	LM
	49	0.5	0.787	0.540	0.852
	49	1.0	0.840	0.583	0.910
	49	4.0	0.883	0.629	0.946
	49	8.0	0.894	0.635	0.952
	81	0.5	0.920	0.705	0.975
	81	1.0	0.949	0.770	0.989
	81	4.0	0.966	0.794	0.997
	81	8.0	0.966	0.797	0.996
	121	0.5	0.986	0.866	0.998
	121	1.0	0.992	0.889	0.999
	121	4.0	0.996	0.914	0.999
	121	8.0	0.998	0.927	1.000
Power Ratio**			0.96	0.78	
$\alpha = 2$					
	49	0.5	0.813	0.625	0.841
	49	1.0	0.827	0.640	0.856
	49	4.0	0.823	0.642	0.845
	49	8.0	0.820	0.638	0.849
	81	0.5	0.906	0.727	0.973
	81	1.0	0.902	0.725	0.972
	81	4.0	0.895	0.726	0.970
	81	8.0	0.902	0.723	0.971
	121	0.5	0.980	0.856	0.997
	121	1.0	0.982	0.864	0.996
	121	4.0	0.982	0.855	0.997
	121	8.0	0.978	0.851	0.996
Power Ratio**			0.96	0.79	

(*) $\lambda = \sigma_u^2 / \sigma_\psi^2$.

(**) The figures in this row are the averages of the ratios of the rejection frequencies of $KR(m_{i.}m_{j.}')$ and $KR(m_{ij} \cup m_{i.}m_{j.}')$ relative to LM.

7.5 Other Processes Involving Heteroskedasticity

For use perhaps in future studies, in this section we introduce three additional disturbance processes in which spatial correlation is intertwined with heteroskedasticity. In two of these processes the heteroskedasticity is 'muted'; in the other it is not. Generalizations and variations of these processes will become evident.

A Spatial Error Component Process with Muted Heteroskedasticity: SECMH

Consider the variation of the process in (7.3)

$$\varepsilon = \mathbf{M}\mathbf{D}^{\alpha} u + \psi,\ \alpha = 1, 2 \tag{7.15}$$

where $\mathbf{M}$, $\mathbf{D}^{\alpha}$, u and ψ are as defined in Section 7.2. For the process in (7.15) the variance-covariance matrix of ε is

$$E[\varepsilon \varepsilon'] = \sigma_u^2 \mathbf{M}\mathbf{D}^{2\alpha}\mathbf{M}' + \sigma_\psi^2 \mathbf{I}$$

which implies that the covariance between the i-th and j-th elements of ε is

$$E[\varepsilon_i \varepsilon_j] = \sigma_u^2 \sum_{r=1}^{N} m_{ir} x_r^{2\alpha} m_{jr}, i < j \tag{7.16}$$

If the i-th and j-th units have many 'neighbors in common' the covariance described in (7.16) will involve a weighted sum of all of the corresponding values of the exogenous variable. The variation of that weighted sum will, under typical specifications, be only 'weakly' related to any given value, or small set of values, of the regressor (e.g., to $x_i x_i$). One would therefore assume that a standard version of the Kelejian-Robinson test will have weak power against the SECMH alternative. The interpretation of this model is that the 10 spill-over component of the error term is heteroskedastic. This model could relate to cross sectional units which have varying degrees of interactions.

A Spatial Autoregressive Process with Muted Heteroskedastic: SARMH

Consider the auto-regressive process

$$\varepsilon = \rho \mathbf{M}\varepsilon + \mathbf{D}^{\alpha}\psi \qquad (7.17)$$

where ρ is a scalar parameter, and **M**, $\mathbf{D}^{\alpha}$ and ψ are defined as in Section 7.2. Assuming that (**I**-ρ**M**) is nonsingular, in this case the variance-covariance matrix of ε is

$$E[\varepsilon \varepsilon'] = \sigma_{\psi}^{2}(I-\rho\mathbf{M})^{-1}\mathbf{D}^{2\alpha}(I-\rho\mathbf{M}')^{-1} \qquad (7.18)$$

Let $c_{i\cdot}$ be the i-th row of $(\mathbf{I}-\rho\mathbf{M})^{-1}$. It then follows from (7.18) that the covariance between ε_i and ε_j is

$$E[\varepsilon_i \varepsilon_j] = \sigma_{\psi}^{2}\sum_{r=1}^{N} c_{ir} x_r^{2\alpha} c_{jr}, i < j$$

so that the heteroskedasticity is 'muted' in the sense described above. Again, one would expect the standard Kelejian-Robinson test to have only weak power against a SARMH process. The interpretation of this model is that the innovation of the auto-regressive process is heteroskedastic. This model could relate cross -ectional units which differ non-negligibly in size.

A Spatial Autoregressive Process with Heteroskedasticity: SARH

Consider the process

$$\mathbf{D}^{-\alpha}\varepsilon = \rho \mathbf{M}\,\mathbf{D}^{-\alpha}\varepsilon + \psi \qquad (7.19)$$

where ρ, $\mathbf{D}^{-\alpha} = (\mathbf{D}^{\alpha})^{-1}$, **M**, and ψ are defined as above. In this case

$$\varepsilon = \mathbf{D}^{\alpha}(\mathbf{I}-\rho\mathbf{M})^{-1}\psi$$

so that the variance-covariance matrix is

$$E[\varepsilon \varepsilon'] = \sigma_{\psi}^{2}\mathbf{D}^{\alpha}(I-\rho M)^{-1}(I-\rho M')^{-1}D^{\alpha} \qquad (7.20)$$

Again, let $c_{i\cdot}$ be the i-th row of $(\mathbf{I}\text{-}\rho\mathbf{M})^{-1}$. Then, in this case the covariance between ε_i and ε_j is

$$E[\varepsilon_i \varepsilon_j] = \sigma_{\psi}^2 x_i^{\alpha} x_j^{\alpha} \sum_{r=1}^{N} c_{ir} c_{jr}$$

so that, by our definition, the heteroskedasticity is not 'muted'. The interpretation of this process is that the heteroskedasticity relates directly to the disturbance vector rather than to the innovation. Again, this model could relate to cross sectional units which differ non-negligibly in size. Clearly, the difference between the SARH and SARMH models is the nature of the heteroskedasticity.

7.6 Summary and a Suggestion for Further Research

The force of this chapter is three-fold. First, we suggested that the Monte Carlo studies which have considered the power of the Kelejian-Robinson test for spatial correlation were done so in a manner that is misleading. The reason for this is that the Kelejian-Robinson test relates to the case in which spatial correlation and heteroskedasticity relating to the exogenous variables are intertwined. The studies in the literature did not considered the heteroskedastic component in their models.

Second, we considered such a disturbance process and found that the power of the Kelejian-Robinson test is much higher than that reported in the literature. Indeed, for the cases considered, its power was 96% of that of the Lagrangian multiplier test which was derived in Anselin (2000) and studied in Anselin and Moreno (2000). We noted two issues concerning this comparison. First, the Kelejian-Robinson test does not require full knowledge of the weighting matrix, while the Lagrangian multiplier test does. Second, the Lagrangian multiplier test was derived under the assumption that the error terms do not have a heteroskedastic component which relates to the exogenous variables. This, in turn, suggests a certain robustness of that Lagrangian multiplier test which has not been noted in the literature.

Third, we introduced four disturbance processes in which spatial correlation is intertwined with heteroskedasticity that relates to the exogenous variables. Two of these processes are such that the heteroskedasticity is 'muted', while in the other two it is not. We defined a 'muted' heteroskedastic process as one in which the heteroskedasticity relates to a weighted sum of, possibly, all of the sample values of the exogenous regressors. Such disturbance processes should be of interest to researchers in the field.

We conclude by suggesting further research. We do this by returning to the issue of qualifications relating to the power comparisons in Anselin and Moreno (2000) of the Lagrangian multiplier test and Kelejian-Robinson test. As indicated earlier, unlike the Lagrangian multiplier test, the Kelejian-Robinson test does not require a detailed knowledge of the weighting matrix. This should be considered when evaluating the power comparisons given in Anselin and Moreno (2000). On the other hand, the Kelejian-Robinson test implicitly approximates all of the non-zero elements in the weighting matrix by a single constant. Therefore, the Kelejian-Robinson test is actually based on an approximation to the weighting matrix. If, in practice, the 'true' weighting matrix is not known, the weighting matrix which is assumed in the Lagrangian multiplier test can also be viewed as an approximation. The relevant issue then would be which approximation is best for the purpose at hand, namely testing for spatial correlation? One would clearly expect the answer to depend upon the nature of the 'true' weighting matrix, as well as the weighting matrix which is assumed in the Lagrangian multiplier test. A somewhat similar issue arises with respect to the possible existence of heteroskedasticity. Specifically, the Lagrangian multiplier test is based on the assumption that heteroskedasticity involving the exogenous variables is absent. Thus in a sense, the Lagrangian multiplier test is based on a homoskedasticity approximation to heteroskedasticity. The Kelejian-Robinson test assumes the existence of such heteroskedasticity and approximates it in terms of the cross products of the regressors corresponding to neighbouring units. In practice the 'true' pattern of heteroskedasticity may not be known. Therefore, the relevant issue again relates to which approximation is best for the purpose at hand. Again , one would expect the answer to depend upon the particulars involved. One way of sorting out the issues raised in this paragraph, as well as issues relating to the R^2 statistic, would be to conduct a wide variety and a large number of Monte Carlo experiments. If those experiments are specified in a way that facilitates the estimation of response functions, the results of such a Monte Carlo study could be extrapolated to scenarios not considered in the study. We leave such an undertaking to future research.

Acknowledgements

Harry Kelejian gratefully acknowledges financial support from the National Science Foundation through grant SES-0001780.

Endnotes

[1]For general background of the literature on spatial econometric issues see Cliff and Ord (1973, 1981), Anselin (1988), and Cressie (1993).
[2]The spatial AR model implies heteroskedasticty for most reasonable specifications of the weighting matrix. The formulation considered by Kelejian and Robinson implies that, perhaps in addition to this, there is a component of the heteroskedasticty which involves exogenous variables.
[3]R^2 values corresponding to the various experiments considered in Anselin and Moreno (2000) were not reported
[4]Ironically, although these disturbance term processes are a by-product of this chapter, they may turn out to be its most important contribution!
[5]At first we also considered the *MI* test but, as in Anselin and Moreno (2000), found that its power was quite low with respect to spatial error component processes. Therefore to better focus our attention on the issues at hand, and to conserve space, we do not report results relating to the *MI* test.
[6]An obvious generalization of **D** in the context of a regression model with *K* regressors, say $z_{i.}$, would be $diag_{i=1}^{N}[f(z_{.i},\gamma)]$, where $f(.)$ is a 'well' behaved function, and γ is a vector of parameters. Let $m_{i.}$ be the *i*-th row of **M**. One scenario in which such a model could arise is the case in which

$$\varepsilon_i = \rho_i m_{i.} u + \psi_i$$

$$\rho_i = f(z_{i.}\gamma), i = 1,\ldots,N$$

where u is defined in (7.3) and ψ_i is the *i*-th element of ψ. Clearly, in the context of (7.2), (7.3) would correspond to the case in which $\rho_i = x_i^{\alpha}$.
[7]Recall that the sample averages of y_t and $\hat{y}_t$ are equal for the case considered.
[8]It is not difficult to show that $[R^2 - R_*^2] \xrightarrow{P} 0$, as $N \to \infty$
[9]A specific model was not considered in Kelejian and Robinson (1992) and so they suggested all combinations of $i<j$ such that $m_{ij} \neq 0$.
[10]There is a typographical error in Anselin and Moreno (2000) in that T_1 should be in the numerator of their expression (7.18).

8 Some Proposals for Discriminating Between Spatial Processes

F. Javier Trivez
Jesús Mur
University of Zaragoza

Abstract

The chapter deals with the problem of selecting a spatial dependence structure for a series with indications of cross-sectional autocorrelation. We consider two models, the autoregressive and the moving average structures and the discussion is limited to series centred on zero. In a generic approach of non-nested models, three possible strategies are examined. The results indicate that the King point optimal test has certain difficulties of instability and asymmetry, while the procedure based on the Wald test lacks power. Finally, the criterion called *minimum sampling variance* offers greater guarantees.

8.1 Introduction

The problem with which we deal in this chapter refers to the selection of a spatial structure for a series with indications of cross-sectional autocorrelation. There are several ways to analyse the hypothesis of spatial independence in a series (Anselin, 1988). However, the following stages in the specification of a spatial stochastic process are not very well structured. We focus on the problem of what to do when the null hypothesis of independence has been rejected. It is obvious that a process of dependencies should be proposed, but which?

Our discussion is carried out for centred variables although it could be generalised. In the second section we propose a previous filtering of the series in order to simplify the analysis. The third section is focussed explicitly on the discrimination problem mentioned above. The results obtained are evaluated with the simulation carried out in the fourth section. The work ends with some conclusions.

8.2 Structures of Spatial Dependence and Series Filtering

In general terms, the Data Generating Process (DGP) of a spatial series can be represented by the following equation:

$$y = \mathbf{A}z \tag{8.1}$$

where y is the $(Rx1)$ vector of observed outputs, z an impulse vector and $\mathbf{A}$ an (RxR) diffusion matrix. The impulse vector may contain systematic elements as well as purely random ones. For simplicity, let us suppose that $z = u \sim \mathrm{N}(0, \sigma^2 \mathbf{I})$. The diffusion matrix may have a moving average structure, SMA: $\mathbf{A}=(\mathbf{I}+\theta\mathbf{W})$, or autoregressive, SAR: $\mathbf{A} = (\mathbf{I} - \rho\mathbf{W})^{-1}$, where $\mathbf{W}$ is a binary contiguity matrix and θ and ρ are parameters.

Expression (8.1) is an isomorphism in a vectorial space of dimension R. In this space we can choose the base and, for example, use that composed by the eigenvectors of $\mathbf{A}$ ($\vec{q}$) instead of the original canonical base ($\vec{e}$). If we order the first set of vectors in matrix $\mathbf{Q}$ in such a way that each vector of base $\vec{q}$ corresponds to a column of this matrix, the transition matrix from $\vec{e}$ to $\vec{q}$ will be $\mathbf{Q}'$ which, used in (8.1), allows us to write:

$$\mathbf{Q}'y=\mathbf{Q}'\mathbf{A}u=\mathbf{Q}'\mathbf{Q}\Delta\mathbf{Q}'u \Rightarrow y^*=\Delta u^* \tag{8.2}$$

where $y^*=\mathbf{Q}'y$ and $u^*= \mathbf{Q}'u$ are the co-ordinates of the vectors y and u in the new base $\vec{q}$. The input vector has the same characteristics as the original: $u^* \sim \mathrm{N}(0,\sigma^2\mathbf{I})$, but the diffusion matrix $\mathbf{A}$ transforms into a diagonal matrix, Δ, made up of the eigenvalues of $\mathbf{A}$. Specifically, if the diffusion matrix was of an SMA type then: $\Delta=(\mathbf{I}+\theta\Lambda)$ and if it was an SAR $\Delta = (\mathbf{I} - \rho\Lambda)^{-1}$, where Λ is the eigenvalue matrix of $\mathbf{W}$. Moreover, if in the canonical base the original series had a certain spatial dynamic, this will disappear in the $\vec{q}$ base, although the variance remains heteroskedastic in both cases:

ORIGINAL SERIES	*FILTERED SERIES*	
$E[y]=\mathbf{A}E[u]=0$	$E[y^*]=\Delta E[u^*]=0$	(8.3)
$V[y]=\mathbf{A}V[u]\mathbf{A}=\sigma^2\mathbf{A}^2$	$V[y^*]=\Delta V[u^*]\Delta=\sigma^2\Delta^2$	

In view of the last expression, we can say that spatial autocorrelation implies heteroskedasticity in the filtered series. Moreover, the skedastic functions in the two cases, SMA or SAR, are:

$$\left.\begin{array}{l} V\left[y_r^*\right]_{SMA} = \sigma_r^2 = \sigma^2(1+\theta\lambda_r)^2 \\ V\left[y_r^*\right]_{SAR} = \sigma_r^2 = \sigma^2(1-\rho\lambda_r)^{-2} \end{array}\right\} \tag{8.4}$$

These results lead us to consider using a heteroskedasticity test as a way of analysing the hypothesis of spatial independence. This test should explicitly incorporate the dependence of the variance of the transformed series y^* in relation to the eigenvalues of **W** (λ_r, r=1, 2,..., R). The results obtained with the Goldfeld-Quant statistic seem satisfactory (Mur and Trivez, 2000). On the other hand, expression (8.4) allows us to advance towards another different test of spatial autocorrelation. For a spatially autocorrelated series it will be true that:

$$\left.\begin{array}{l} E\left[\left|y_r^*\right|\right] = c\sigma_r = \mu_r \\ V\left[\left|y_r^*\right|\right] = (1-c^2)\sigma_r^2 \end{array}\right\} \tag{8.5}$$

with $c = \sqrt{2/\pi}$. Also we can write that:

$$\left|y_r^*\right| = E\left[\left|y_r^*\right|\right] + v_r = \mu_r + v_r \tag{8.6}$$

where $v_r = \left|y_r^*\right| - E\left[\left|y_r^*\right|\right]$. The error term is centred on zero ($E[v_r] = 0; \forall r$); it is uncorrelated ($E[v_r v_s] = 0; \forall r \neq s$) although it will remain heteroskedastic ($E[v_r^2] = (1-c^2)\sigma_r^2$). For the SMA case we obtain:

$$\left|y_r^*\right| = \mu_r + v_r = c\sigma(1+\theta\lambda_r) + v_r = c\sigma + c\sigma\theta\lambda_r + v_r \Rightarrow \left|y_r^*\right| = \theta_0 + \theta_1\lambda_r + v_r \tag{8.7}$$

with $\theta_0 = c\sigma$ and $\theta_1 = c\sigma\theta$. In the SAR case the result is :

$$\left|y_r^*\right| = \frac{\alpha_0}{1-\alpha_1\lambda_r} + v_r \tag{8.8}$$

with $\alpha_0 = c\sigma$ and $\alpha_1 = \rho$. This quotient in the *stability interval* (where $|\delta\lambda_r|<1$, with $\delta=\theta$ or ρ) may be approximated by:

$$\left|y_r^*\right| \cong \gamma_0 + \gamma_1\lambda_r + \gamma_2\lambda_r^2 + ... + \gamma_p\lambda_r^p + n_r \tag{8.9}$$

with $\gamma_j = \gamma_1^j / \gamma_0^{j-1}$ for $j \geq 2$, $\gamma_0 = c\sigma$, $\gamma_1 = \delta\lambda_r$ and n_r the sum of v_r and the approximation error. Expression (8.9) could be used as the basis for a simple test of autocorrelation. Spatial independence implies that $\delta=0$, in which case the auxiliary equation (8.9) reduces to:

$$\left|y_r^*\right| = \gamma_0 + u_r^* \tag{8.10}$$

When there are SAR or SMA dependencies, the inclusion of the eigenvalues of **W** on the right-hand side of (8.9) will prove useful. In order to construct the test we can specify the equation:

$$\left|y_r^*\right| = \gamma_0 + \gamma_1 \lambda_r + m_r \tag{8.11}$$

with $m_r = v_r$ in (8.7) or $m_r = v_r + \sum_{j=2} \gamma_j \lambda_r^j$ in (8.8). LS estimators in (8.11) will be unbiased and consistent if the series was an SMA, although they will be biased and inconsistent in the case of SAR. However, even in the latter situation, the coefficient γ_1 should be statistically strong.

On the other hand, the error term of (8.11), m_r, will be white noise in the case of non-correlation, but it will be heteroskedastic under the alternative hypothesis. In the SAR case it will also include the specification error from the functional form. This range of possibilities can be dealt with using the procedure of White (1980). If by **X** we denote the $(Rx2)$ matrix of observations of the regressors of (8.11), that is $\mathbf{X}=[1,\lambda_r]$, a consistent estimate of the covariance matrix for the LS estimators of the vector $\beta=[\gamma_0, \gamma_1]'$ can be obtained as:

$$\left.\begin{aligned} \hat{V}[\beta] &= [\mathbf{X}'\mathbf{X}]^{-1}\hat{\mathbf{C}}_{\mathbf{R}}[\mathbf{X}'\mathbf{X}]^{-1} \\ \hat{\mathbf{C}}_{\mathbf{R}} &= \sum_{R=1}^{R} \frac{\hat{u}_r^2 x_r x_r'}{d_r} \end{aligned}\right\} \tag{8.12}$$

with $\hat{u}_r$ the rth LS residual of (8.11), x_r' the rth row of **X**, and d_r the standardisation factor discussed by Davidson and MacKinnon (1993). It is important to point out, as White indicates (1980, p. 824), that this matrix *'is a consistent estimator of the parameter covariance matrix even when the model is misspecified'* (see also White, 1982).

In conclusion, the test of spatial independence can be solved through a simple t-ratio obtained from the LS estimation of (8.11). The null hypothesis will be that the parameter γ_1 is zero, that is to say:

$$\left.\begin{array}{l} H_0 : \delta = 0 \\ H_A : \delta \neq 0 \end{array}\right\} \Leftrightarrow \left.\begin{array}{l} H_0 : \gamma_1 = 0 \\ H_A : \gamma_1 \neq 0 \end{array}\right\} \Rightarrow t_{\hat{\gamma}_1} = \frac{\hat{\gamma}_1}{\sigma_{\hat{\gamma}_1}} \underset{as}{\sim} N(0;1) \qquad (8.13)$$

with $\sigma_{\hat{\gamma}_1}$ being the estimated standard deviation of $\hat{\gamma}_1$ according to (8.12).

8.3 The Problem of Discriminating Between SAR and SMA Processes

When the null hypothesis of spatial independence has been rejected, the usual practice is to specify an SAR process. This decision is generally taken automatically and can result in specification errors with important consequences. For example, the nature of the series will be misunderstood and the estimation of the (wrongly) proposed model will be inconsistent. However, this discussion has not received very much attention in the literature. Haining (1979) proposes an LR test in which the SMA model is the null hypothesis, though the regions must form a perfectly regular grid. Cliff and Ord (1981) propose using the correlogram and the partial correlogram of the series, though their properties are not well known. Hooper and Hewings (1981) are particularly sceptical about this strategy. More recently, Hepple (1995) introduces Bayesian methods into the discussion, which is still analytically complex. The set of LM tests proposed by Mur (1999) is badly behaved in small samples and produces non-conclusive situations. The discussion about Common Factors is evidently related to the problem at hand (see Burridge, 1981, Bivand, 1984, or Blommestein and Nijkamp, 1986), although this approach is not easy to implement in a case like this where there are no exogenous variables.

Below we analyse several new proposals, all of them based on a generic approach of non-nested models.

The Wald Test

We consider just two rival DGPs, SAR or SMA, represented in equations (8.7) and (8.8), both of which are nested in the artificial model of (8.9). As a consequence, and given that the series is spatially correlated, we might expect to obtain a significant relationship between the absolute value of the filtered series and the eigenvalues of **W**. Without additional information it seems reasonable to propose a lineal relationship, as in (8.9). Then, if in this equation we cannot reject the restrictions $\gamma_2 = \gamma_3 = ... = \gamma_p = 0$, the evidence favours an SMA structure. On the other hand, if we cannot reject the restrictions $\gamma_j \gamma_0^{j-1} = \gamma_1^j$ for $j \geq 2$ the evidence favours SAR structures. In conclusion, the stages that we are considering in this procedure are the following:

(1)– Unrestricted LS estimation of (8.9) using a moderate value of p. Our findings tend to support a maximum value of 3 for this variable.

(2)– Analysis of the SMA alternative. The null hypothesis of the test is:

$$\left.\begin{array}{c} H_0 : \gamma_2 = \gamma_3 = ... = \gamma_p = 0 \\ H_A : No \quad H_0 \end{array}\right\} \qquad (8.14)$$

which can be tested through a Wald test:

$$WALD_{SMA} = R\left[\mathbf{S}\beta - b\right]\left[\mathbf{S}\hat{\mathbf{V}}(\hat{\boldsymbol{\beta}})\mathbf{S}'\right]^{-1}\left[\mathbf{S}\beta - b\right] \underset{as}{\sim} \chi^2(p-1) \qquad (8.15)$$

with β of order $(p+1)x1$, the LS estimation of β in (9). **S** is the matrix $(p-1)x(p+1)$ containing the linear restrictions of (8.14) and b its value under the null hypothesis:

$$\mathbf{S} = \begin{bmatrix} 0 & 0 & 1 & 0 & 0 & \dots & 0 & 0 \\ 0 & 0 & 0 & 1 & 0 & \dots & 0 & 0 \\ 0 & 0 & 0 & 0 & 1 & \dots & 0 & 0 \\ \dots & \dots & \dots & \dots & \dots & \dots & \dots & \dots \\ 0 & 0 & 0 & 0 & 0 & \dots & 1 & 0 \\ 0 & 0 & 0 & 0 & 0 & \dots & 0 & 1 \end{bmatrix} \qquad b = \begin{bmatrix} 0 \\ 0 \\ 0 \\ \dots \\ 0 \\ 0 \end{bmatrix} \qquad (8.16)$$

Finally, $\hat{\mathbf{V}}(\hat{\boldsymbol{\beta}})$ is a consistent estimation of the covariance matrix of β.

(3)– Analysis of the SAR alternative. In this situation, the hypothesis of the test consists of $(p-1)$ non-linear restrictions, gathered in the vector $h(\beta)$:

$$\left.\begin{matrix} H_0 : \gamma_j \gamma_0^{j-1} - \gamma_1^j = 0; \forall j \geq 2 \\ H_A : No \quad H_0 \end{matrix}\right\} \Leftrightarrow \left.\begin{matrix} H_0 : h(\beta) = 0 \\ H_A : h(\beta) \neq 0 \end{matrix}\right\} \qquad (8.17)$$

The test statistic may also be a Wald test:

$$WALD_{SAR} = R\left[h(\hat{\gamma})\right]'\left[\mathbf{H}(\hat{\gamma})\hat{\mathbf{V}}(\hat{\gamma})\mathbf{H}(\hat{\gamma})'\right]^{-1}\left[h(\hat{\gamma})\right] \underset{as}{\sim} \chi^2(p-1) \qquad (8.18)$$

with:

$$\mathbf{H}(\boldsymbol{\beta}) = \frac{\partial h(\beta)}{\partial \beta}_{|\beta} = \begin{bmatrix} \hat{\gamma}_2 & -2\hat{\gamma}_1 & \hat{\gamma}_0 & 0 & 0 & \dots & 0 & 0 \\ 2\hat{\gamma}_0\hat{\gamma}_3 & -3\hat{\gamma}_1^2 & 0 & \hat{\gamma}_0^2 & 0 & \dots & 0 & 0 \\ 3\hat{\gamma}_0^2\hat{\gamma}_4 & -4\hat{\gamma}_1^3 & 0 & 0 & \hat{\gamma}_0^3 & \dots & 0 & 0 \\ \dots & \dots & \dots & \dots & \dots & \dots & \dots & \dots \\ (p-2)\hat{\gamma}_0^{p-3}\hat{\gamma}_{p-1} & -(p-1)\hat{\gamma}_1^{p-2} & 0 & 0 & 0 & \dots & \hat{\gamma}_0^{p-2} & 0 \\ (p-1)\hat{\gamma}_0^{p-2}\hat{\gamma}_p & -p\hat{\gamma}_1^{p-1} & 0 & 0 & 0 & \dots & 0 & \hat{\gamma}_0^{p-1} \end{bmatrix}$$

(8.19)

(4)– Interpretation of the results. A positive identification of the series will only be attained when the relationship between both tests is one of Acceptance–Rejection or Rejection–Acceptance. The relationships Acceptance–Acceptance and Rejection–Rejection do not allow a decision to be taken, and thus the analysis will be inconclusive.

In the fourth section it will be shown that the performance of this strategy is far from satisfactory, which led us to explore other approaches.

The King Point Optimal Test

The procedure proposed by King (1983) fits the present situation. Concretely, we have a spatially correlated series and the question is whether it would be better to introduce an SAR or an SMA process. Based on these terms, it is about resolving the following test:

$$\left.\begin{aligned} H_0 &: \; y \sim \mathrm{N}(0, \boldsymbol{\Omega}(\rho)) \\ H_A &: \; y \sim \mathrm{N}(0, \boldsymbol{\Sigma}(\theta)) \end{aligned}\right\} \Leftrightarrow \left.\begin{aligned} H_0 &: \; y \sim \mathrm{SAR}(1) \\ H_A &: \; y \sim \mathrm{SMA}(1) \end{aligned}\right\} \tag{8.20}$$

where $\boldsymbol{\Omega}(\rho)$ is the covariance matrix corresponding to the SAR process and $\boldsymbol{\Sigma}(\theta)$ that associated with the SMA. The recommendation is *to 'reject H_0 for small values of:*

$$g(\theta_0, \rho_0) = \frac{y'\boldsymbol{\Sigma}^{-1}(\theta_0)y}{y'\boldsymbol{\Omega}^{-1}(\rho_0)y} \tag{8.21}$$

assuming covariance matrices $\boldsymbol{\Sigma}(\theta_0)$ *and* $\boldsymbol{\Omega}(\rho_0)$ *respectively'* (King, 1983, p. 42). The problem in (8.21) is the determination of the true α level critical value, c_g, which fulfils:

$$Pr\left[g(\theta_0, \rho_0) < c_g / H_0\right] = \alpha \tag{8.22}$$

or:

$$Pr\left[\sum_{r=1}^{R} \omega_r \zeta_r^2 < 0 \mid \zeta \sim N(0, \mathrm{I})\right] \le \alpha \tag{8.23}$$

where the scalars $\{\varpi_1, \varpi_2, \ldots, \varpi_R\}$ are the characteristic roots of the matrix $\boldsymbol{\Gamma}^{-2}(\rho_0)\boldsymbol{\Gamma}^{-2}(\theta_0) - c_g\mathbf{I}$, with $\boldsymbol{\Gamma}(\rho_0)=\mathbf{I}-\rho_0\boldsymbol{\Lambda}$, $\boldsymbol{\Gamma}(\theta_0)=\mathbf{I}+\theta_0\boldsymbol{\Lambda}$ $\mathbf{I}$ the identity matrix and $\boldsymbol{\Lambda}$ the eigenvalue matrix of $\mathbf{W}$. The probability of (23) for each pair (θ_0, ρ_0) can be solved using, for example, Farebrother's (1980) PAN procedure. Furthermore, the decision about which model should be placed in the null hypothesis and which in the alternative is not determinate, given that the procedure can be adapted to either case.

King's proposal constitutes a simple procedure, appropriate when the number of alternatives is only two but it is useless when this number increases.

The Minimum Variance Criterion

The third procedure we wish to analyze is based the differences observed in terms of the skedastic function of each process, as expressed in equations (8.7) and (8.8). The elements that make up this proposal are the following:

(1)- The series has been identified as spatially autocorrelated using a contiguity matrix.

(2)- If the series has been generated by an SMA process, equation (8.7) will be the right one and the LS estimators of θ_0 and of θ_1 (in other words, of σ and of θ) will be consistent. On the other hand, if the DGP is of the SAR type, the correct equation will be (8.8) which should be estimated by non-linear methods.

(3)- The imposition of an SAR structure on an SMA series, or vice versa, results in inconsistent estimations and in an increase in the sum of squared residuals. In conclusion, the DGP with the least sampling variance should be selected. In the Appendix we discuss the consistency of this criterion.

Figure 8.1: Right Decision. SMA Processes. P=2

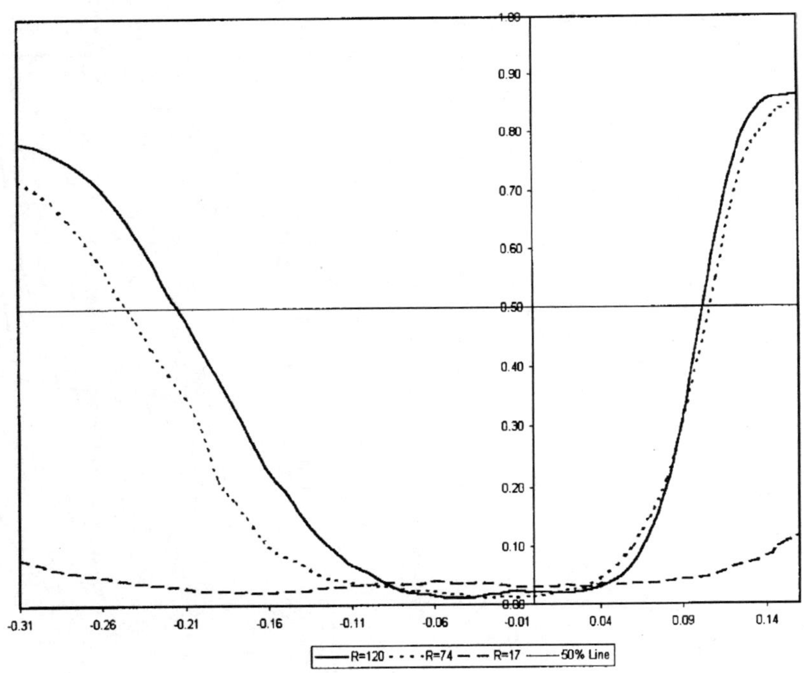

Before finishing the section we would like to add that even a fourth method could be used in the present framework which consists of introducing the Box-Cox elasticities developed by van Gastel and Paelinck (1998). The Box-Cox approach deals mainly with the problem of non-linearities, which we have also found in our research, and which is under consideration.

8.4 A Simulation Exercise

In this section we will comment on the main results of a simulation exercise in relation to the discriminatory capacity of the three methods: the Wald tests of (8.15) and (8.18), the King point optimal test of (8.21) and the criterion of minimum variance developed in the Appendix. The most relevant aspects of the simulation are the following:

- Three sampling sizes have been simulated, with 17, 74 and 120 observations.
- In each case we have employed a binary contiguity matrix based on the geographical adjacency criterion between the regions. The 120 sampling size corresponds, roughly, to the NUTS II European regional system of 15 Member States. The other two sampling sizes are smaller pieces of the former.

Figure 8.2: Right Decision. SAR Processes. P=2

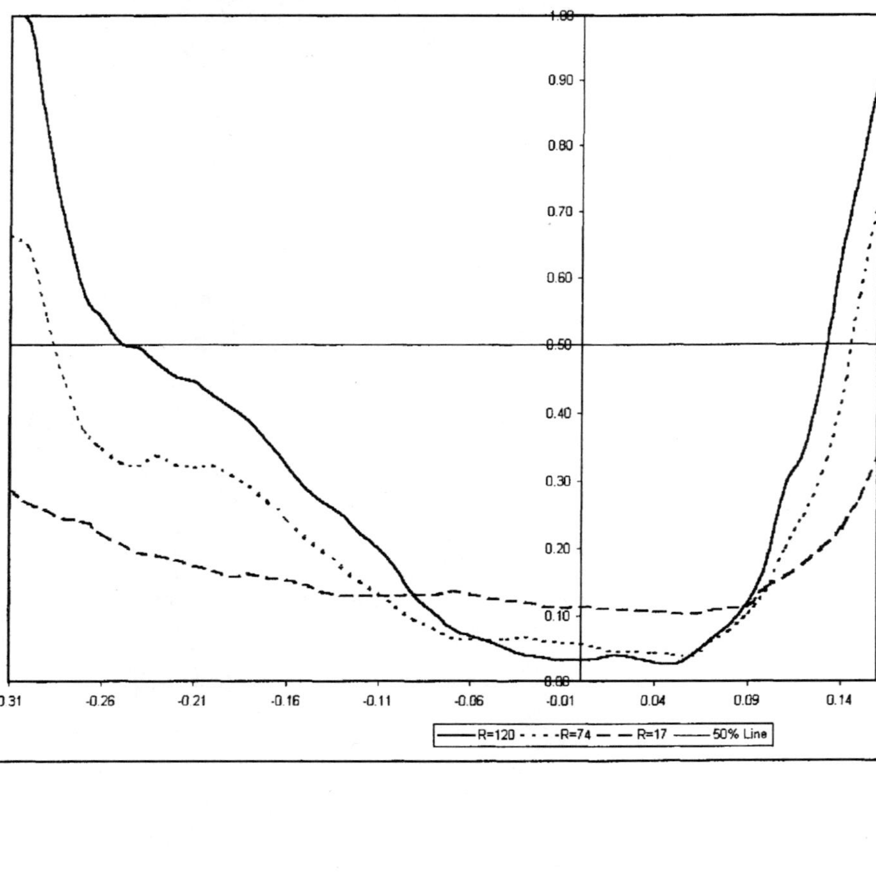

- In each case we have employed a binary contiguity matrix based on the geographical adjacency criterion between the regions. The 120 sampling size corresponds, roughly, to the NUTS II European regional system of 15 Member States. The other two sampling sizes are smaller pieces of the former.
- Series of random numbers have been obtained from an N(0,1) distribution, which have later been transformed into SAR or SMA processes using (1). In each case 1000 simulations have been resolved.
- Only values inside the stability interval for each matrix ($|\delta\lambda_r|<1$) have been employed, although we recognise that values outside this interval are also acceptable.

Figure 8.3: Right Decision. SMA Processes. P=3

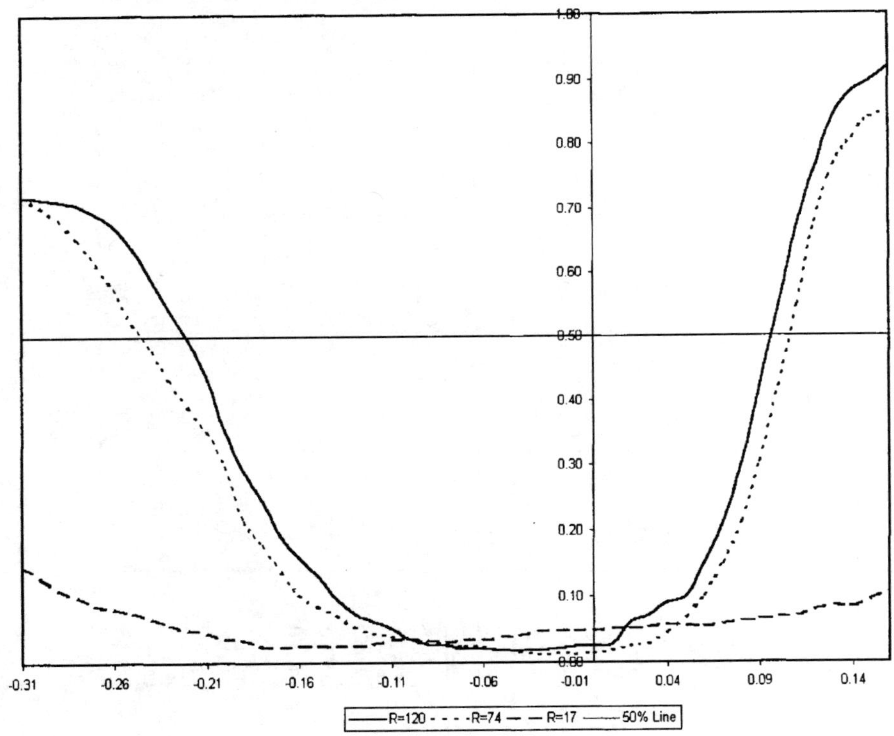

Table 8.1: The Wald Test. Selected points of the estimated discrimatory function

p=2	R=120			R=74			R=17		
SMA	*Right*	*Wrong*	*Unsolved*	*Right*	*Wrong*	*Unsolved*	*Right*	*Wrong*	*Unsolved*
-0.30	0.78	0.02	0.21	0.70	0.03	0.26	0.07	0.22	0.71
-0.25	0.67	0.04	0.29	0.52	0.06	0.42	0.05	0.21	0.74
-0.20	0.43	0.06	0.51	0.30	0.06	0.64	0.02	0.19	0.79
-0.15	0.20	0.06	0.75	0.08	0.06	0.86	0.02	0.15	0.83
-0.10	0.06	0.04	0.90	0.03	0.05	0.92	0.04	0.12	0.84
-0.05	0.01	0.03	0.96	0.02	0.04	0.94	0.04	0.11	0.86
0.00	0.02	0.03	0.95	0.01	0.06	0.93	0.03	0.12	0.85
0.05	0.05	0.04	0.91	0.07	0.07	0.87	0.03	0.13	0.84
0.10	0.47	0.04	0.49	0.42	0.05	0.53	0.05	0.12	0.84
0.15	0.86	0.00	0.14	0.84	0.00	0.16	0.10	0.10	0.80
p=2	R=120			R=74			R=17		
SAR	*Right*	*Wrong*	*Unsolved*	*Right*	*Wrong*	*Unsolved*	*Wrong*	*Right*	*Unsolved*
-0.30	0.99	0.00	0.01	0.64	0.00	0.36	0.27	0.03	0.70
-0.25	0.50	0.01	0.49	0.33	0.05	0.63	0.21	0.07	0.73
-0.20	0.43	0.12	0.46	0.32	0.21	0.47	0.17	0.07	0.76
-0.15	0.29	0.30	0.41	0.22	0.22	0.56	0.15	0.06	0.80
-0.10	0.17	0.15	0.68	0.11	0.10	0.79	0.13	0.04	0.83
-0.05	0.06	0.03	0.91	0.06	0.03	0.91	0.13	0.03	0.84
0.00	0.03	0.02	0.95	0.06	0.01	0.93	0.12	0.03	0.85
0.05	0.03	0.02	0.95	0.04	0.02	0.94	0.10	0.04	0.85
0.10	0.18	0.03	0.79	0.14	0.02	0.84	0.14	0.04	0.82
0.15	0.76	0.02	0.22	0.57	0.03	0.40	0.27	0.02	0.71

Table 8.1: The Wald Test. Selected points of the estimated discrimatory function (continued)

p=3	R=120			R=74			R=17		
SMA	*Right*	*Wrong*	*Unsolved*	*Right*	*Wrong*	*Unsolved*	*Right*	*Wrong*	*Unsolved*
-0.30	0.72	0.00	0.28	0.70	0.03	0.26	0.13	0.19	0.68
-0.25	0.64	0.01	0.35	0.52	0.06	0.42	0.08	0.20	0.73
-0.20	0.35	0.05	0.60	0.30	0.06	0.64	0.04	0.20	0.76
-0.15	0.13	0.10	0.77	0.08	0.06	0.86	0.03	0.21	0.77
-0.10	0.04	0.09	0.87	0.03	0.05	0.92	0.03	0.20	0.76
-0.05	0.02	0.08	0.90	0.02	0.04	0.94	0.04	0.17	0.79
0.00	0.03	0.07	0.91	0.01	0.06	0.93	0.05	0.15	0.81
0.05	0.10	0.06	0.84	0.07	0.07	0.87	0.05	0.13	0.82
0.10	0.54	0.02	0.43	0.42	0.05	0.53	0.07	0.08	0.85
0.15	0.90	0.00	0.10	0.84	0.00	0.16	0.09	0.06	0.85

p=3	R=120			R=74			R=17		
SAR	*Right*	*Wrong*	*Unsolved*	*Right*	*Wrong*	*Unsolved*	*Right*	*Wrong*	*Unsolved*
-0.30	0.74	0.00	0.26	0.64	0.02	0.33	0.25	0.07	0.69
-0.25	0.42	0.00	0.58	0.33	0.02	0.65	0.16	0.06	0.78
-0.20	0.37	0.01	0.62	0.32	0.02	0.65	0.11	0.06	0.83
-0.15	0.25	0.02	0.73	0.22	0.02	0.76	0.09	0.06	0.85
-0.10	0.13	0.03	0.84	0.11	0.03	0.86	0.10	0.07	0.83
-0.05	0.06	0.03	0.91	0.06	0.02	0.92	0.12	0.06	0.83
0.00	0.07	0.03	0.90	0.06	0.01	0.93	0.15	0.05	0.81
0.05	0.07	0.02	0.92	0.04	0.02	0.94	0.18	0.04	0.78
0.10	0.25	0.02	0.73	0.14	0.02	0.84	0.21	0.04	0.76
0.15	0.75	0.03	0.22	0.57	0.03	0.40	0.26	0.03	0.72

The Wald Test

A summary of the results obtained for the Wald test appears in Table 8.1 and in Figures 8.1 to 8.4. It is clear from this data that adding on powers of the eigenvalues of W in equation (8.9) does not improve the performance of the method. On the contrary, the inclusion of higher powers translates into an increase of non-conclusive situations, to the detriment of the percentage of right decisions.

The procedure offers certain guarantees when the autocorrelation parameter is at the extremes of the stability interval. However, non-conclusive situations clearly dominate for intermediate and low values of this parameter. Besides, the performance of the Wald method is more regular in the SMA case than in SAR processes. The functions shown in Figures 8.2 and 8.4 tend to be asymmetrical and even non-monotonous in certain areas of the parametric space. Another aspect that merits some comments is the disastrous behaviour of the procedure in small samples. The percentage of right decisions is always below the 30% threshold using a sample with only 17 observations. Moreover, for this sampling size the number of erroneous decisions is greater than the number of correct ones in the case of SMA processes. In conclusion, the experiment has shown that a discriminatory strategy based solely on the Wald test is not very trustworthy.

Figure 8.4: Right Decision. SAR Processes. P=3

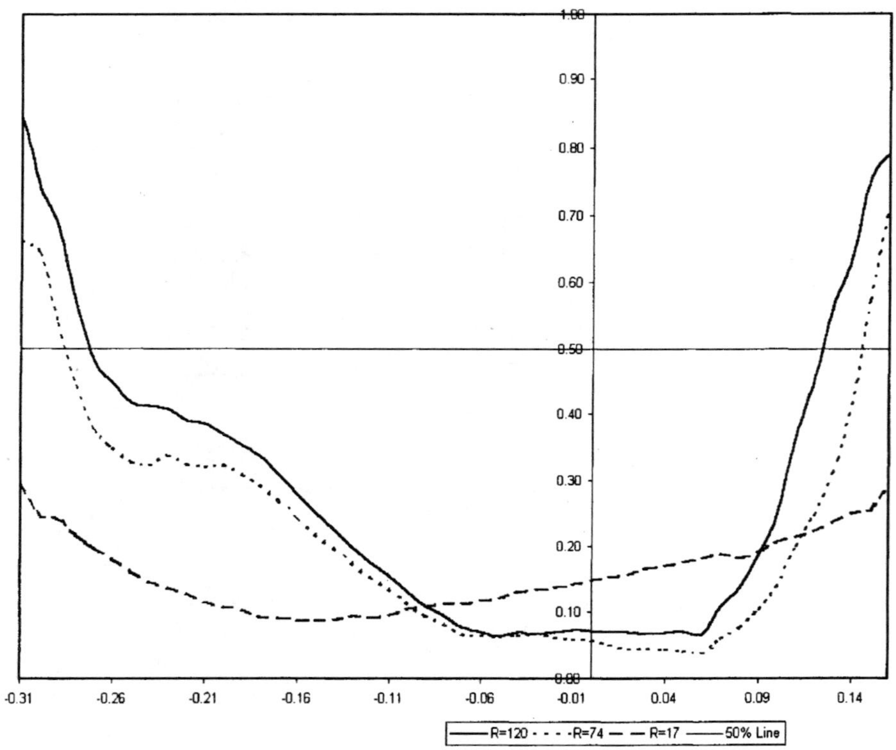

The King Point Optimal Test

King's test was very popular some years ago in a time series context, although the evolution of econometric methodology has left it out of the mainstream. That is surprising to some extent because it seems to be a good test. King and McAleer, for example, indicated (1987, p.662) that '*it always has the correct size and is the most powerful test. Its high power is obtained through the use of the Neyman-Pearson lemma without estimating* ρ *and* θ, *and by taking account of the one-sided nature of the testing problem*'. Also, the simulations carried out by Burke et al. (1990), support this position.

The main problem with this test is that the testing procedure is not standard. King and McAleer (1987, p.655) state that '*suitable values of* $\{\theta_0, \rho_0\}$ *must be chosen without reference to* y', for which we have to obtain their corresponding α level critical values. In a time series context and with positive autocorrelation, the pair (0.75,0.75) seems to be the optimum. In our case, we will restrict our attention to the stability interval of each contiguity (non row-standardised) matrix considering the pairs of the Table 8.2.

The first column for each matrix contains the pair of values applied when negative autocorrelation was detected, whereas the pairs in the second column were applied in the case of positive autocorrelation. Finally, the optimum pairs for each case were the following:

- R=17 and Positive autocorrelation: (0.20; 0.20) with critical value 1.578.
- R =17 and Negative autocorrelation: (-0.20; -0.20) with critical value 1.578.
- R =74 and Positive autocorrelation: (0.15; 0.20) with critical value 1.428.
- R =74 and Negative autocorrelation: (-0.15; -0.20) with critical value 1.428.
- R =120 and Positive autocorrelation: (0.15; 0.20) with critical value 1.418.
- R =120 and Negative autocorrelation: (-0.15; -0.20) with critical value 1.418.

Table 8.2: The King Test. Pairs of parameters analysed

R=17		R=74		R=120	
θ ρ	θ ρ	θ ρ	θ ρ	θ ρ	θ ρ
-0.03;-0.03	0.03;0.03	-0.03;-0.03	0.03;0.03	-0.03;-0.03	0.03;0.03
-0.05;-0.05	0.05;0.05	-0.05;-0.05	0.05;0.05	-0.05;-0.05	0.05;0.05
-0.10;-0.10	0.10;0.10	-0.10;-0.10	0.10;0.10	-0.10;-0.10	0.10;0.10
-0.15;-0.15	0.15;0.15	-0.15;-0.15	0.15;0.15	-0.15;-0.15	0.15;0.15
-0.20;-0.20	0.20;0.20	-0.15;-0.20	0.20;0.15	-0.15;-0.20	0.20;0.15
-0.20;-0.30	0.30;0.20	-0.15;-0.25	0.25;0.15	-0.15;-0.25	0.25;0.15
-0.20;-0.35	0.35;0.20	-0.15;-0.30	0.30;0.15	-0.15;-0.30	0.30;0.15
Stability Interval	- 0.36;0.21	**Stability Interval**	- 0.31;0.17	**Stability Interval**	- 0.30;0.16

Figure 8.5: The King Test. SMA Processes. Positive autocorrelation

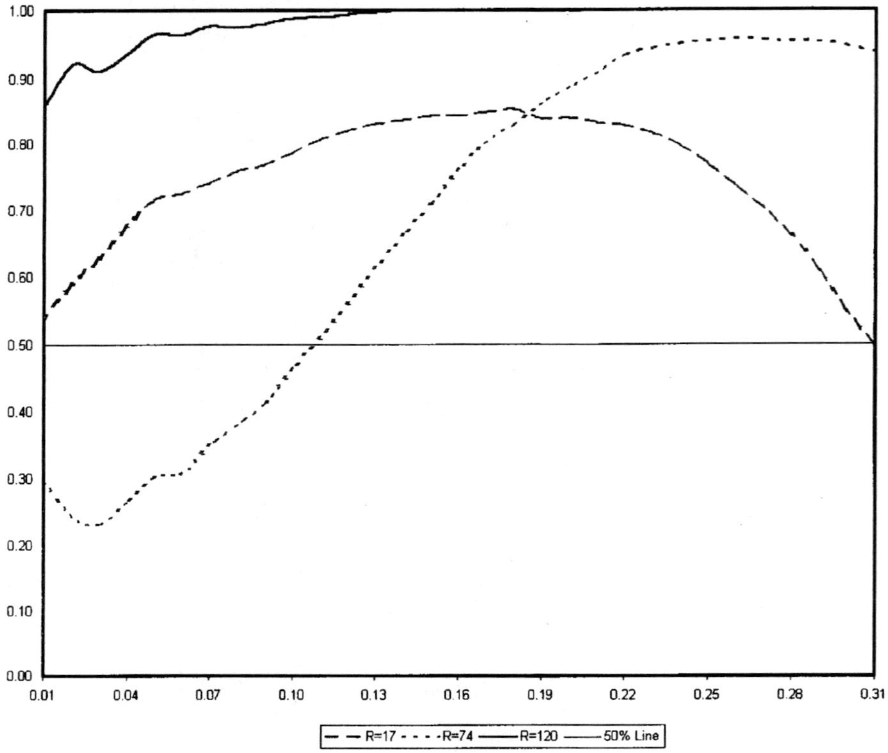

In Figures 8.5 to 8.8 we present the percentage of right decisions taken using these pairs (θ_0, ρ_0) in each case. The set of Figures tends to be encouraging in relation to the King test. This procedure works very well in the case of positive autocorrelation and with high sampling sizes. The estimated probability of committing an error in the selection of the DGP using 120 observations is negligible, even when the parameter of the process is in the neighbourhood of zero. Fewer observations imply less power, especially in the SAR case when, for the 17 observations sample, the test hardly reaches the 20% threshold. The performance of the test worsens slightly when we introduce negative autocorrelation in the DGP. In this case, the sample size effect is less evident as, for example, in Figure 8.7 where the differences between the three sizes are only small. Finally, it is worth underlining the following points in relation to the King method:

- Its performance is quite unstable in the sense that small changes in the pair used result in sharp changes in the estimated power functions.
- The test is very sensitive to the existence of outliers in the series and also to departures from the hypothesis of normality. Both situations weaken its confidence.

Figure 8.6: The King Test. SAR Processes. Positive autocorrelation

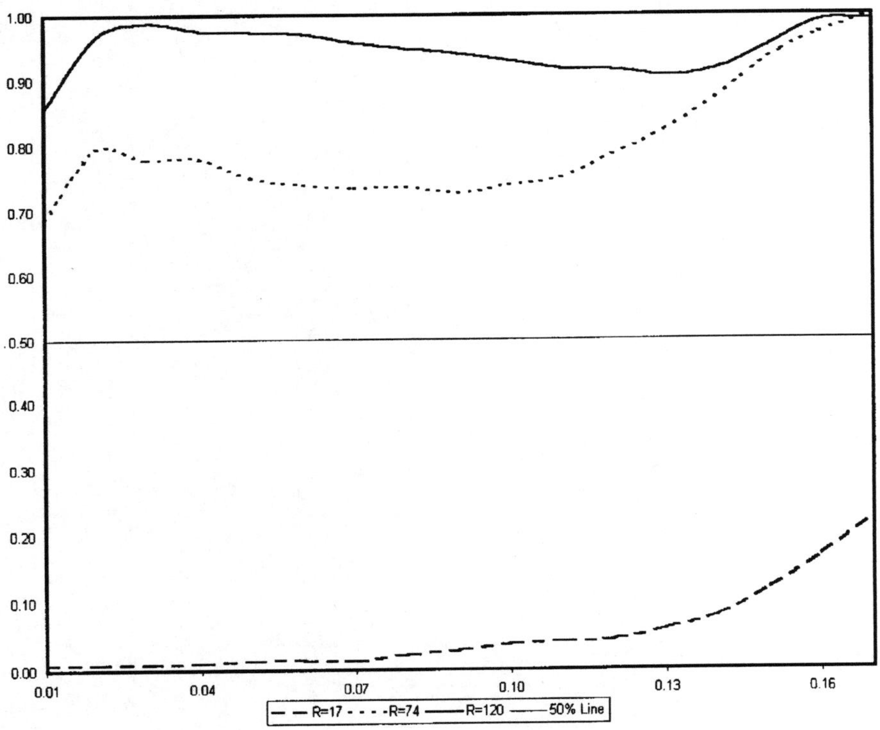

Our results are in line with those obtained in a time series context, since the optimum pairs (θ_0, ρ_0) also lie near the extremes of the stability interval.

- The global functioning of the procedure is strange due to the lack of symmetry between negative and positive autocorrelation. Also, there are sharper than expected differences in the behaviour of the test when applied to SMA or SAR processes.

The Minimum Variance Criterion

The advantages of this proposal with respect to the previous two are that non-conclusive situations will not arise and we do not need any other information, such as critical values. We only have to estimate two equations (8.7 and 8.8) and assign whichever process minimises the estimated variance of the series. The results obtained in terms of percentage of correct identifications are summarised in Figures 8.9 and 8.10.

Figure 8.7: The King Test. SMA Processes. Negative autocorrelation

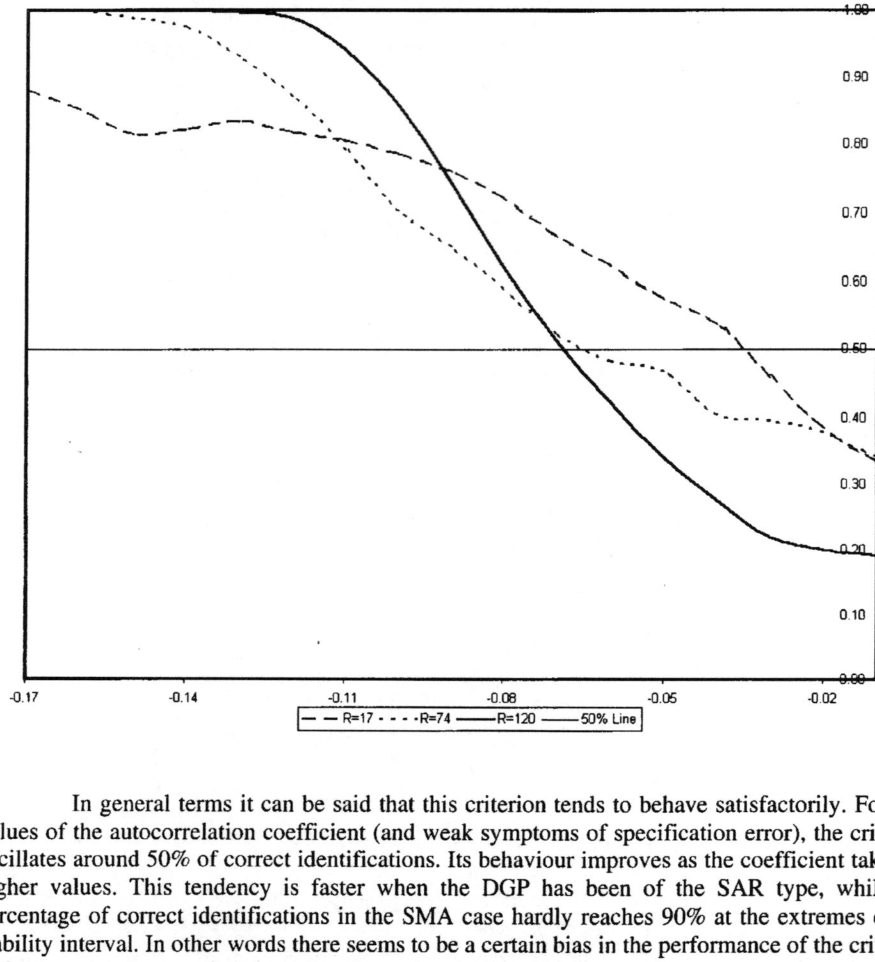

In general terms it can be said that this criterion tends to behave satisfactorily. For low values of the autocorrelation coefficient (and weak symptoms of specification error), the criterion oscillates around 50% of correct identifications. Its behaviour improves as the coefficient takes on higher values. This tendency is faster when the DGP has been of the SAR type, while the percentage of correct identifications in the SMA case hardly reaches 90% at the extremes of the stability interval. In other words there seems to be a certain bias in the performance of the criterion in favour of autoregressive processes. Also, the size effect is more clearly present when the DGP has been of an SAR type. Lastly, and more importantly, the behaviour of the criterion in the case of small samples (R=17) is far from negligible.

Figure 8.8: The King Test. SAR Processes. Negative autocorrelation

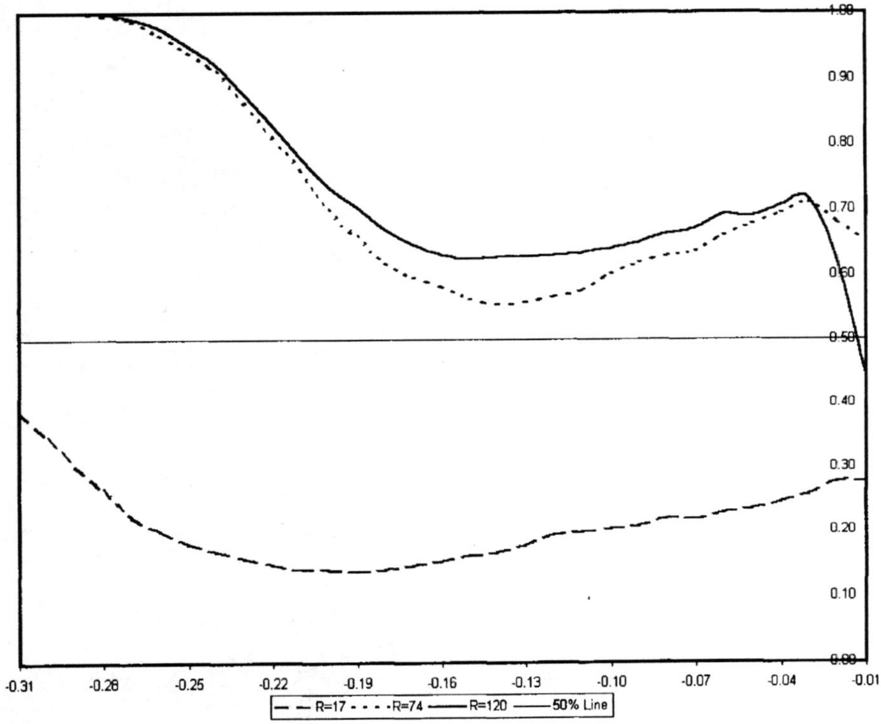

8.5 Conclusion

The problem set out at the beginning of the chapter was that of proposing a structure of dependencies for a spatially autocorrelated series. We restricted the study to the case of a series lacking any systematic structure and centred on zero, as a first step towards a more general approach. Also, the number of alternatives under consideration for modelling the series has been limited to just two: autoregressive and moving average structures. Using a non-nested models approach we have examined three possible strategies for selecting the best alternative.

The most relevant aspect in our results is the deficient behaviour of the Wald test as well as the problems of the King point optimal test when the sample is small. The performance of the latter procedure improves greatly when the sample size grows although as a selection strategy it remains unstable and asymmetric. Lastly, the criterion based on the minimum sampling variance seems to be the most advisable both because of its simplicity and its trustworthiness. The main problem with this criterion is that, because we use the eigenvectors and eigenvalues of the weight matrix, the calculation needs increase sharply with the sample size. Furthermore, this contiguity matrix should be symmetrical for its resolution to be simple.

Acknowledgements

This research has been financed by the CICYT-FEDER 2FD97-1004-C03-03 project for whose collaboration we are sincerely thankful.

Figure 8.9: The estimated variance method. SMA Processes

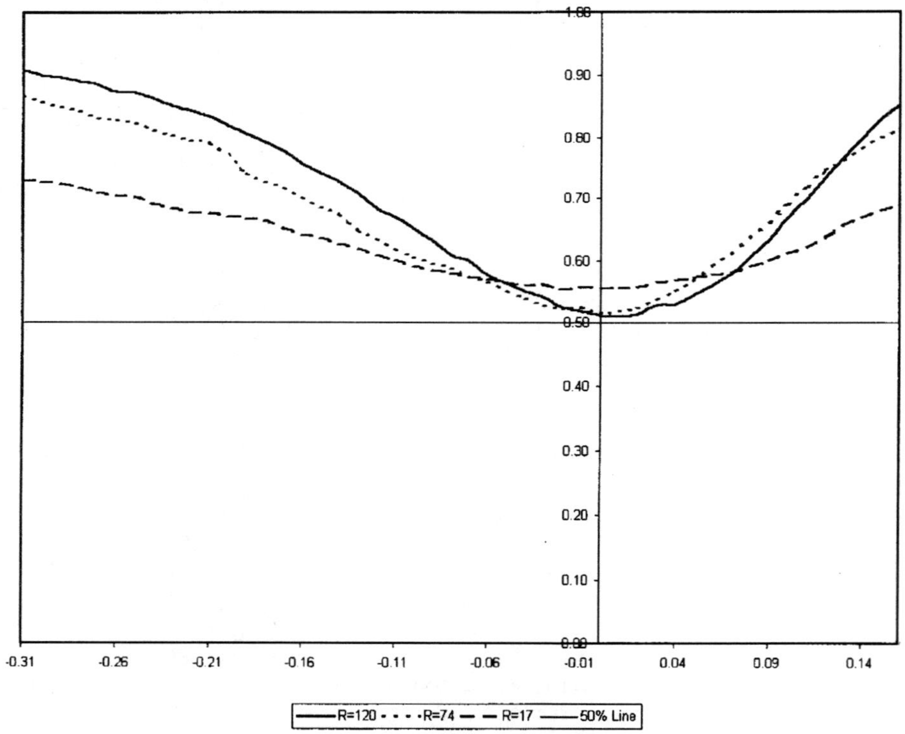

Appendix: Consistency of the criterion of minimum sampling variance

The DGP associated with a spatially autocorrelated series can be written, once it has been filtered and assuming normality, as:

$$y_r^* \sim iidN\left[0, \sigma_r^2\right] \tag{A.1}$$

with $\sigma_r^2 = \sigma^2(1+\theta\lambda_r)^2$ in the SMA case and $\sigma_r^2 = \sigma^2(1-\rho\lambda_r)^{-2}$ in the SAR case. The structure of the variance allows us to put forward the following auxiliary equations:

Figure 8.10: The estimated variance method. SAR Processes

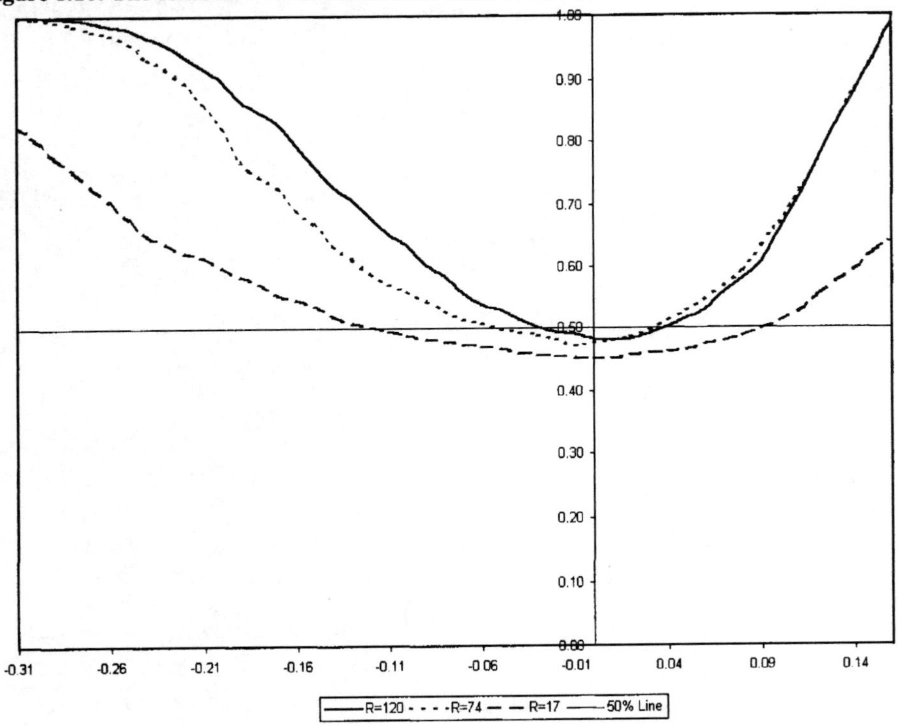

$$SMA \quad CASE \quad \Rightarrow \left|y_r^*\right| = \theta_0 + \theta_1 \lambda_r + v_r \tag{A.2}$$

$$SAR \quad CASE \quad \Rightarrow \left|y_r^*\right| = \frac{\alpha_0}{1-\alpha_1 \lambda_r} + v_r \tag{A.3}$$

which correspond to (8.7) and (8.8) in the text. The error term in both cases is heteroskedastic. The selection strategy proposed in section 8.3 implies considering both equations (A.2) and (A.3) and choosing the model with the least sampling variance. Below we outline the basis for this proposition.

Case I: The DGP is an SMA

If an SMA structure is proposed for the series, the equation that will be specified will be (A.2) which can be estimated by LS. To simplify, let us write this equation in matrix notation:

$$\left|y^*\right| = \mathbf{X}\beta + v \tag{A.4}$$

where **X** is the *(Rx2)* matrix of observations of the exogenous variables, $\beta' = [\theta_0 ; \theta_1]$ and v an error term vector. The LS estimation of $\hat{\beta}$ is consistent:

$$\hat{\beta} = [\mathbf{X'X}]^{-1}\mathbf{X}'\left|y^*\right| = \beta + [\mathbf{X'X}]^{-1}\mathbf{X}'v \Rightarrow plim\hat{\beta} = \beta \tag{A.5}$$

The LS residuals may be obtained in the usual manner:

$$\hat{v} = \left|y^*\right| - \mathbf{X}\hat{\beta} = \left[\mathbf{I} - \mathbf{X}(\mathbf{X'X})^{-1}\mathbf{X'}\right]v = \mathbf{M}v \tag{A.6}$$

where **M** is the projection matrix into the orthogonal space to **X**. Its expected value is zero and the sum of squared residuals (*SR*) could be expressed as a quadratic form of the error vector:

$$SR_{SMA} = \hat{v}'\hat{v} = v'\mathbf{M}v \tag{A.7}$$

where the subindex indicates that the adjusted DGP is an SMA. The expected value of *SR* is:

$$E[\hat{v}'\hat{v}] = \sigma^2(1-c^2)tr\mathbf{M\Omega} \tag{A.8}$$

with:

$$E[vv'] = \sigma^2(1-c^2)\begin{bmatrix} (1+\theta\lambda_1)^2 & 0 & 0 & \dots & 0 \\ 0 & (1+\theta\lambda_2)^2 & 0 & \dots & 0 \\ 0 & 0 & (1+\theta\lambda_3)^2 & \dots & 0 \\ \dots & \dots & \dots & \dots & \dots \\ 0 & 0 & 0 & \dots & (1+\theta\lambda_R)^2 \end{bmatrix} = \sigma^2(1-c^2)\mathbf{\Omega} \tag{A.9}$$

If we standardise the *SR* by the number of terms, we obtain a biased estimator of the variance:

$$\sigma^2_{SMA} = \frac{SR}{R} \Rightarrow E[\sigma^2_{SMA}] = \sigma^2(1-c^2)\frac{tr\mathbf{M\Omega}}{R} \tag{A.10}$$

This estimator will also be inconsistent, although it will converge to a finite constant:

$$plim\,\sigma^2_{SMA} = \sigma^2(1-c^2)\left[plim\frac{tr\mathbf{\Omega}}{R} - plim\frac{tr\,(\mathbf{X'X})^{-1}\mathbf{X'\Omega X}}{R}\right] =$$

$$= \sigma^2(1-c^2)\left[plim\sum_{r=1}^{R}\frac{(1+\theta\lambda_r)^2}{R} - plim\frac{1}{R}tr\left(\frac{\mathbf{X'X}}{R}\right)^{-1}\left(\frac{\mathbf{X'\Omega X}}{R}\right)\right] =$$

$$= \sigma^2(1-c^2)\left[plim\sum_{r=1}^{R}\frac{(1+\theta\lambda_r)^2}{R} - plim\frac{1}{R}tr\,\mathbf{\Sigma_{XX}}^{-1}\mathbf{\Sigma_{X\Omega\Omega}}\right] =$$

$$= \sigma^2(1-c^2)plim\sum_{r=1}^{R}\frac{(1+\theta\lambda_r)^2}{R} = \sigma^2(1-c^2)(1+\theta^2\bar{S}_2) = \sigma^2_{SMA} \tag{A.11}$$

with:

$$plim\frac{\mathbf{X'\Omega X}}{R} = \mathbf{\Sigma_{X\Omega\Omega}} = \begin{bmatrix} 1+\theta^2\bar{S}_2 & \theta(2\bar{S}_2+\theta\bar{S}_3) \\ \theta(2\bar{S}_2+\theta\bar{S}_3) & \bar{S}_2+\theta(2\bar{S}_3+\theta\bar{S}_4) \end{bmatrix}$$

$$plim\frac{\mathbf{X'X}}{R}=\mathbf{\Sigma_{xx}}=\begin{bmatrix}1 & 0\\ 0 & \overline{S}_2\end{bmatrix}$$

where $\overline{S}_j = plim\frac{\sum_{r=1}^{R}\lambda_r^j}{R}$; j=2,3,4 are finite elements that depend on $\mathbf{W}$. That is to say:

$$plim\,\hat{\sigma}^2_{SMA}=\sigma^2_{SMA} \tag{A.12}$$

If a specification error has been committed, imposing an SAR structure on an SMA series, we will specify equation (A.3) which must be estimated by non-linear methods. For simplification, let us rewrite the last equation as:

$$\left|y_r^*\right| = f(\lambda_r)+n_r \tag{A.13}$$

with $f(\lambda_r)=\dfrac{\alpha_0}{1-\alpha_1\lambda_r}$ and $n_r = v_r + \left[x_r'\beta - f(\lambda_r)\right] = v_r + g(\lambda_r)$. The vector of estimators $\hat{\alpha}$ obtained in (A.13) will be inconsistent in relation to β, although it will be true that $plim\hat{\alpha}=\alpha^*\neq\alpha$ where α^* is the point in the parametric space where the objective function reaches its optimum (White, 1982). The lack of consistency is due to the fact that the error term has a non-zero mean:

$$E[n_r]=E[v_r]+E[g(\lambda_r)]=g(\lambda_r) \tag{A.14}$$

This error will inflate the SR of the algorithm:

$$SR=\sum_{r=1}^{R}\left[\left|y_r^*\right|-\hat{f}(\lambda_r)\right]^2=\sum_{r=1}^{R}\left[(x_r'\beta-\hat{f}(\lambda_r))+v_r\right]^2=\sum_{r=1}^{R}g(\lambda_r)^2+\sum_{r=1}^{R}v_r^2+2\sum_{r=1}^{R}g(\lambda_r)v_r \tag{A.15}$$

Standardising once again by the number of terms, we get a biased and inconsistent estimator of the variance:

$$\hat{\sigma}^2_{SAR}=\frac{SR}{R}=\sum_{r=1}^{R}\frac{g(\lambda_r)^2}{R}+\sum_{r=1}^{R}\frac{v_r^2}{R}+2\sum_{r=1}^{R}\frac{g(\lambda_r)v_r}{R} \tag{A.16}$$

To obtain the limit of $\hat{\sigma}^2_{SAR}$ it is necessary to consider that:

$$plim\sum_{r=1}^{R}\frac{v_r^2}{R}=\sigma_{SMA}^2 \quad \text{(A.17)}$$

$$plim\sum_{r=1}^{R}\frac{g(\lambda_r)^2}{R}=\hat{g}^2>0 \quad \text{(A.18)}$$

$$plim\sum_{r=1}^{R}\frac{g(\lambda_r)v_r}{R}=plim\sum_{r=1}^{R}\frac{[(x_r'\beta)v_r]}{R}-plim\sum_{r=1}^{R}\frac{\hat{f}(\lambda_r)v_r}{R}=0 \quad \text{(A.19)}$$

The first plim on the right hand side of (A.19) is zero due to the orthogonality condition between $E\left[y_r^*\right]$ and v_r and the second one on applying the Law of Large Numbers ($plim\sum_{r=1}^{R}\frac{\hat{f}(\lambda_r;\beta)v_r}{R}=plim\sum_{r=1}^{R}\frac{f(\lambda_r;\beta^*)v_r}{R}=0$). The final result is that:

$$plim\,\hat{\sigma}_{SAR}^2=\sigma_{SMA}^2+\hat{g}^2>plim\,\hat{\sigma}_{SMA}^2=\sigma_{SMA}^2 \quad \text{(A.20)}$$

Case II: The DGP is an SAR

In this case the correct equation will be (A.3) which generates consistent estimators, with an associated SR:

$$SR=\sum_{r=1}^{R}\hat{v}_r^2=\sum_{r=1}^{R}\left[\left(f(\lambda_r)-\hat{f}(\lambda_r)\right)+v_r\right]^2=$$
$$\sum_{r=1}^{R}\left[\left(f(\lambda_r)-\hat{f}(\lambda_r)\right)\right]^2+\sum_{r=1}^{R}v_r^2+2\sum_{r=1}^{R}\left[f(\lambda_r)-\hat{f}(\lambda_r)\right]v_r \quad \text{(A.21)}$$

Standardising and using limits, we obtain:

$$plim\,\hat{\sigma}_{SAR}^2=plim\frac{\sum_{r=1}^{R}\left[\left(f(\lambda_r)-\hat{f}(\lambda_r)\right)\right]^2}{R}+plim\frac{\sum_{r=1}^{R}v_r^2}{R}+2plim\frac{\sum_{r=1}^{R}\left[f(\lambda_r)-\hat{f}(\lambda_r)\right]v_r}{R}$$

(A.22)

The first and the third terms on the right-hand side of (A.22) become zero given that the estimation is consistent, so that:

$$plim\,\hat{\sigma}^2_{SAR} = \sigma^2_{SAR} \qquad \text{(A.23)}$$

where:

$$\sigma^2_{SAR} = \sigma^2(1-c^2)\,plim\sum_{r=1}^{R}\frac{(1-\rho\lambda_r)^{-2}}{R} \qquad \text{(A.24)}$$

If an error has been committed in the specification of the functional form, assigning a moving average structure to the series, the equation proposed will be:

$$\left|y_r^*\right| = x_r'\beta + n_r \qquad \text{(A.25)}$$

with $n_r = v_r + \left[f(\lambda_r) - x_r'\beta\right]$ and $x_r' = [1, \lambda_r]$. The LS estimators in this equation are biased and inconsistent:

$$\begin{aligned} \hat{\beta} &= (\mathbf{X'X})^{-1}\mathbf{X'}\left|y^*\right| = \beta + (\mathbf{X'X})^{-1}\mathbf{X'}n \\ \Rightarrow E[\hat{\beta}] &= \beta + (\mathbf{X'X})^{-1}\mathbf{X'}[f(\lambda) - \mathbf{X}\beta] = (\mathbf{X'X})^{-1}\mathbf{X'}f(\lambda) \end{aligned} \qquad \text{(A.26)}$$

with $f(\lambda)$ the $(Rx1)$ vector with typical elements $\{f(\lambda_r),\ r = 1, 2, \ldots, R\}$. These anomalies will affect the residuals:

$$\hat{n} = \left|y^*\right| - \mathbf{X}\hat{\beta} = \mathbf{M}[v + f(\lambda)] \qquad \text{(A.27)}$$

resulting in a sum of squared residuals:

$$SR = \hat{n}'\hat{n} = [v + f(\lambda)]'\mathbf{M}[v + f(\lambda)] \qquad \text{(A.28)}$$

Once again standardising by the number of terms and using limits, we obtain:

$$plim\,\hat{\sigma}^2_{SMA} = \sigma^2(1-c^2)[N_1 + N_2 + N_3 + N_4] \qquad \text{(A.29)}$$

with:

$$N_1 = plim \frac{tr\,\mathbf{\Omega}^{-1}}{R}$$

$$N_2 = plim \frac{f(\lambda)' f(\lambda)}{R}$$

$$N_3 = plim \frac{1}{R} tr\left(\frac{\mathbf{X'X}}{R}\right)^{-1}\left(\frac{\mathbf{X'\Omega^{-1}X}}{R}\right) = plim \frac{tr\,\mathbf{\Sigma}_{\mathbf{xx}}^{-1}\mathbf{\Sigma}_{\mathbf{x\Omega^{-1}x}}}{R} = 0$$

$$N_4 = plim \frac{1}{R} tr\left(\frac{\mathbf{X'X}}{R}\right)^{-1}\left(\frac{\mathbf{X'F(\lambda(\lambda}}{R}\right) = plim \frac{tr\,\mathbf{\Sigma}_{\mathbf{xx}}^{-1}\mathbf{\Sigma}_{\mathbf{xFx}}}{R} = 0$$

where $\mathbf{F(\lambda)}=f(\lambda)f(\lambda)'$ and $plim\left(\frac{\mathbf{X'FX}}{R}\right) = \mathbf{\Sigma}_{\mathbf{xFx}}$. That is:

$$plim\,\sigma^2_{SMA} = plim \frac{tr\,\mathbf{\Omega}^{-1}}{R} + plim \frac{f(\lambda)' f(\lambda)}{R} = \sigma^2_{SAR} + \bar{f}^2 \qquad \text{(A.30)}$$

The two cases considered are summarised in Table 8.3.

Table 8.3: Specification errors and estimated variance. Comparison of results

	MODEL PROPOSED	
PGD	**SMA**	**SAR**
SMA	$plim\,\sigma^2_{SMA} = \sigma^2_{SMA}$	$plim\,\sigma^2_{SAR} = \sigma^2_{SMA} + \hat{g}^2$
SAR	$plim\,\sigma^2_{SMA} = \sigma^2_{SAR} + \bar{f}^2$	$plim\,\sigma^2_{SAR} = \sigma^2_{SAR}$

9 Serial and Spatial Error Dependence in Space-Time Models

J. Paul Elhorst
University of Groningen

Abstract

Spatial scientists have recently paid a great deal of attention to the problem of spatial error dependence, but neglected offering equal attention to the problem of serial error dependence. This chapter regenerates the problem of serial error dependence; it delineates an iterative two-step procedure to obtain the full maximum likelihood estimates of the parameters in a space-time model with first-order serial and spatial autocorrelation. It turns out that this model is *not* just the sum of the first-order serial and the first-order spatial autocorrelation model. Especially, the transformation on the cross-section of first observations complicates the analysis.

9.1 Introduction

It is frequently observed that the null hypotheses of no serial and no spatial autocorrelation of the error term require rejection when estimating a relationship on space-time data. This finding is problematic, since failing to account for serial and spatial autocorrelation when present causes the commonly used estimators, mostly OLS, to lose their property of efficiency. One remedial reaction could be to re-estimate the model using methods that assume that the errors are generated by a first-order (or higher order) serial and spatial autoregressive process. Leading textbooks still present this approach as an appropriate reaction, however, separately for either serial autocorrelation (Greene, 1993, Ch.15; Griffiths et al., 1993, Ch.16) or spatial autocorrelation (Anselin, 1988, Ch.8; Odland, 1988). Surprisingly, not one single textbook or article considers first-order serial and spatial autocorrelation within one common framework, while it seems such a natural combination when using space-time data. The objective of this chapter follows on from this remark and is to study a space-time model with first-order serial and spatial autocorrelation. In section 9.2 the space-time model is specified and its estimation procedure delineated; we discuss an iterative two-step procedure to obtain the full maximum likelihood estimates of the parameters involved. In section 9.3 we sketch the conditions under which the model is stationary. In section 9.4, a Phillips-curve model and a labor force participation rate equation are estimated, on the one hand as applications on real-world data and on the other hand as the basis of a Monte Carlo simulation experiment to assess the validity of the proposed estimation procedure.

It should be noted that space-time models with serial and spatial autocorrelation are not entirely absent from the literature. Bronars and Jansen (1987) specified a space-time model with serial and spatial autocorrelation by beginning with lattice data, the two-dimensional analogue of a time series, but this type of data is rather unusual in that it oversimplifies the spatial arrangement of the spatial units. In space, there is typically no natural order for arranging sample data. Earlier textbooks on spatial econometrics also dealt with autoregressive specifications in which the regressand is lagged in both space and time, but on further observation, not together in estimation[1] (see Cliff et al., 1975; Bartels, 1979; Bennett, 1979; Wrigley and Bennett, 1981). Finally, Pfeifer and Deutsch (1980) introduced regression equations in which the regressand, as well as the error term, are lagged in both space and time. To estimate this model, also known as a space time autoregressive moving average model (STARIMA), they used maximum likelihood, but they struggled with the initial values of lagged error terms, which they simply set at zero (ibid, 1980: 41).

In addition, it should be noted that first-order serial and spatial autocorrelation is not the only way to model serial and spatial error dependence. Other models suggested in the literature, not considered in this chapter, are the typical panel data model with both space and time effects (Anselin, 1988, 150-152), and a variant of the Zellner model (Hordijk and Nijkamp, 1977; Anselin, 1988, 37-38). In these models, serial and spatial error dependence are not expressed in terms of a particular parameterized function, but left unspecified as a general covariance.

Finally, it should be noted that the practice of correcting for autocorrelation has also been criticized. In section 9.5, we summarize the core criticism on the practice of correcting for serial autocorrelation put forward in the time series literature and show that these comments also apply to the practice of correcting for spatial autocorrelation and for serial and spatial autocorrelation simultaneously. It is shown that in a spatio-temporal context more dependence structures can exist than first-order serial and spatial autocorrelation. This section is meant to give a more balanced view of the importance of first-order serial and spatial autocorrelation.

In section 9.6 we present our final conclusion.

9.2 Model Specification and the Estimation Procedure

The space-time model with first-order serial and spatial autocorrelation is considered in vector form for a cross-section of observations at time t

$$Y_t = \mathbf{X}_t \beta + \varepsilon_t \tag{9.1a}$$

$$\varepsilon_t = \delta \mathbf{W} \varepsilon_t + \tau \varepsilon_t + u_t \tag{9.1b}$$

$$u_t \sim N(0, \sigma^2 \mathbf{I}_n), \tag{9.1c}$$

where Y_t denotes a $n \times 1$ vector consisting of one observation for every spatial unit (i=1,...,n) of the dependent variable in the tth time period (t=1,...,T); $\mathbf{X}_t$ denotes a $n \times k$ matrix of explanatory variables in the tth time period; β is a $k \times 1$ vector of response parameters; ε_t and u_t are $n \times 1$ vectors containing the error terms; the normality assumption with respect to u_t is to make ML estimation possible computationally; τ is the serial autoregressive coefficient; δ is the spatial autoregressive coefficient; and $\mathbf{W}$ denotes a $n \times n$ weight matrix describing the spatial arrangement of the spatial units. It is assumed that $\mathbf{W}$ is a matrix of known constants and that all of its diagonal elements are zero, since no spatial unit may be simultaneously viewed as its own neighbor. Moreover, it is assumed that the characteristic roots of $\mathbf{W}$, $\{\omega_1, ..., \omega_n\}$ are known. This assumption is needed to ensure that the log-likelihood function of the model can be computed and presupposes that the characteristic roots can be computed accurately based on computing technology available to empirical researchers. Kelejian and Prucha (1999) have pointed out that this might be problematic even for moderate sample sizes (n=400).

Means of estimating the first-order serial and the first-order spatial autocorrelation model are widely discussed in the literature. The common procedure of estimating the first-order serial autocorrelation model is an iterative two-step procedure in which β, as well as σ^2 and τ, are alternately estimated until convergence occurs. β and σ^2, given τ, are estimated using feasible GLS, while for τ, given β and σ^2, any estimator suffices so long as it is consistent (see for example Greene, 1993, Ch.15). The common procedure of estimating the first-order spatial autocorrelation model is almost similar (Anselin, 1988, 181-182; Anselin and Hudak, 1992). The two differ in that a wide range of consistent estimators exists for τ in the first-order serial autocorrelation model, whereas no such estimator exists for δ in the first-order spatial autocorrelation model (Anselin, 1988, 59). The most commonly suggested method to overcome this problem is to estimate the model by ML (see Anselin, 1988, 181-182; Anselin and Hudak, 1992). For that reason, we also choose ML to estimate the combined first-order serial and spatial autocorrelation model.

In order to obtain the log-likelihood function of the model in (9.1), we make use of

$$f(Y_1,...,Y_T) = f(Y_1) f(Y_2 \mid Y_1) ... f(Y_T \mid Y_{T-1}) \tag{9.2}$$

where f denotes the normal density function. Apart from the vector of first observations, the regression model for the vectors of subsequent observations (t>1) may be rewritten as

$$Y_t \mid Y_{t-1} = \mathbf{X_t}\beta + \tau\mathbf{B}^{-1}Y_{t-1} - \tau\mathbf{B}^{-1}\mathbf{X_{t-1}}\beta + \mathbf{B}^{-1}u_t \quad \text{with } \mathbf{B} = \mathbf{I_n} - \delta\mathbf{W} \quad (9.3)$$

This means that

$$f(Y_t \mid Y_{t-1}) = \mid \mathbf{B} \mid f_{\mathrm{u_t}}[\mathbf{B}Y_t - \tau Y_{t-1} - \mathbf{B}\mathbf{X_t}\beta + \tau\mathbf{X_{t-1}}\beta] \quad t>1 \quad (9.4)$$

To obtain the probability function of Y_1, the vector of first observations in the sample, we rewrite (9.1b) to get

$$\mathbf{B}\varepsilon_t = \tau\varepsilon_t + u_t \quad (9.5)$$

To proceed, we repeatedly lag this equation by one period. For ε_{t-m} ($m \geq 1$) we get

$$\mathbf{B}\varepsilon_{t-m} = \tau\varepsilon_{t-(m+1)} + u_{t-m} \quad \text{or } \varepsilon_{t-m} = \tau\mathbf{B}^{-1}\varepsilon_{t-(m+1)} + \mathbf{B}^{-1}u_{t-m} \quad (9.6)$$

Then, by substitution of ε_{t-1} into (9.5), next ε_{t-2} into (9.5) up to $\varepsilon_{t-(m-1)}$ into (9.5), we have

$$\mathbf{B}\varepsilon_t = \tau^m\mathbf{B}^{-(\mathbf{m-1})}\varepsilon_{t-m} + u_t + \tau\mathbf{B}^{-1}u_{t-1} + \ldots + \tau^{m-1}\mathbf{B}^{-(\mathbf{m-1})}u_{t-(m-1)} \quad (9.7)$$

If $|\tau\mathbf{B}^{-1}|<1$, the term $\tau^m\mathbf{B}^{-(\mathbf{m-1})}$ of the first right-hand side variable approaches zero as m approaches infinity. Thus, if we assume that $|\tau\mathbf{B}^{-1}|<1$ and that the process started long ago,

$$E(\mathbf{B}\varepsilon_t) = E(u_t + \tau\mathbf{B}^{-1}u_{t-1} + \tau^2\mathbf{B}^{-2}u_{t-1} + \ldots) = 0 \quad (9.8)$$

which implies that $E(\varepsilon_t)=0$. Since the successive values of u_t are uncorrelated, the variance of $\mathbf{B}\varepsilon_t$ under these assumptions can be written as

$$\begin{aligned} Var(\mathbf{B}\varepsilon_t) &= E(u_t u_t') + \tau^2\mathbf{B}^{-1}E(u_{t-1}u_{t-1}')\mathbf{B}'^{-1} + \tau^4\mathbf{B}^{-2}E(u_{t-2}u_{t-2}')\mathbf{B}'^{-2} + \ldots = \\ &= \sigma^2\mathbf{I_n} + \tau^2\mathbf{B}^{-1}\sigma^2\mathbf{I_n}\mathbf{B}'^{-1} + \tau^4\mathbf{B}^{-2}\sigma^2\mathbf{I_n}\mathbf{B}'^{-2} + \ldots = \\ &= \sigma^2(\mathbf{I_n} + \tau^2\mathbf{B}^{-1}\mathbf{B}'^{-1} + \tau^4\mathbf{B}^{-2}\mathbf{B}'^{-2} + \ldots) \end{aligned} \quad (9.9a)$$

which implies that $Var(\varepsilon_t)=\sigma^2\mathbf{B}^{-1}(\mathbf{I_n}-\tau^2(\mathbf{B'B})^{-1})^{-1}\mathbf{B}'^{-1}$. Provided that ε_m is not correlated with any subsequent u_t for which $t>m$, it further follows that

$$Cov(\varepsilon_t,\varepsilon_{t-m})=E(\varepsilon_t\varepsilon_{t-m}^{'})=\sigma^2\tau^m\mathbf{B}^{-m}\mathbf{B}^{-1}\left(\mathbf{I_n}-\tau^2(\mathbf{B'B})^{-1}\right)^{-1}\mathbf{B}'^{-1} \quad (9.9b)$$

Thus the vector process ε has a constant unconditional mean and a constant unconditional variance, independent of time, while the autocovariances depend only on the length of time separating the observations, and not on the date of observation itself. A process satisfying these conditions is said to be covariance stationary.

Under the model assumptions, the regression model for the vector of first observations may be written as

$$Y_1=\mathbf{X}_1\beta+\left(\mathbf{P}_1^{1/2}\mathbf{B}\right)^{-1}u_1, \text{ with } \mathbf{P}_1=\mathbf{I}_n-\tau^2(\mathbf{B'B})^{-1} \quad (9.10)$$

This means that

$$f(Y_1)=|\mathbf{P}_1^{1/2}||\mathbf{B}|f_{u_1}(\mathbf{P}_1^{1/2}\mathbf{B}(Y_1-\mathbf{X}_1\beta)), \quad (9.11)$$

The determinants $|\mathbf{B}|=|\mathbf{I_n}-\delta\mathbf{W}|$ in equation (9.4) and (9.11) and $|\mathbf{P_1}|=|\mathbf{I_n}-\tau^2(\mathbf{B'B})^{-1}|$ in equation (9.11) can be expressed as the product of the characteristic roots of the spatial weight matrix **W**, i.e., $\prod_{i=1}^n 1-\delta\varpi_i$ for the matrix **B** and $\prod_{i=1}^n 1-\tau^2/(1-\delta\varpi_i)^2$ for the matrix $\mathbf{P_1}$ (see Griffith, 1988, 44, table 3.1).

Collecting terms, the log-likelihood function of the complete sample of size T is

$$\begin{aligned}LogL=&-\frac{nT}{2}ln(2\pi\sigma^2)+\frac{1}{2}\sum_{i=1}^{n}ln(1-\frac{\tau^2}{(1-\delta\varpi_i)^2})+T\sum_{i=1}^{n}ln(1-\delta\varpi_i)\\&-\frac{1}{2\sigma^2}(Y_1-\mathbf{X}_1\beta)'\mathbf{B}'(\mathbf{I_n}-\tau^2(\mathbf{B'B})^{-1})\mathbf{B}(Y_1-\mathbf{X}_1\beta)\\&-\frac{1}{2\sigma^2}\sum_{t=2}^{T}(\mathbf{B}Y_t-\tau Y_{t-1}-\mathbf{B}\mathbf{X_t}\beta+\tau\mathbf{X_{t-1}}\beta)'(\mathbf{B}Y_t-\tau Y_{t-1}-\mathbf{B}\mathbf{X_t}\beta+\tau\mathbf{X_{t-1}}\beta)\end{aligned}$$

(9.12)

The first order conditions for the ML estimators are obtained by taking the partial derivatives of the log-likelihood function with respect to β, σ^2, τ and δ. Starting with β and σ^2, we get

$$\frac{\partial \log L}{\partial \beta} = \frac{1}{\sigma^2}\mathbf{X}_1^{'}\mathbf{B}'(\mathbf{I}_n - \tau^2(\mathbf{B}'\mathbf{B})^{-1})\mathbf{B}(Y_1 - \mathbf{X}_1\beta) + \\ + \frac{1}{\sigma^2}\sum_{t=2}^{T}(\mathbf{B}\,\mathbf{X}_t - \tau\mathbf{X}_{t-1})'(\mathbf{B}\,Y_t - \tau Y_{t-1} - \mathbf{B}\,\mathbf{X}_t\beta + \tau\mathbf{X}_{t-1}\beta) \tag{9.13a}$$

$$\frac{\partial \log L}{\partial \sigma^2} = -\frac{nT}{2\sigma^2} + \frac{1}{2\sigma^4}(Y_1 - \mathbf{X}_1\beta)'\mathbf{B}'(\mathbf{I}_n - \tau^2(\mathbf{B}'\mathbf{B})^{-1})\mathbf{B}(Y_1 - \mathbf{X}_1\beta) \\ + \frac{1}{2\sigma^4}\sum_{t=2}^{T}(\mathbf{B}\,Y_t - \tau Y_{t-1} - \mathbf{B}\,\mathbf{X}_t\beta + \tau\mathbf{X}_{t-1}\beta)'(\mathbf{B}\,Y_t - \tau Y_{t-1} - \mathbf{B}\,\mathbf{X}_t\beta + \tau\mathbf{X}_{t-1}\beta) \tag{9.13b}$$

When applying the following transformations to the data

$$Y_1^* = P_1^{1/2} B Y_1 \qquad X_1^* = P_1^{1/2} B X_1 \\ Y_t^* = B Y_t - \tau Y_{t-1}, \qquad X_t^* = B X_t - \tau X_{t-1} \qquad \text{for } t > 1, \tag{9.14}$$

these formulas reduce to

$$\frac{\partial \log L}{\partial \beta} = \frac{1}{\sigma^2}\sum_{t=1}^{T}\mathbf{X}_t^{*'}(Y_t^* - \mathbf{X}_t^*\beta), \tag{9.13a'}$$

$$\frac{\partial \log L}{\partial \sigma^2} = -\frac{nT}{2\sigma^2} + \frac{1}{2\sigma^4}\sum_{t=1}^{T}(Y_t^* - \mathbf{X}_t^*\beta)'(Y_t^* - \mathbf{X}_t^*\beta) \tag{9.13b'}$$

So if τ and δ are known, the solution for β and σ^2 is least squares on the transformed data or generalized least squares on the original data.

For τ and δ we find that

$$\frac{\partial \log L}{\partial \tau} = -\sum_{i=1}^{n}\frac{\tau}{(1-\delta\varpi_i)^2 - \tau^2} + \frac{\tau}{\sigma^2}e_1^{'}\mathbf{B}'(\mathbf{B}'\mathbf{B})^{-1}\mathbf{B}\,e_1 + \frac{1}{\sigma^2}\sum_{t=2}^{T}e_{t-1}^{'}(\mathbf{B}\,e_t - \tau e_{t-1}), \tag{9.13c}$$

$$\frac{\partial logL}{\partial\delta}=-\sum_{i=1}^{n}\frac{\tau^2\varpi_i}{((1-\delta\varpi_i)^2-\tau^2)(1-\delta\varpi_i)}-\sum_{i=1}^{n}\frac{T\varpi_i}{1-\delta\varpi_i}+\frac{1}{\sigma^2}(\mathbf{W}e_1)'(e_1-\delta\mathbf{W}e_1)$$

$$+\frac{\tau^2}{2\sigma^2}e_1'[\mathbf{W}'(\mathbf{B}'\mathbf{B})^{-1}\mathbf{B}+\mathbf{B}'(\mathbf{B}'\mathbf{B})^{-1}\mathbf{W}-\mathbf{B}'(\mathbf{B}'\mathbf{B})^{-1}(\mathbf{B}'\mathbf{W}+\mathbf{W}'\mathbf{B})(\mathbf{B}'\mathbf{B})^{-1}\mathbf{B}]e_1$$

$$+\frac{1}{\sigma^2}\sum_{t=2}^{T}(\mathbf{W}e_t)'(e_t-\delta\mathbf{W}e_t-\tau e_{t-1}) \tag{9.13d}$$

where $e_t=Y_t-\mathbf{X}_t\beta$. Note that if the matrix **W**, and thus the matrix **B** are symmetric, the second term at the right-hand side of (9.13c) reduces to $\tau/\sigma^2\times e_1'e_1$ and that the fourth term at the right-hand side of (9.13d) reduces to zero. But even then a closed-form solution for δ and τ does not exist. For this reason, δ and τ must be solved by numerical methods. In view of this, we propose the following iterative two-step estimation procedure. First, an initial estimate of β and σ^2 are needed, which can be obtained by estimating equation (9.1) by OLS. Estimation stops if the null hypotheses of no spatial or serial autocorrelation need not be rejected. Upon inserting the maximizing values of β and σ^2 into the log-likelihood function, the concentrated log-likelihood function of τ and δ is obtained, which then is maximized. To obtain the maximum likelihood estimates of β and σ^2 given τ and δ, the observations may be transformed with the help of equation (9.14) to apply feasible GLS. Finally, these successive steps must be repeated until convergence occurs.

The above procedure is based on Amemiya (1985, 125-127) and Magnus (1978), who have proved that an iterative procedure of this type converges to a consistent root of the first-order maximizing conditions and that this root is a ML estimator. In the next section we discuss some additional restrictions to be placed on the model related to stationarity and stability. Prior to this, two other issues require further explanation. The first issue is the area of τ and δ on which the (concentrated) log-likelihood function of τ and δ is to be maximized. Let ω_{min} denote the smallest and ω_{max} the largest characteristic root of the spatial weight matrix **W** ($\omega_{min}<0$, $\omega_{max}>0$). Then the log-likelihood function in equation (9.12) is defined only for $|\tau|<1-\delta\varpi_{max}$ if $\delta\geq 0$ and $|\tau|<1-\delta\varpi_{min}$ if $\delta<0$, which is in line with the stationarity condition $|\tau\mathbf{B}^{-1}|<1$. Figure 9.1 shows the total area of τ and δ on which the (concentrated) log-likelihood function is defined. If $\delta = 0$, these conditions reduce to $|\tau| < 1$, which is the stationarity condition on the first-order serial autocorrelation model which can be found in most econometric textbooks. If $\tau = 0$, these conditions reduce to $1/\varpi_{min}<\delta<1/\varpi_{max}$, which is the standard condition on the first-order spatial autocorrelation model pointed out by Ord (1981) and Anselin (1982). Figure 9.1 shows that there is a trade-off between serial and spatial autocorrelation. From equation (9.7) it can be seen that the disturbance term of spatial unit i at time t, ε_{it}, embodies the entire past history of the u's, not only of its own, but also of the other spatial units. In other words, spatial dependence is not restricted to a functional relationship between two points in space, but is also transmitted through time. Consequently, if spatial autocorrelation is strong, serial autocorrelation has to be weak, and vice-versa.

The second issue requiring further explanation is whether the solution found for τ and δ, when maximizing the concentrated log-likelihood function of τ and δ, is the global maximum and not a local maximum. If the concentrated log-likelihood function were concave and its domain convex, the solution of the first-order maximizing conditions is the global maximum, as a result of which it would be sufficient to find this solution. Since every pair of τ and δ can be joined by a line segment that lies entirely within the domain of the concentrated log-likelihood function of τ and δ (see figure 9.1), its domain is convex. In section 9.4 we will show that the concentrated log-likelihood function of τ and δ is also concave.

Another issue how to implement the proposed estimation procedure in commercial econometric software is explained in the Appendix.

Figure 9.1 The area on which the (concentrated) log-likelihood function is defined for τ and δ

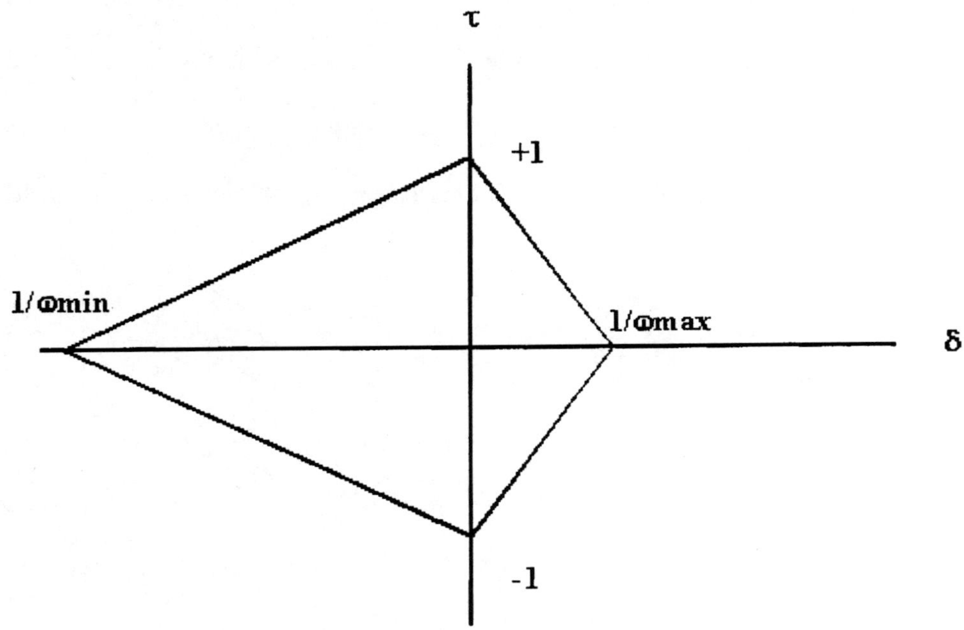

9.3 Consistency

One problem related to the proposed estimation procedure is whether the feasible GLS estimator of β is asymptotically equivalent to the true GLS estimator of β. Two general conditions have appeared in the literature that indicate this may be true. Let $\mathbf{N} \equiv \mathbf{n} \times \mathbf{T}$ denote the number of observations and $\mathbf{\Omega}$ the general covariance structure. Then $1/N \times \mathbf{X}'\mathbf{\Omega}(\hat{\mathbf{s}})\mathbf{X}$ has to converge to $1/N \times \mathbf{X}'\mathbf{\Omega}(\mathbf{s})\mathbf{X}$ and $1/\sqrt{N} \times \mathbf{X}'\mathbf{\Omega}(\hat{\mathbf{s}})^{-1} u_N$ has to converge to $1/\sqrt{N} \times \mathbf{X}'\mathbf{\Omega}(\mathbf{s})^{-1} u_N$, where s denotes a set of unknown parameters independent of β such that $\mathbf{\Omega}=\mathbf{\Omega}(\mathbf{s})$. Magnus (1978) has shown that these conditions are met in most familiar settings (see also Greene, 1993, 366). By contrast, Mandy and Martins-Filho (1994, 972) have shown that these conditions are insufficient for the covariance structure in Magnus (1978) unless further restrictions are placed upon the model. The general covariance structure considered by Magnus (1978) should also meet the three following conditions so that the feasible GLS estimator of β be asymptotically equivalent to the GLS estimator:

[I] $\mathbf{\Omega}^{-1}(\mathbf{s})$ has at most $C_1<\infty$ distinct nonzero elements for every N. In other words, there are N^2-C_1 elements that are either zero or duplicates of other nonzero elements in $\mathbf{\Omega}^{-1}$.

[II] The number of nonzero elements in each row (and column) of $\mathbf{\Omega}^{-1}(\mathbf{s})$ is bounded uniformly by $C_2<\infty$ as $N\rightarrow\infty$.

[III] The row and column sums of $\mathbf{\Omega}(\mathbf{s})$ are bounded uniformly in absolute value as $N\rightarrow\infty$.

The covariance structure of the space-time model with first-order serial and spatial autocorrelation is

$$E\varepsilon\varepsilon' = \sigma^2\mathbf{\Omega} = \sigma^2 \begin{bmatrix} \mathbf{Z}_n & \mathbf{Z}_n\tau\mathbf{B}'^{-1} & \cdot & Z_n\tau^{T-1}\mathbf{B}'^{-(T-1)} \\ \tau\mathbf{B}^{-1}\mathbf{Z}_n & \mathbf{Z}_n & \cdot & \mathbf{Z}_n\tau^{T-2}\mathbf{B}'^{-(T-2)} \\ \cdot & \cdot & \cdot & \cdot \\ \tau^{T-1}\mathbf{B}^{-(T-1)}\mathbf{Z}_n & \tau^{T-2}\mathbf{B}^{-(T-2)}\mathbf{Z}_n & \cdot & \mathbf{Z}_n \end{bmatrix} \tag{9.14}$$

where $\mathbf{Z_n} = \mathbf{B}^{-1}(\mathbf{I_n} - \tau^2(\mathbf{B'B})^{-1})^{-1}\mathbf{B'}^{-1}$. The inverse of $\mathbf{\Omega}$, $\mathbf{\Omega}^{-1}$, is[2]

$$\mathbf{\Omega}^{-1} = \begin{bmatrix} \mathbf{Z_n}^{-1} + \tau^2\mathbf{I_n} & -\tau\mathbf{B} & 0 & . & 0 & 0 \\ -\tau\mathbf{B'} & \mathbf{B'B} + \tau^2\mathbf{I_n} & -\tau\mathbf{B} & . & 0 & 0 \\ 0 & -\tau\mathbf{B'} & \mathbf{B'B} + \tau^2\mathbf{I_n} & . & 0 & 0 \\ . & . & . & . & . & . \\ 0 & 0 & 0 & . & \mathbf{B'B} + \tau^2\mathbf{I_n} & -\tau\mathbf{B} \\ 0 & 0 & 0 & . & -\tau\mathbf{B'} & \mathbf{B'B} \end{bmatrix} \tag{9.15}$$

Since a space-time data set has two dimensions, it is possible to consider asymptotic behavior as $n \to \infty$, $T \to \infty$, or both.

First, these additional conditions are checked when the number of observations in the time domain tends to infinity, $T \to \infty$, and the number of spatial units denoted by n are fixed. It can be seen that each block-row and block-column of the matrix $\mathbf{\Omega}^{-1}$ contains at most three non-zero blocks verifying [II] with $C_2=3n$. Since $\mathbf{\Omega}^{-1}$ has five different blocks and these blocks are independent of T, $\mathbf{\Omega}^{-1}$ has at most $C_1=5n^2$ distinct elements for every number of T. This verifies [I]. The sum of every block-row or block-column ($m=1,\ldots,T$) of the covariance matrix $\mathbf{\Omega}$ in expression (9.14) is

$$\sigma^2[(\tau\mathbf{B}^{-1} - \tau^m\mathbf{B}^{-m})\mathbf{Z_n}(\mathbf{I_n} - \tau\mathbf{B}^{-1})^{-1} + \mathbf{Z_n}(\mathbf{I_n} - \tau^{m+1-m}\mathbf{B}'^{-(T+1-m)})(\mathbf{I_n} - \tau\mathbf{B}'^{-1})^{-1}] \tag{9.16}$$

These sums are bounded uniformly when T tends to infinity in case (i) the matrices **B**, **B'**, **B'B**, $\mathbf{I_n} - \tau^2(\mathbf{B'B})^{-1}$, $\mathbf{I_n} - \tau\mathbf{B}^{-1}$, and $\mathbf{I_n} - \tau\mathbf{B}'^{-1}$ are nonsingular and (ii) $|\tau\mathbf{B}^{-1}| < 1$. Condition (i) holds if condition (ii) holds. As condition (ii) is simply the stationarity condition, this verifies [III].

In contrast to stationarity in time, stationarity in space is quite difficult to consider. Quite recently, Griffith and Lagona (1998) pointed out that the correlation between two spatial observations should converge to zero as the distance separating them increases to infinity, while Kelejian and Prucha (1998, 1999) formulated one necessary condition that must be satisfied: the row and column sums of the spatial weight matrix **W** must be bounded uniformly in absolute value as $n \to \infty$. When the spatial weight matrix is a binary contiguity matrix, this condition is satisfied. Normally speaking, no spatial unit is assumed to be a neighbor to more than a given number, say q, of other spatial units. When the spatial weight matrix is an inverse distance matrix, this condition need not automatically be satisfied. Consider an infinite number of spatial units that are linearly arranged. The distance of each spatial unit to its first left and right neighbor is 1, to its second left and right neighbor is 2, and so forth. When **W** is an inverse distance matrix, the off-diagonal elements of **W** being of the form $1/d_{ab}^{\alpha}$ with $\alpha > 1$ or of the form $exp(-\alpha d_{ab})$, where d_{ab} is the distance between two spatial units a and b, then the sum of each row of **W** would be finite. By contrast, if the off-diagonal elements of **W** would be of the form $1/d_{ab}$, the sum of each row of **W** is $\sum 2/d$, a series which is not finite[3].

In sum, when n is fixed and $T \rightarrow \infty$, the vector process ε is stationary, but when $n \rightarrow \infty$ and T is fixed the weights formulated as a function of distance between spatial units must decline rapidly; otherwise large sample theory may not be applied. In addition, it should be realized that a spatial process, though stable, may not be stationary (see Haining, 1990: 82; Kelejian and Robinson, 1995, 78). This is a separate problem beyond the scope of this chapter (for further details Lahari, 1996; Griffith and Lagona, 1998).

9.4 Numerical Illustrations

In this section we first estimate a Phillips-curve model and a relationship between the labor force participation rate and the unemployment rate as applications based on real-world data, and then we use these models as the basis of a Monte Carlo simulation experiment to check whether the solution produced by the proposed estimation procedure is the global maximum for different values of τ and δ. It is to be noted that these models are selected primarily to illustrate spatial and serial effects, and not to make a contribution to the substantive understanding of the space-time dynamics of labor market adjustments.

The Phillips-curve model is estimated on observations for 25 South-Western Ohio counties in two time periods, 1981 and 1983. It relates changes in wage rates (WAGE) to the inverse unemployment rate (UN), a net-migration rate (NMR), and a dummy variable (SMSA) which assumes a value of 1 for counties that are part of a metropolitan area. The spatial weight matrix is a binary contiguity matrix. The data, the model as well as the spatial weight matrix are taken from Anselin (1988, 203-215). The results are reported in table 9.1.

The relationship between the labor force participation rate (LFPR) and the unemployment rate (UR) is estimated on observations for 22 regions in France, 31 regions in Germany and 11 regions in the UK over the period 1983-1993. These data are derived from Eurostat. The spatial weight matrix used in these estimations is a normalized inverse (linear) distance matrix. The results are reported in table 9.2.

To stress the differences between both applications, we repeat that the first is based on American data, a binary contiguity matrix and two time periods, while the second is based on European data, an inverse distance matrix and 11 time periods. In both cases, we first estimated a static regression equation by OLS and tested for serial and spatial autocorrelation and then we estimated the static regression equation corrected for both serial and spatial autocorrelation.

Table 9.1: Estimation results of the Phillips curve model

Independent variables	Dependent variable WAGE			
	Static model		Serial and spatial autocorrelation	
Intercept	1.013 (42.76)		1.063 (70.43)	
UN	0.584 (2.31)		0.030 (0.20)	
NMR	0.954 (2.37)		0.657 (2.38)	
SMSA	-0.010 (-0.52)		0.003 (0.44)	
τ	-		-0.802 (-11.07)	
δ	-		0.018 (0.18)	
R^2	0.23		0.99	
LM test for serial aut.	8.66			
LM test for spatial aut.		0.07		
# spatial units	25		ω_{min}	-0.508
# time periods	2		ω_{max}	1.000

T-values between parentheses

Table 9.2: Estimation results of the labor force participation rate equation

Independent variables	Dependent variable LFPR		
	France	Germany	The UK
STATIC MODEL			
Intercept	0.742 (102.29)	0.745 (141.45)	0.820 (100.95)
UR	-0.863 (-12.72)	-1.202 (-16.07)	-0.917 (-12.42)
LM test for serial aut.	118.5	180.2	43.4
LM test for spatial aut	43.5	110.5	32.3
R^2	0.56	0.42	0.67
SERIAL AND SPATIAL AUTOCORRELATION			
Intercept	0.617 (38.07)	0.782 (55.69)	0.751 (40.51)
UR	-0.443 (-11.22)	-0.529 (-9.20)	-0.490 (-5.09)
τ	0.772 (28.73)	0.803 (35.24)	0.834 (17.39)
δ	0.227 (8.35)	0.196 (8.73)	0.161 (3.33)
R^2	0.86	0.87	0.95
# spatial units	22	31	11
# time periods	11	11	11
ω_{min}	-0.225	-0.218	-0.355
ω_{max}	1.000	1.000	1.000

T-values between parentheses

When the Phillips-curve model is estimated in static form, there appears to be an indication of the Phillips-curve effect; the coefficient of UN is positive and significant. The coefficient of NMR is also positive and significant. The null hypothesis whether the model is acceptable on the data (or, alternatively, must be extended with serial autocorrelation) has to be rejected; the value of the Lagrange Multiplier test, which has a chi-squared distribution with 1 degree of freedom, appeared to be greater than the 5 percent critical value of this distribution. By contrast, a similar test for spatial autocorrelation indicated that the null hypothesis of no spatial autocorrelation need not be rejected. When estimating the first-order serial and spatial autocorrelation model, these test results reappear. Whereas the serial autocorrelation coefficient appears to be significant, the spatial autocorrelation coefficient does not. In other words, serial effects are more important than spatial effects in this application. Looking at the other estimation results, we see that the main change is in the coefficient of UN. Although still positive, it has lost its significance and to a great extent also its relevance. By contrast, the positive impact of NMR appears to be quite robust. Although its coefficient has fallen slightly, it is again significant.

In the labor force participation rate equation, the absence of serial autocorrelation and of spatial autocorrelation in the static equation must be rejected too. When estimating the first-order serial and spatial autocorrelation model, not only the serial but also the spatial autocorrelation coefficient appears to be significant, though again serial effects appear to be more important than spatial effects. Looking at the coefficient of the unemployment rate, we see that its effect on the participation rate tends to be highly overestimated when serial and spatial autocorrelation are not taken into account.

Although these are just two examples, we may conclude that under normal circumstances (say during periods without wars or climatic catastrophes in which socio-economic variables observed at spatial units show little if any change over time), it is not unreasonable to suspect serial effects being present in a greater class of space-time models. Consequently, the investigator should better not be focussed on spatial effects only.

The simulation experiment of checking whether the solution found by the proposed estimation procedure is really the global maximum is based on the estimated relationships. The data generating process reads as

$$\begin{aligned} &Y_t = \mathbf{X}_1\beta + (\mathbf{P}_1^{1/2}\mathbf{B})^{-1}u && \text{for } t=1 \\ &Y_t \mid Y_{t-1} = \mathbf{X}_t\beta + \tau\mathbf{B}^{-1}(Y_{t-1} - \mathbf{X}_{t-1}\beta) + \mathbf{B}^{-1}u && \text{for } t>1 \qquad (9.17) \\ &u \sim N(0, \sigma^2\mathbf{I}_n) \end{aligned}$$

In this experiment Y_t, $\mathbf{X}_t$ as well as β and σ^2 are replaced successively by the variables and the parameter values of the Phillips-curve model in table 9.1 or of the labor force participation rate equation in table 9.2. In addition, 6 different combinations are considered for τ and δ. In case of the Phillips curve model, the original estimates and the combinations (δ,τ)=(0.1,0.1), (0.45,0.45), (0.45,-0.45), (-1.5,0.45), and (-1.5,-0.45), and in

Figure 9.2 The concave shape of the concentrated log-likelihood in function of τ and δ

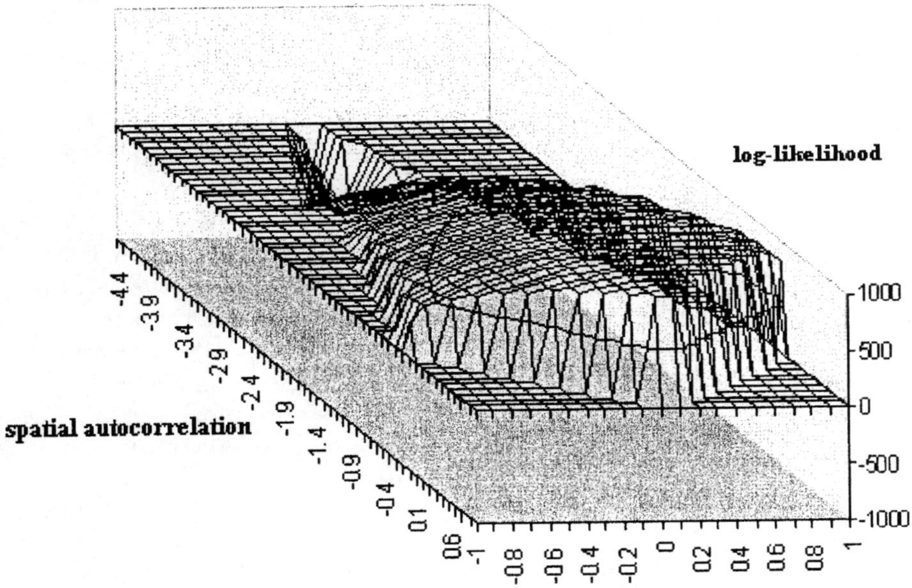

case of the labor force participation rate equation, the original estimates (for France, Germany and the UK, respectively) and the combinations (δ,τ)=(0.1,0.1), (0.45,0.45), (0.45,-0.45), (-2,0.45), and (-2,-0.45). All these combinations are part of the area of τ and δ on which the (concentrated) log-likelihood function is defined[4].

These simulation experiments have demonstrated that the concentrated log-likelihood function of τ and δ is concave on its whole domain of τ and δ. Its concave shape is illustrated in figure 9.2 for the true combination of (δ,τ)=(0.45,0.45) in the German labor force participation rate equation and one particular draw of the disturbance term from the normal distribution. The results obtained for the other combinations of τ and δ, the other countries, and for the Phillips curve model showed similar patterns but are not reported for reasons of space.

9.5 Criticism

It should be stressed that the practice of correcting for residual autocorrelation has also been criticized. Hendry and Mizon (1978) were among the first to point out that correcting for serial autocorrelation cannot be considered a serious effort to find the 'correct' equation (see also Gilbert (1986) on professor Hendry's econometric methodology and Mizon (1995) for a recent update). Instead of improving an initial econometric model when it appears to be unsatisfactory, one can better start with a more general model containing, nested within it as special cases, a series of simpler models that ideally represents all the alternative economic hypotheses requiring consideration. The general model Hendry and Mizon have recommended, as a generalization of the first-order serial autocorrelation model, is the first-order serial autoregressive distributed lag model; a linear dynamic regression model in which Y_t is regressed on Y_{t-1}, $\mathbf{X}_t$ and $\mathbf{X}_{t-1}$ and the error term is Gaussian white noise.

Hendry and Mizon's model is a typical time series model. Its counterpart in the spatial regression literature, the first-order spatial autoregressive distributed lag model covering the first-order spatial autocorrelation model as a special case, has been discussed by Burridge (1981), Bivand (1984), Blommestein and Nijkamp (1986), Anselin (1988, 226-230), and Florax and Folmer (1992). Below, a first-order autoregressive distributed lag model in both space and time is introduced to further explain Hendry and Mizon's objection.

Consider a stable linear regression equation in vector form relating Y_t to its own serially and spatially lagged value and to a set of regressors $\mathbf{X}_t$, as well as their serially and spatially lagged values

$$Y_t = \tau Y_{t-1} + \delta \mathbf{W} Y_t + \mathbf{X}_t \beta_1 + \mathbf{X}_{t-1} \beta_M + \mathbf{W} \mathbf{X}_t \beta_W + u_t \tag{9.18}$$

where $u_t \sim N(0, \sigma^2 \mathbf{I_n})$ and is serially and spatially independent. It follows that this regression equation changes into

$$[\mathbf{I_n} - \tau L - \delta \mathbf{W}] Y_t = [-\tau L - \delta \mathbf{W}] \mathbf{X}_t \beta_1 + u_t \tag{9.19}$$

when $\beta_M = -\tau \beta_1$ (L is defined such that $LY_t = Y_{t-1}$) and $\beta_W = -\delta \beta_1$, so that the terms involving Y_t and X_t have a common factor of $(\mathbf{I_n} - \tau L - \delta \mathbf{W})$. Multiplying both sides of this equation by the inverse of this common factor yields

$$Y_t = \mathbf{X}_t \beta_1 + [\mathbf{I_n} - \tau L - \delta \mathbf{W}]^{-1} u_t = \mathbf{X}_t \beta_t + \varepsilon_t \tag{9.20}$$

where $\varepsilon_t = \tau \varepsilon_{t-1} + \delta \mathbf{W} \varepsilon_t + u_t$, from which it is seen that ε_t is generated by a first-order serial and spatial autoregressive process. Equation (9.18) is obviously more general than equation (9.20). Practically, the variables $\tau Y_{t-1} + \delta \mathbf{W} Y_t + \mathbf{X}_{t-1} \beta_M + \mathbf{W} \mathbf{X}_t \beta_W$ are omitted from the regression equation and are being approximated by the error autoregression $\tau \varepsilon_{t-1} + \delta \mathbf{W} \varepsilon_t$. However, an investigator who allows for first-order serial and spatial autocorrelation in case the errors of a regression of Y_t on $\mathbf{X}_t$ break the condition of Gaussian white noise, does not know whether the common factor restriction is satisfied. Under these circumstances, if equation (9.18) would be the true model, equation (9.20) could still be a poor approximation. In this respect, a better strategy would have been to estimate the general model and to investigate whether it could be simplified. According to Mizon (1995), a general-to-specific modeling strategy, which starts from a general congruent model and then tries to narrow it down by looking for simplications that are acceptable on the data, is to be preferred vis-à-vis a specific-to-general modeling strategy. On the other hand, there may be reservations held on whether one can know a priori that (9.18) is the true model and that this model cannot be estimated when $\mathbf{X}_t$ is constant over time or between spatial units (see Maddala, 1993, 393-432).

9.6 Conclusion

Space-time data are usually more oriented toward cross-section analysis, as they are wide but typically short. Heterogeneity across units is an integral part and often the central focus of the analysis. By contrast, time series modeling is somewhat unusual in the analysis of space-time data. This is conspicuous since it is well-known that not only the presence of spatial autocorrelation, but also the presence of serial autocorrelation in the errors of a linear regression equation, induces problems for inference based solely on OLS methods. Hence, it is important in econometric modeling to use residuals to test for the absence of both serial and spatial autocorrelation in the model errors.

On the other hand, if the null hypotheses of no serial and no spatial autocorrelation require rejection, there is no uniquely alternative model to adopt, since all the test results have established is that the linear model estimated by OLS is inadequate. Consequently, the practice of correcting for first-order serial and spatial correlation might be inadequate to characterize the available data. Hence, the model discussed in this chapter is no more, but also no less, than a mixture of the well-known first-order serial autocorrelation model and first-order spatial autocorrelation model that was missing from the literature to date.

Appendix: Practical problems using commercial econometric software

Anselin and Hudak (1992) have illustrated how the basic elements of a spatial econometric methodology can be implemented in commercial econometric software packages. In most commercial econometric software, the built-in maximum likelihood procedure expresses the log-likelihood function in terms that correspond to each observation and the overall log-likelihood as the sum of these terms: $LogL = \sum_t \sum_i Log\ L_{ti}$. From equation (9.12) it can be seen that the characteristic roots of the matrices $\mathbf{B} = \mathbf{I_n} - \delta\mathbf{W}$ and $\mathbf{P_1} = \mathbf{I_n} - \tau^2(\mathbf{B'B})^{-1}$ can be expressed in this form by introducing them as additional pseudo variables; the order in which these characteristic roots are listed being irrelevant, since they appear separately from the other variables in the log-likelihood (see Anselin and Hudak, 1992, 532). By contrast, the transformation on the vector of first observations by the operator $\mathbf{P_1^{1/2}B}$ cannot be expressed in this form. There are two solutions for this problem. The *first solution* can only be used when the spatial weight matrix is symmetric. If W is symmetric, **W=W'**, we get **B'=B**, which in turn means that $\mathbf{P_1^{1/2}} = (\mathbf{I_n} - \tau^2(\mathbf{B'B})^{-1})^{1/2} = (\mathbf{I_n} - \tau^2\mathbf{B}^{-2})^{1/2} = (\mathbf{I_n} - \tau\mathbf{B})^{1/2}(\mathbf{I_n} + \tau\mathbf{B})^{1/2}$.

One important property of this matrix is that it has the same characteristic vectors as the matrix **W** (see Griffith, 44, table 3.1, and Greene, 1993, 34-36). Let **C** be a $n \times n$ matrix whose i^{th} column is the characteristic vector c_i of the spatial weight matrix **W**, **C**=[c_1 c_2 ... c_n], and let **R** be a diagonal matrix whose i^{th} diagonal element is r_i, where r_i is the ith chracteristic root of $\mathbf{P_1^{1/2}}$ that corresponds to c_i, i.e., r_i is the square root of $1-\tau^2/(1-\delta\varpi_i)^2$. Then we may also use the transformation matrix **RC'** instead of $\mathbf{P_1^{1/2}}$ This is because the determinant of **RC'** equals the determinant of $\mathbf{P_1^{1/2}}$ and $\mathbf{P_1^{1/2'}P_1^{1/2}} = (\mathbf{RC'})'(\mathbf{RC'}) = \mathbf{CR^2C'} = \mathbf{I_n} - \tau^2\mathbf{B}^{-2}$ and the log-likelihood function and the first derivatives do not change as a result. Since the matrix of characteristic vectors **C** is independent of τ and δ, part of the transformation **RC'** on the vector of first observations can be computed in advance by determining the variables $\mathbf{C'}Y_1$, $\mathbf{C'W}Y_1$, $\mathbf{C'X_1}$ and $\mathbf{C'WX_1}$. When computing the concentrated log-likelihood function of τ and δ, these variables then only need to be multiplied by the matrix **R**. Since **R** is a diagonal matrix, the ith row of $\mathbf{C'}Y_1$, $\mathbf{C'W}Y_1$, $\mathbf{C'X_1}$ and $\mathbf{C'WX_1}$ must in fact be multiplied by r_i, which can be programmed for each observation separately. The advantage of this solution is that once more we can use one of the built-in maximum likelihood procedures of commercial econometric software. The disadvantage, however, is that we require the characteristic vectors of the spatial weight matrix **W**, which might be problematic when **W** is large.

The *second solution* can be used when the spatial weight matrix is asymmetric or when it is not possible to determine the characteristic vectors of **W**. In this case, the characteristic vectors of the matrix $\mathbf{P_1}$ are different for different values of τ and δ, making it impossible to use one of the built-in maximum likelihood procedures of commercial econometric software. Since the concentrated log-likelihood function of τ and δ is concave and both τ and δ are upper and lower bounded, we instead suggest a bisection search algorithm as described in Anselin and Hudak (1992, 527-530). At each iteration, the partial derivative of the concentrated log-likelihood function is evaluated for τ and δ at the midpoint between its lower and upper bound. If this derivative is positive, the midpoint becomes the new lower bound, whereas if it is is negative, the midpoint becomes the new upper bound.

Endnotes

[1]Excepting OLS, but as has been said this estimator is inefficient.

[2]On verifying $\mathbf{\Omega} \times \mathbf{\Omega}^{-1} = \mathbf{I}$, the following expressions are of help $\mathbf{B'}^{-1} = \mathbf{B}^{-1'}$, $\tau\mathbf{B}^{-1}\mathbf{Z_n} = (\mathbf{Z_n}\tau\mathbf{B'}^{-1})'$, $\mathbf{B} = \mathbf{B'}^{-1}(\mathbf{B'B})$, $\mathbf{Z_n} - \tau^2\mathbf{B}^{-1}\mathbf{Z_n}\mathbf{B}^{'-1} = (\mathbf{B}^{'}\mathbf{B})^{-1}$, $\mathbf{B'} = (\mathbf{B'B})\mathbf{B}^{-1}$, and $\mathbf{Z_n} = \mathbf{B}^{-1}(\mathbf{I_n} + \tau^2\mathbf{B}^{-1}\mathbf{B}^{'-1} + \tau^4\mathbf{B}^{-2}\mathbf{B}^{'-2} + ...)\mathbf{B}^{'-1}$

[3]In this example, the sum of the elements of a particular row of **W** both left and right from the diagonal element is: $1+1/2+1/3+1/4+......$ This means that each row total equals $\sum_{d=1}^{\infty}\frac{2}{d}$.

[4]The interval $1/\varpi_{min} < \delta < 1/\varpi_{max}$ can be computed using ω_{min} and ω_{max} in table 9.1 and 9.2.

10 Spatial Dependence in the Evolution of Regional Income Distributions

Sergio J. Rey
San Diego State University

Abstract

This chapter introduces three measures of spatial dependence for use in the analysis of regional income distributions and their evolution. The first builds upon the notion of regional conditioning (Quah, 1993), and is derived as a trace statistic from a modified Markov transition matrix. The remaining two statistics are intended for use in a dynamic context and measure the degree of spatial clustering and regional cohesion in income rank mobility. All three measures are applied in an empirical analysis of per capita income patterns in the lower 48 United States over the 1929-99 period.

10.1 Introduction

The question of regional economic convergence continues to attract much attention (Cuadrado-Roura et al., 1999; Kim, 1998; Nijkamp and Poot, 1998; Tsionas, 2000). This rich and growing literature contains a number of fascinating debates surrounding the theories of spatial economic growth and change, as well as the appropriate empirical strategies that can be used to confront these theories (Sala-i- Martin, 1996).

On a theoretical front, the main competitors appear to be the traditional neoclassical growth paradigm (e.g., Barro and Sala-i-Martin, 1991) and the more recent arrival, endogenous growth theory (e.g., Aghion and Howitt, 1998). When applied to the question of regional growth and change, the two perspectives generate radically different policy implications. The neoclassical approach predicts eventual convergence due to the presence of diminishing returns to capital. Accordingly, because the convergence process will operate to reduce initial income differentials, policy interventions to correct the latter are viewed as unnecessary. Endogenous growth theory tells a very different story. The presence of increasing returns to scale leads to the possibility of persistent and even widening levels of regional income disparities. While this would appear to make a strong case for policy interventions, endogenous growth theorists are quick to point out the enormous problems associated with information, coordination and the question of the proper geographical scale to use in designing appropriate policies (Bröcker, 1997). Beyond the policy debates, much interest centers on the convergence hypothesis as a vehicle to differentiate between these two competing approaches towards economic growth (Sala-i-Martin, 1996).

While both theories do touch on the spatial dimensions of regional growth processes, a number of authors have argued that much more needs to be accomplished in this regard before the promise of the theories is fully realized (Nijkamp and Poot, 1998; Oosterhaven, 1997). At the same time, the issues associated with spatially referenced data that are at the heart of empirical analyses of the convergence hypotheses are also beginning to attract increased attention (Kelejian et al., 1997; López-Bazo et al., 1998; Rey and Montouri, 1999; Magalhães et al., 2000).

This chapter takes up the latter area of concern and examines the role of spatial dependence in a class of models for regional income convergence that have been recently suggested in the literature. It introduces several new measures for spatial dependence in regional income distributions and their dynamics. These measures are extensions of recent work reported in Rey (2001) for the analysis of spatial dependence in regional growth and change. While the substantive focus is on the question of regional income convergence, the methodological issues examined appear to have relevance for the study of a wide class of phenomena that have spatial and temporal dimensions.

This chapter is organized as follows. The next section provides a brief overview of the substantive issues surrounding convergence of regional income distributions. This is followed by a discussion of the role of spatial dependence in the analysis of regional income distributional dynamics. Section 10.4 introduces a number of new empirical measures designed to provide insights as to the role of spatial dependence in regional income growth and distributions. These methods are then applied in an empirical analysis of U.S. regional income growth patterns over the 1929-99 period, in section 10.5. The chapter concludes with a summary of main findings and some directions for future research.

10.2 Distributional Dynamics

A number of recent studies have focused on the evolution of cross-sectional distributions in addressing the question of regional convergence (Ioannides and Overmann, 2000; Magrini, 1999; Overman and Ioannides, 1999; Quah, 1996b; Tsionas, 2000). The interest on distributional dynamics centers on several dimensions of regional evolutions. The first concerns the overall shape characteristics of the regional income distribution, primarily spread and skewness, and the evolution of these characteristics over time. The second dimension relates to the amount of internal mixing or rank mobility that occurs within these same distributions over time.

10.2.1 Morphological Characteristics

The morphology of the regional income distribution can reveal important insights as to the current situation regarding regional income disparities. When viewed in a dynamic context, changes in the shape characteristics of these distributions can illuminate aspects of the regional growth processes. For example, Figure 10.1 summarizes the evolution of the state income distributions at 10-year intervals beginning in 1929. The densities are generated for the lower 48 coterminous states using a normal kernel estimator.

Inspection of the densities reveals several important changes over this 71-year period. Perhaps most pronounced is the overall decline in the level of dispersion of the regional income distributions. The drop in dispersion is greatest between the 1939-1949 period and each additional decade yields a further reduction in spread with the exception of the 1979-89 period in which the situation was reversed. However, after 1989 the collapsing of the distribution appears to have continued.

Figure 10.1: US Regional Per-Capita Income Distribution 1929-99

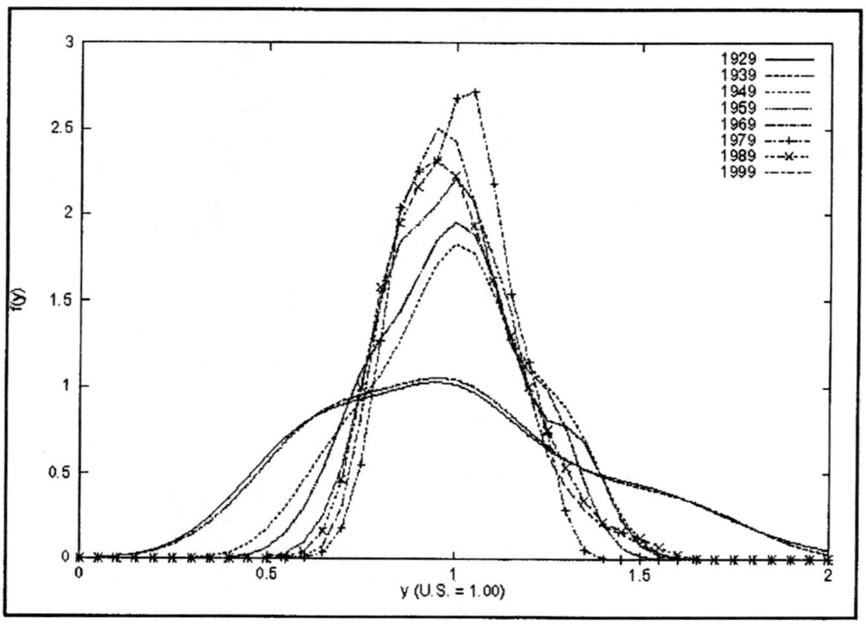

It would also appear that several other characteristics of the distribution have changed over these decades. The first two densities display a higher degree of positive skewness than the later densities, although this is complicated by the radically different levels of variance over time. At the same time, a casual inspection suggests that the relative kurtosis in the distributions has changed over time, but here again the comparison is complicated by non-constant variances.

Figures 10.2 and 10.3 provide a more detailed annual view of these individual characteristics. The long term trend of so called σ-convergence is evident in Figure 10.2, where the standard deviation of the relative incomes had steadily declined up until the late 1970s when the reversal noted above is reflected. While this increase in dispersion appeared to end by the late 1980s (see Figure 10.1), there is some evidence of another shift towards increasing dispersion (divergence) towards the very end of the sample period.

Figure 10.3 plots measures of skewness and kurtosis for the distributions from each year. Because these measures are standardized against a normal distribution with the same variance of the distribution in question, the confounding effect of changing variance is controlled for here. These annual views reveal that the trends in these characteristics are much more mixed than is the case for dispersion. For example, skewness is positive and fluctuates around 0.40 until the mid 1940s at which point it begins a rapid decline that bottoms out in 1949. This period of decline in skewness coincides with the rapid decline in dispersion (convergence) from Figure 10.2. In the middle portion of the study period the skewness is positive but generally lower than in the pre-1940 period. However, the middle 1970s through 1990 is a period in which the distribution becomes more positively skewed. Here again, this coincides with the turn around towards increased divergence in Figure 10.2. Also during this period the kurtosis displays a strong increase above what was a negative value for most of the study period[1].

Given the interest in the convergence hypothesis, much of the focus has been on how the measures of dispersion have changed over time. The evidence presented above, however, suggests that not only are measures of dispersion sensitive to regional convergence/divergence but higher order moments of the distribution may also signal changes in regional growth processes. In this regard, the use of kernel densities to trace out the evolution of the income distributions provides a powerful visualization framework, providing a flexible exploratory analysis perspective that complements more formal, theory based, econometric approaches that have dominated the convergence literature.

Figure 10.2: Dispersion:US Regional Per-Capita Income Distribution 1929-99

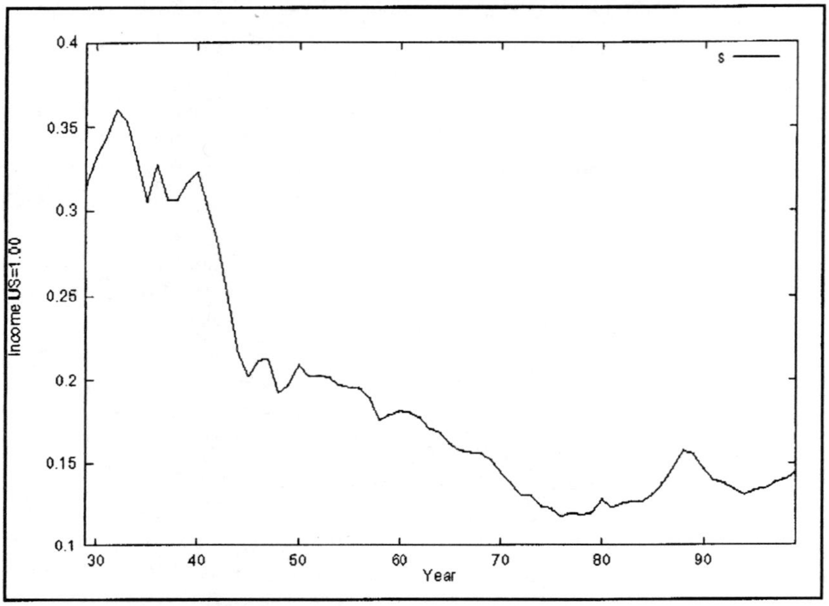

Figure 10.3: Skewness and Kurtosis: US Regional Per-Capita Income Distribution 1929-99

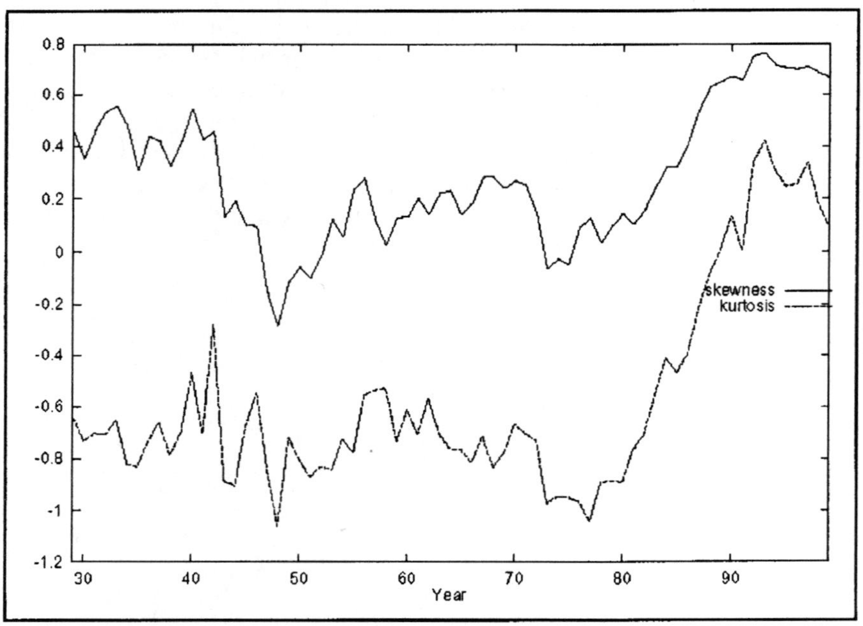

10.2.2 Class Mobility

Recent work on regional income dynamics has adopted Markov chain approaches to study the evolution of the regional income distribution over time (Quah, 1993, 1996a; Fingleton, 1999). At the core of these approaches is an empirical transition probability matrix:

$$\mathbf{M}_{t,t+s} = \begin{pmatrix} m_{11} & m_{12} & \cdots & m_{1k} \\ m_{21} & m_{22} & \cdots & m_{2k} \\ \cdots & \cdots & \cdots & \cdots \\ m_{k1} & m_{k2} & \cdots & m_{kk} \end{pmatrix} \qquad (10.1)$$

where m_{ij} are the probabilities of a region making the transition from income class i to class j over the period t to $t + s$. The classes serve to discretize the space of income values into fixed classes that are typically held constant over the study period. Estimates of the transition probability matrices for interval s are obtained by accumulating the empirical transitions over each s length interval within the larger study period and normalizing by the number of regions that begin a period within an income class[2].

The empirical transition probability matrix provides rich insights as to the evolution of the distribution over time. For example, strong convergence in regional incomes over time would be reflected in the distribution concentrating in the middle income classes. Conversely, divergence would result in a more multi-model, or twin-peaks (Quah, 1996b), distribution.

Table 10.1: Estimated Annual Transition Matrix, US 1929-99

	Class				
n	0.59	0.79	0.89	1.10	∞
158	0.87	0.13	0.00	0.00	0.00
565	0.02	0.88	0.09	0.01	0.00
635	0.00	0.07	0.82	0.11	0.00
1409	0.00	0.00	0.05	0.93	0.02
593	0.00	0.00	0.00	0.06	0.94

The transition probability matrix maps the income distribution from one period into the distribution for the next period:

$$P_{t+s} = P_t \mathbf{M}_{t,t+s} \qquad (10.2)$$

where P_t is a $1 \times k$ frequency vector that summarizes the distribution for period t. Table 10.1 reports the empirical transition matrix for the case of the US over the 1929-99 period. The estimates are based on annual transitions between five income classes, the upper bounds of which correspond to the quintiles of the relative (US=1.00) incomes in 1929.

There are a number of indices that can be calculated from the transition matrix that provide summaries of the amount of mobility in the distribution. Shorrock (1978) introduced the following index:

$$SI = \frac{k - Tr(\mathbf{M}_{t,t+s})}{k-1} \tag{10.3}$$

where Tr indicates the trace operator. With $k = 5$, Shorrock's index is bounded on the [0,1.25] with interval lower values reflecting less mobility across income classes over time. The value of this index for the annual transitions from table 10.1 is 0.141. While this is towards the low end of the mobility spectrum it is important to recall that the probabilities are estimated for annual transitions. Even at this interval, the relatively higher mobility rates for the first three classes work to have the final marginal distribution such that there are no states remaining in the first class, with 5, 11, 26 and 6, in classes 2-5. This is evidence of very strong upward convergence.

10.2.3 Internal Mixing

While Shorrock's measures of mobility obtained from a transition matrix provides one view of the evolution of regional income distributions it does not necessarily speak to the question of relative mobility. Indeed, mobility across income classes need not imply changes in the relative positions of economies in the distribution. By the same token, low measures of income mobility, as measured by the Shorrock's index, could also be associated with a high degree of positional flux.

There are several mobility measures that are sensitive to changes in the ordinal ranking of economies in the income distribution. Kendall's τ statistic considers the degree of concordance in the rankings of all pairs of observations for two variables. In the context of income mobility, the first variable would pertain to the regional incomes for the initial year, while the second would be the incomes in the end year of the interval period. If two regions have the same relative rankings in both periods that pair is said to be concordant. However, if the relative rankings of the two switch over the interval, then the pair is discordant. With n observations there are $(n^2-n)/2$ pairwise comparisons to be made. Kendall's τ is given as:

$$\tau = \frac{N_c - N_d}{\left(n^2 - n\right)/2} \tag{10.4}$$

where N_c is the number of concordant pairs, and N_d the number of discordant pairs. If all pairs are concordant, then $N_c = (n^2 - n)/2$, $N_d = 0$ and $\tau = 1$, while if all pairs are discordant, then $\tau = -1$.

As a non-parametric test, τ is useful in cases where the normality assumption is tenuous. At the same time, however, when applied to spatially referenced bi-variate distributions, the independence assumption may also be violated. In this chapter, the properties of τ in the presence of spatial autocorrelation are examined. Extensions of τ to take spatial dependence into account are also presented and evaluated.

10.3 Spatial Dimensions

It seems desirable to consider the different dimensions (morphology, mobility and mixing) of regional income distributions together when drawing inference about the nature of regional economic growth processes. Limiting attention to only one of these dimensions may result in misguided inferences and only a partial understanding of growth dynamics. For example, an income distribution that retains the same shape over long periods of time may actually mask a great deal of internal mixing as individual economies move up and down the income distribution. At the same time a high degree of internal mixing might also occur over periods where the morphology of the distribution changes dramatically.

While all these dimensions are clearly important, the methods that have been adopted thus far to study these issues essentially ignore the spatial characteristics of the data and the role that regional context may play in shaping economic growth. There are a number of reasons why the spatial dimension needs to be incorporated more fully in analyses of regional income distributions.

From a theoretical perspective, some extensions to the work on endogenous growth theory have explored the role of technology diffusion as a mechanism to actually generate convergence even in the presence of increasing returns to scale (Barro and Sala-i-Martin, 1995, Ch 8). In these models, poor economies are late adopters of technology which has been previously developed in the wealthier economies. To the extent that adoption costs are lower than innovation costs, the impact of the adopted technology works over a much shorter time-span in the poor economies compared to the wealthier regions. In this case the late adopters should experience higher growth rates following adoption.

Curiously absent in these diffusion-based endogenous growth models is any discussion of how the diffusion process may operate over space. Essentially these models consider only the relative technology gap separating economies, and it is that distance that determines the relative growth rates. But if diffusion operates as a spatial process, as a very rich geographic literature suggests it does, then some interesting spatial dynamics may lie beneath the regional income patterns examined in the literature. Extending these diffusion-led endogenous growth models to incorporate spatial structure would appear to be a promising direction to pursue.

A related question is the extent to which there are spill-overs in the regional growth process, and what forces might be driving any spill-overs that exist. This question touches on the debate surrounding cooperative versus competitive regional growth. For example, it is not clear if the ascendency of a region of states, such as the southern US states (Crown and Wheat, 1995), is driven by competition or cooperation between the member states. Of course both forces could be operating simultaneously within a group of states. There is also the issue of scale and the possibility that the nature of growth processes within groups of states is different from that across groups of states. For example, it could be that regional growth is collaborative within a region of states (i.e., Florida, Georgia, South Carolina) but competitive when viewed from a wider perspective (i.e., the southern states versus the northeastern states).

Another possibility is that growth is competitive, but that the nature of that competition varies depending upon whether the states are members of the same region or different regions. States within a given region may see their neighbouring states as more direct competitors and focus their economic development strategies accordingly but give less attention to the activities of more distant states. A related question is whether intraregional competition could work to strengthen the region's competitive posture vis-à-vis other regions. Put another way, are the benefits of competition bounded by some regional context?

All of these issues surround the notion of regional cohesion and the extent to which economic growth is articulated in a heterogenous fashion across space. In other words, when studying the question of regional economic convergence, does the relevant unit of analysis become an individual state or should the focus be on groups of states and how these different groups move against one another?

While these are theoretical motivations for focusing on space in convergence analysis, what is also needed are new empirical strategies that can shed light on these issues. The recent focus on intra-distributional mobility has yielded important insights on regional growth dynamics, however, the possibility that moves of individual economies in the income distribution may be dependent upon those of their regional neighbours has largely been overlooked[3].

10.4 Methods for Spatial Dependence in Distributional Dynamics

10.4.1 Regional Conditioning

In the initial application of Markov models to the question of regional convergence Quah (1993) was cognizant of the possibility for regional clustering of incomes. He suggested the notion of 'regional conditioning' as a way to represent this phenomenon. This is implemented using two different income distributions for a given point in time. The first is the distribution of incomes standardized to the national average, or what is referred to as the nationally conditioned distribution. This is compared to the regionally conditioned income distribution in which each economy's per capita income is expressed relative to the average of the incomes for its regional neighbours, defined as first order contiguous. From these two observations a 'transition' matrix can be constructed to summarize the degree of correspondence between the positions of an observation in each conditional distribution:

$$\mathbf{C_{R,N,t}} = \begin{pmatrix} c_{R1,N1,t} & c_{R1,N2,t} & \cdots & c_{R1,Nk,t} \\ c_{R2,N1,t} & c_{R2,N2,t} & \cdots & c_{R2,Nk,t} \\ \cdots & \cdots & \cdots & \cdots \\ c_{Rk,N1,t} & c_{Rk,N2,t} & \cdots & c_{Rk,Nk,t} \end{pmatrix} \qquad (10.5)$$

where $c_{R2,Nk,t}$ is the empirical proportion of economies that were in class 2 in the regionally (R) conditioned distribution in time t that fell in the kth income class of the nationally (N) conditioned distribution during the same period. Quah suggested that if incomes were randomly distributed in space, the transition matrix in (10.5) would be largely diagonal. Intuitively, this is akin to randomly sampling for the neighbours of each economy and then constructing the regionally conditioned distribution. In other words, individual economies would occupy similar positions in the two different conditional distributions if geography did not matter.

Curiously, Quah (1993) did not extend this reasoning to develop a statistic to formalize these intuitive notions. This is the task that we take up here. First we consider the construction of a statistic that could be used to evaluate the extent to which the regional income distributions depart from spatial randomness. Next we consider the distributional properties of this statistic.

One way to proceed is to treat each of the standardized (relative) per capita income distributions as marginal distributions and consider whether their joint distribution departs from the independence assumption. Under the independence assumption, an estimate of the expected joint probability that an economy is in class 2 of the regionally conditioned distribution and class 3 of the nationally conditioned distribution would be:

$$\hat{c}_{2,3,t} = \frac{n_{2,R,t}}{\sum_{l}^{k} n_{l,R,t}} \times \frac{n_{3,N,t}}{\sum_{l}^{k} n_{j,N,t}} \qquad (10.6)$$

where $n_{2,R,t}$ ($n_{3,R,t}$) is the number of observations in class 2 (3) of the regionally (nationally) conditioned distribution.

The difficulty here is that the substantive hypothesis we are interested in would be reflected in the joint probability matrix being non-diagonal. A non-diagonal joint probability matrix (i.e., what our 'transition matrix' is being viewed as) could very well be consistent with the hypothesis that the two relative incomes are independent variates, but it is not the only outcome which is consistent with this. Put another way, spatial clustering (non-diagonality of the transition matrix) may or may not be consistent with the variate independence assumption (in a joint distribution sense).

A more direct way to implement a test for spatial clustering would be to examine whether the joint probability matrix departs from diagonality:

$$\xi_t = 1 - \sum_{l}^{k} c_{l,l,t} \tag{10.7}$$

where $c_{l,l,t} = n_{l,l,t}/n$ with $n_{l,l,t}$ being the number of observations that are simultaneously members of class l in both relative distributions. As $0 \le \xi \le 1$, lower values point to spatial randomness, and higher values indicate clustering[4].

This statistic is similar to those used to test for non-diagonality of error covariance matrices in seemingly unrelated regressions (Breusch and Pagan, 1980). Of course the distributional properties of the latter tests are likely to be different from those for a test of the diagonality of a joint probability matrix. We will revisit this issue in the empirical analysis later in the chapter.

10.4.2 Spatial Decomposition of Mobility

Quantifying the amount of regional income mobility is clearly an important task, which has attracted much recent attention. In this section, two new measures are presented which provide insights on the extent to which spatial dependence does or does not play a role in the overall level of regional income mobility.

A measure that allows for the decomposition of distributional mobility to analyze the amount of spatial clustering can be obtained as follows. Let $\theta_{i,t}$ represent the rank of state i's per-capita income in year t. Then between any two periods, say $t0$ and $t1$, a scalar measure of spatial clustering of distributional transitions is:

$$\Theta_{t1-t0} = \frac{\sum_R \left| \sum_{i \in R} \theta_{i,t1} - \theta_{i,t0} \right|}{\sum_i \left| \theta_{i,t1} - \theta_{i,t0} \right|} \tag{10.8}$$

where R is one of a set of exhaustive and mutually exclusive groups of states[5].

The denominator of this measure is simply the sum of the absolute rank changes over the period. To the extent that movements in the income distribution are cohesive within regions this measure should be closer to 1. The extreme case would be all states from one region increasing their ranks at the expense of states belonging to another region. In this case growth could be said to be competitive across the regions, or inter-regionally competitive. With less cohesion, the measure approaches 0. This is because the absolute value of the sum of the intraregional rank changes should be smaller when the states in question are moving in different directions in the distribution. In this case growth is competitive irrespective of geography.

A second approach towards studying the role of spatial dependence in income mobility is to apply a spatially modified rank correlation statistic. This focuses more on the question of internal regional cohesion, or the extent to which relative positions in the income distribution remain fixed, or not fixed, depending upon whether two economies are regional neighbours. The concordant and discordant count measures from (10.4) can be further decomposed to consider which pairs involved observations from the same region and which considered observations from different regions. More specifically the number of concordant pairs can be decomposed into:

$$N_c = N_{c,r} + N_{c,o} \tag{10.9}$$

where N_c is the number of concordant pairs involving locations belonging to the same region, and $N_{c,o}$ are the number of concordant pairs for observations belonging to different regions. A similar decomposition holds for the discordant pairs N_d. To develop a spatial version of the τ statistic from equation (10.4), define:

$$\omega = (n^2 - n)/2 \tag{10.10}$$

which is the denominator in the original form of the statistic. This is also equal to the number of elements above (or below) the main diagonal in the spatial weights matrix. Each element in that matrix specifies whether observations i and j are considered members of the same region (or more generally neighbours). The sum of these elements can also be decomposed:

$$\omega = \omega_r - \omega_o \tag{10.11}$$

where ω_r is the number of pairs of observations that are members of the same region. From this, a spatial version of the rank mobility measure is[6]

$$\tau_r = \frac{N_{c,r} + N_{d,r}}{\omega_r} \tag{10.12}$$

A similar measure can be derived for the non-neighbour pairs of observations τ_o. The original measure of rank mobility can then be decomposed as follows:

$$\tau = \psi \tau_r + (1 - \psi) \tau_0 \tag{10.13}$$

with $\psi = \omega_r / \omega$.

This decomposition opens up some interesting possibilities. For example, we can ask if the discordant counts are mainly due to changes in the rankings involving pairs of economies from different regions (i.e., interregional) or from within the same region (i.e., intraregional). Even when $\tau = 0$, in which case the number of concordant and discordant pairs are equal, this question still is an interesting one since it would shed light on the nature of regional cohesion.

The decomposition can also speak to the question of whether growth is competitive within regions as well as across regions. Interregional growth is characterized by one group of states moving ahead of another group in the distribution. The spatial decomposition would allow us to examine whether the relative positions between the states in a group moving up (or down) the distribution remain invariant to the movement of the group as a whole.

Rank correlation tests are non-parametric but still rest on the random sampling assumption. Consequently, when applied to georeferenced data, this assumption becomes questionable. As such, the tests are initially treated as indices of spatial dependence in distributional mobility and are applied in an exploratory mode. An initial investigation into the properties of these tests is also carried out using empirically generated sampling distributions based on Monte Carlo techniques.

10.5 Empirical Illustration

To provide some initial insight as to the properties of these new measures, a series of simulations was carried out using income data for the U.S. over the 1929-99 period.

10.5.1 Regional Conditioning

The first analysis focuses on the trace statistic ζ from (10.7) for measuring the degree of spatial autocorrelation in regional incomes in a given year. The results are presented in figure 10.4 which plots the trace statistic together with the value of Moran's *I* for each year in the sample. The trace statistic in Figure 10.4 is implemented with five classes where the class boundaries are expressed in terms of incomes as a percentage of the national average. The relative incomes associated with the quintiles for the first year are used to define the cut-offs for the classes in the remaining years. The same row-standardized first order contiguity matrix is used to implement both the trace statistic and Moran's *I*.

The two statistics display similar behaviour with respect to generally declining trends in the level of spatial clustering, although the trend is clearly more pronounced for Moran's *I* than for the trace statistic. Both tests also display sharp increases in value over the 1980's, followed by a resumption of the long term decline towards the end of the century. It is tempting to conclude that Moran's *I* displays more power relative to the trace statistic since it takes on a larger value for all except a few of the years. Because the two statistics have different scales, however, a direct comparison of their values as indicators of spatial clustering would be misleading. Instead, the relative strength of the signal each statistic yields regarding spatial autocorrelation needs to be viewed from an inferential perspective.

Figure10.4: Spatial Autocorrelation of Per-Capita Income Distribution 1929-99

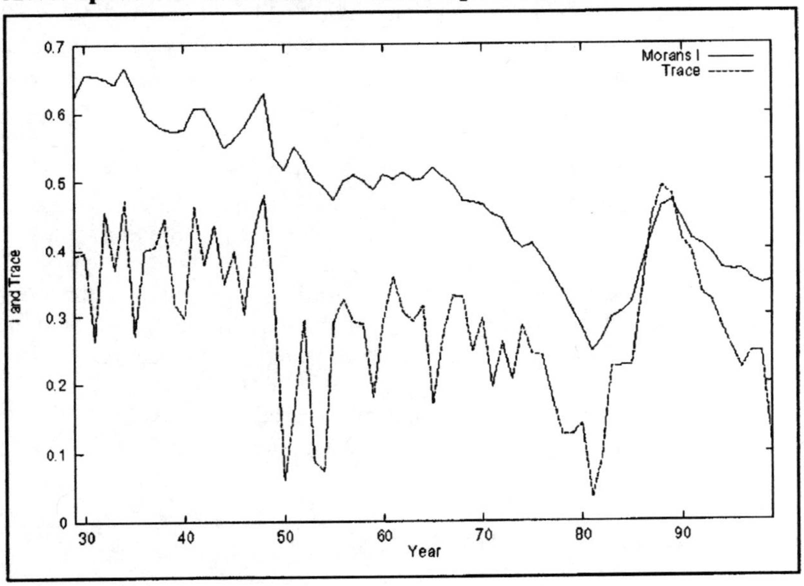

As a formal evaluation of the properties of the trace statistic have not yet been carried out, inference is based on 10000 random permutations of the 48 incomes for each year in order to develop an empirical sampling distribution for the statistic. Pseudo *p*-values for the both the trace statistic and Moran's *I* are derived by considering the percentage of the simulated values that were more extreme than the value for the actual map of income values for that year. For each year of the study, the empirical value for Moran's *I* was significant at the 0.10, 0.05 and 0.01 levels, while the trace statistic was found to be significant in 68, 62 and 53 of the years, respectively. In all years the *p*-values are higher for the trace statistic compared to those for the Moran's *I*. Intuitively, because the trace statistic is based on an ordinal metric while Moran's *I* is based on a ratio metric, one could expect the former to display lower power against spatial dependence. This may account for the pattern displayed in Figure 10.4, although a more detailed evaluation of this issue would require a formal analysis of the properties of the trace statistic.

10.5.2 Regional Cohesion of Rank Mobility

Figure 10.5 summarizes the results of applying the regional cohesion measure to the annual rank changes in state per capita incomes. The value of Θ from (10.8) for the realized income changes is plotted against the average value from 1000 random permutations of the income changes for each year. In 56 of the 70 annual changes, the statistic takes on a value that is larger than the value that could be expected if the rank changes were randomly distributed in space. In 23 of the 70 cases the values are significantly different from their expected values, at the 0.05 level, while 12 and 27 values were significant at the 0.01 and 0.10 levels, respectively.

The value of Θ increases towards the end of the sample, while its expected value remains fairly constant over the entire period. Using 1968 as a split point confirms this as the average value of Θ before this year was 0.54, but for the period following 1968 the average value was 0.63. The values obtained from the randomized maps averaged 0.445 before 1968 and 0.451 after the break.

These findings suggest that in over a third of the years external regional cohesion is present and that there has been a slight increase in the amount of such cohesion since 1968. An interesting possibility to consider is if this increase is related to the short run turn-around toward divergence over the 1979-89 period. Elsewhere, Rey and Montouri (1999) have found that spatial autocorrelation in regional incomes has strengthened during periods of divergence during this century.

10.5.3 Spatial Decomposition of Rank Dynamics

Figures 10.6-10.9 display the traditional and spatial versions of the τ statistic for rank mobility over four different interval periods. Several findings emerge from a visual inspection of these figures. First, as expected, the amount of mobility tends to increase with the length of the interval under consideration[7]. The second regularity is that the mobility measure for the non-contiguous pairs is much closer to that of the overall-pattern than is the case of mobility defined for contiguous pairs of states.

The main differences between the mobility rates for the contiguous and noncontiguous state are reflected in the generally higher rates for the contiguous pairs in the early part of the study, but lower mobility towards the end of the period. This is found for each of the four different interval lengths as well.

To test whether the differences in the rank mobility rates displayed in these series were significant, a randomization strategy was adopted. More specifically, 1000 permutations were carried out, in each of which a state's time series was allocated to a potentially different location. For each permutation, the rank correlation statistics were calculated for all the different length intervals. Values for these measures from the permutations were then used to develop an empirical sampling distribution against which the actual differences in mobility measures were compared. This allows for a consideration of the differences in the mobility measures for different length intervals as well as for specific years within these intervals.

Table 10.2 reports the specific intervals in which significant differences were found between the τ statistics from the neighbouring and non-neighbour pairs. Using plus or minus two standard deviations from the simulated rank differences for each interval, 29 of the 243 intervals generated significant differences in mobility between the groups. In eighteen of these intervals, rank mobility was greater within the set of neighbouring pairs ($\tau_r < \tau_o$) than in the case of non-neighbouring pairs of states. The majority of these cases were in the shorter (1 and 5-year intervals) and in the beginning of the study period (i.e., 1935-53).

The eleven intervals in which rank mobility was significantly weaker for the set of neighbouring states versus the non-neighbouring set tend to occur in the latter portion of the study period and involve longer lengthed intervals. These two patterns emerge during periods in which the rate and direction of convergence were different. One possible explanation for the higher mobility rates displayed by the contiguous pairs in the earlier years is that the overall levels of income dispersion were larger then, as was the level of spatial dependence. Consequently, the differences in incomes between neighbouring pairs of states was likely to be smaller than income gaps separating non-contiguous states, and discordance in rank pairs was more likely for the former than latter group. In any event, the patterns suggest a possible relationship between the level of overall convergence and the strength of spatial autocorrelation in regional incomes.

Figure 10.5: Spatial Cohesion Index: 1-year interval

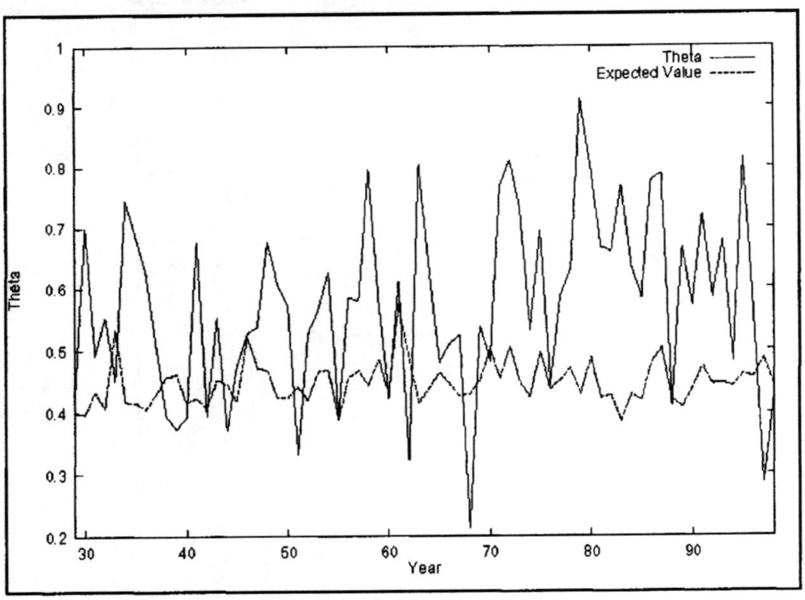

Figure 10.6: Spatial Rank Mobility Index: 1-year interval

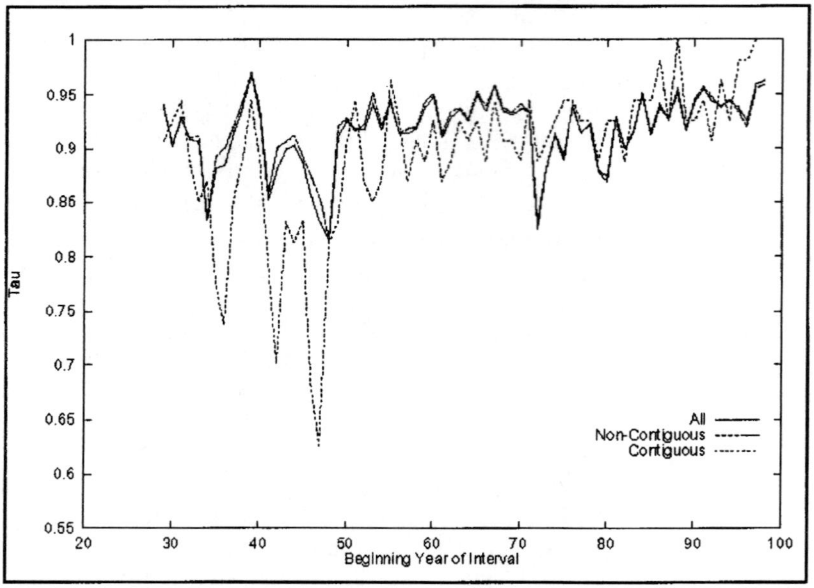

Figure 10.7: Spatial Rank Mobility Index: 5-year interval

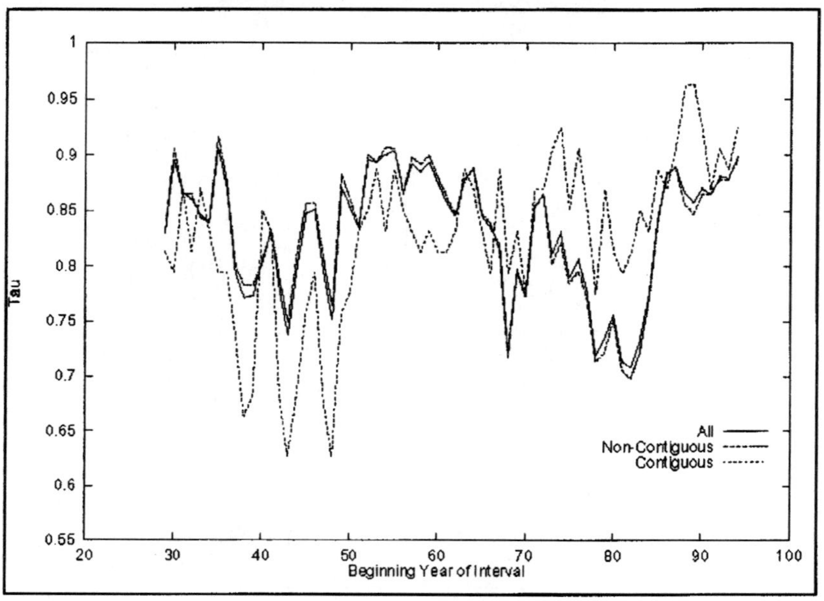

Figure 10.8: Spatial Rank Mobility Index: 10-year interval

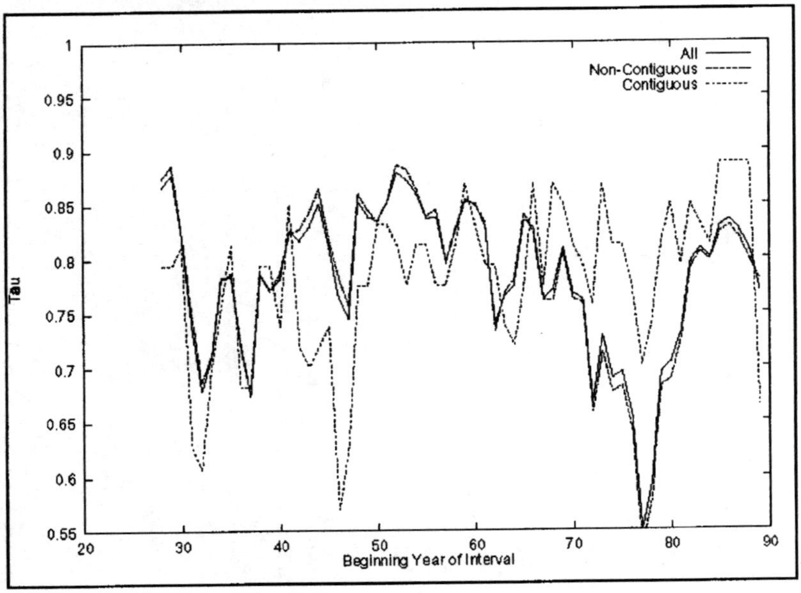

Figure 10.9: Spatial Rank Mobility Index: 25-year interval

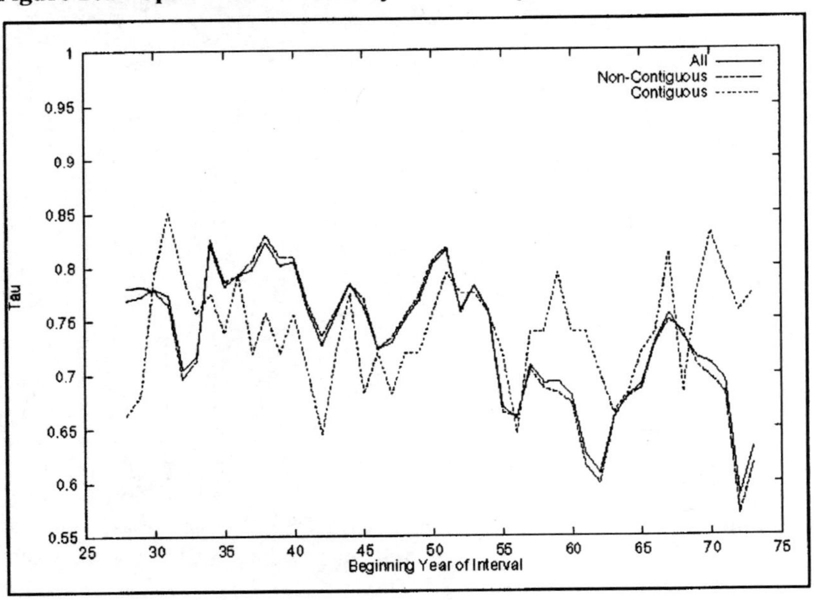

Table 10.2: Differences in Regional Mobility

Start Year	Interval Length	τ_r-τ_o	Critical Value[a]
35	1	-0.117	0.097
36	1	-0.162	0.092
42	1	-0.199	0.093
44	1	-0.099	0.087
46	1	-0.190	0.100
47	1	-0.227	0.109
49	1	-0.088	0.084
53	1	-0.101	0.067
30	5	-0.112	0.090
35	5	-0.123	0.086
44	5	-0.130	0.118
47	5	-0.124	0.119
48	5	-0.139	0.130
49	5	-0.125	0.094
79	5	0.147	0.135
83	5	0.129	0.128
88	5	0.106	0.099
89	5	0.115	0.099
44	10	-0.142	0.105
45	10	-0.145	0.104
47	10	-0.214	0.123
54	10	-0.107	0.097
74	10	0.155	0.128
78	10	0.167	0.160
79	10	0.159	0.156
81	10	0.162	0.143
71	25	0.135	0.131
73	25	0.188	0.151
74	25	0.160	0.147

[a] 1.96 empirical standard deviations.

10.6 Conclusion

This study has introduced three new measures for the study of spatial dependence in the context of regional income distributions and their dynamics. A trace statistic for spatial dependence was found to be in general agreement with the inferences based on the conventional Moran's *I* statistic. Because it is based on a ordinal metric the new test appears to possess somewhat less power than Moran's I, but a definitive conclusion on this awaits more detailed analysis of the test's formal properties.

Application of the two new measures of spatial clustering in rank mobility indicates that changes in ordinal positioning in the income distribution do not occur randomly in space. Interestingly, these statistics also appear to be sensitive to changes in the direction of the convergence path. This suggests that the relationship between spatial structure and regional income convergence may be worth further exploration in future studies.

While these new statistics show promise, much more work needs to be done on their theoretical properties and a number of implementation issues also need to be further investigated. Chief among these is the definition of the spatial weights matrix used to operationalize the mobility statistics and whether that matrix should be based on the notion of regional membership or more conventional distance and contiguity criteria.

Acknowledgements

A previous version of this chapter was presented in the Applied Econometric Association conference held in conjunction with the Western Regional Science Association Fortieth Anniversary Annual meeting in Palm Springs, California, February 2001.

Endnotes

[1]The increase moves the kurtosis of the distribution towards that displayed by a normal distribution.

[2]The values in the estimated transition matrix will be sensitive to the length of the interval s.

[3]The work on estimation of kernel densities (Quah, 1996a) and mixture models (Tsionas, 2000) to study regional income distributions has ignored the complications that spatial dependence may pose for these approaches.

[4]Recall that the transition matrix is row-standardized, so technically speaking it is not the same form as a joint probability matrix since the latter is typically doubly stochastic.

[5]A careful referee pointed out that the denominator in (10.8) could be zero. This would only occur in cases were there was no ordinal mobility in the regional income distribution during the interval under consideration. In these cases, however, the question of whether there was any spatial dependence in the income transitions becomes moot. Therefore, this measure should only be applied when there is rank mobility in the income distribution.

[6]The spatial τ statistic presented here can be viewed as a dynamic extension of the rank adjacency statistic recently presented in Ekwaru and Walter (2001a,b). Future work will explore the relationships between these two statistics.

[7]Higher values for τ reflect lower rates of mobility.

Part IV

Spatial Patterns and Local Spatial Analysis

11 On Connectropy

Johan F. Kaashoek
Erasmus University Rotterdam

Jean H.P. Paelinck
Erasmus University Rotterdam

Henry G. Zoller
Université Catholique de Louvain

Abstract

The present chapter introduces an axiomatic concept of proximity, and develops a method to operationalise it. The method is applied to the case of Dutch and Belgian provinces (in both the aspects of their contiguity structure and their economic interrelations) and conclusions are compared to prior knowledge about the spatial economic structure of Belgium and the Netherlands.

11.1 Introduction

The present study attempts to formalise some previous research on the structuring of connectivity matrices using linear assignment (Paelinck, 2000a and b).
The central idea was to isolate groups of elements (agents, sectors, institutions,...) on the basis of the overall strength of their mutual connections; in the conclusions this idea will be further elaborated on and justified.

In what follows a more formal rationale is developed in terms of a 'nearness' concept, which in turn forms the basis for the definition of a 'connectropy' measure, akin to that of entropy. The concept is then quantified through linear assignment, the technique being applied to two Dutch and two Belgian matrices. Conclusions and references follow.

11.2 Connectropy

One will start with an axiomatic definition of 'nearness' or 'proximity', along the lines of that of a distance measure or metric (see Blumenthal, 1970, p. 329, where 'nearer than' is defined in terms of a metric, but no other formal 'nearness' axiomatics is presented; on metrics, see Paelinck, 1983, pp.44-50).

Take any set of elements, *S*, and consider three of those elements, *x*, *y* and *z*. Define on two of them a numerical function, $n(.,.)$, which obeys the following properties:

$$n(x,y) = 1,\ x=y \tag{11.1}$$
$$0 < n(x,y) < 1,\ x \neq y \tag{11.2}$$
$$n(x,y) = n(y,x) \tag{11.3}$$
$$n(x,y)\, n(y,z) \leq n(x,z) \tag{11.4}$$

Function $n(.,.)$ will then be called a 'nearness' measure; the similarity with an axiomatic definition of a metric is obvious, it being stressed that the well-known triangular inequality is now stated in terms of a product and a $\leq$-relation.

A possible and useful member of the family of nearness (or closeness) measures is the following:

$$n(i,j) = exp(-d_{i,j}) \tag{11.5}$$

where d_{ij} is any proper distance measure between entities *i* and *j*, e.g. the degree of contiguity, which is indeed a proper distance measure, and which will be used later on; one can easily check that (11.5) is a proper closeness measure, given the exponential additivity property and the triangular inequality property of metrics.

Consider now the function:

$$c_{ij} = d_{ij}\, exp(-d_{ij}) \equiv -n(i,j)\, ln\, n(i,j) \tag{11.6}$$

of which the similarity with an entropy measure is again obvious (on entropy and probability, see Wilson, 1972, paper 7; on entropy and information, Theil 1971, pp.639-641); indeed read 'probability' for variable *n*, and the formal likeness is complete.

But also from the intrinsic point of view this is true; equation (11.6) can indeed be given an informational content. The occurrence of a long distance relation is relatively rare, so its observation contains a high degree of information, which is measured here by d_{ij}; on the other hand, it is known from distance interaction that the effective intensity of interaction declines with distance (in (11.6) in a negative exponential way, 'radioactive decay'), and as entropy measures the expected information, so (11.6), which we will call 'connectropy', measures the expected intensity.

11.3 Technique

Suppose one avails of a matrix of connectropy measures; one could then try to extract groups with certain values of total connectropy, $\Sigma_{i,j}c(i,j)$ (to be compared to total entropy), e.g. declining levels from a maximum to a minimum; this amounts in fact to applying linear assignment to the matrix, as will be shown now.

Let **A**, of order *nxn*, be that matrix; note that if the diagonal is not void, its terms will be put equal to zero in all the exercises to come (section 11.4).

One could try and generate several paths, possibly generating cyclical groups (or circuits), in the sense that the last entity connects with the first one (Roy, 1969, pp.162-169), each maximising the sum total of the interrelationships present, the objective being to isolate *several* groups with declining maximal total internal intensity, and study the resulting interconnections.

The *initial* mathematical program can then be written as:

$$\max_{x} \varphi \overset{\Delta}{=} [vec(\mathbf{A}')]\, x \qquad (11.7)$$

subject to:

$$\mathbf{J}x \leq i \qquad (11.8)$$

$$\hat{\mathrm{x}}\, x = x \qquad (11.9)$$

i being a vector of ones.

The following definitions and interpretations should be given:

- *vec*(**A'**) is the vectorisation of the transpose of **A**, all diagonal elements omitted, so of order *n(n-1)x 1*;
- *x* is a vector of order *n(n-1)x1* of binary variables (conditions (11.9), where $\hat{\mathbf{X}}$ is the diagonal matrix constructed from *x*);
- conditions (11.8) are weakened assignment conditions, matrix **J** being binary and of order *2nxn(n-1)*; it is the so-called assignment matrix, assigning a unique place to each of the units taken up in the solution. The following example illustrates this for the case of three entities :

$$\mathbf{J} = \begin{bmatrix} 1 & 1 & 0 & 0 & 0 & 0 \\ 0 & 0 & 1 & 1 & 0 & 0 \\ 0 & 0 & 0 & 0 & 1 & 1 \\ 1 & 0 & 0 & 0 & 0 & 1 \\ 0 & 1 & 0 & 1 & 0 & 0 \\ 0 & 0 & 1 & 0 & 1 & 0 \end{bmatrix} \tag{11.10}$$

If the weak inequalities were to be replaced by equalities, exactly *n directed* relations would be selected, each agent appearing twice; relaxation so allows of generating *incomplete groups*.

In fact, the mathematical program (11.7) through (11.9) is the inverse (a maximum instead of a minimum) of the traveling salesman problem, but here the traveling salesman is in no hurry at all, and not obliged to pass through all the potential nodes, and moreover he is allowed to go back and forth and to travel along disconnected paths! One will come back to those aspects.

Once the solution to the mathematical program has been obtained (by simple linear programming: see Murty , 1976, p.413, but the extensions to be discussed further down would not necessarily have that characteristic), one has obtained a first group with maximal internal cohesion, meaning here total connectropy; one then cancels the corresponding entries of **A**, leading up to a matrix **A*** to be treated in the same manner, and so on until all the entries have been exhausted.

Symmetry of matrix **A** leads to symmetrical flows in the optimal solution, which can be prevented by introducing the conditions:

$$x_{ij} + x_{ji} \leq 1, \ \forall i<j \tag{11.11}$$

but they would still not guarantee complete connectivity; such a requirement would in fact lead to a quadratic assignment problem (Kaashoek and Paelinck, 2000).

11.4 Applications

11.4.1. The Netherlands

The technique of section 11.3 will first be applied to the Dutch matrix of provincial contiguity degrees (source: van Gastel and Paelinck, 1995, p.152); table 11.1 hereafter pictures that matrix, table 11.2 defining the entries and map in Figure 11.1 visualising the whole.

Table 11.1: contiguity degrees of the Dutch provinces

Provinces	**1**	**2**	**3**	**4**	**5**	**6**	**7**	**8**	**9**	**10**	**11**	**12**
1	0	1	1	2	2	3	3	3	4	5	4	4
2	1	0	1	1	1	2	2	2	3	4	3	3
3	1	1	0	1	2	2	3	3	3	4	3	3
4	2	1	1	0	1	1	2	2	2	3	2	2
5	2	1	2	1	0	1	1	1	2	3	2	2
6	3	2	2	1	1	0	1	2	1	2	1	1
7	3	2	3	2	1	1	0	1	1	2	2	2
8	3	2	3	2	1	2	1	0	1	2	2	3
9	4	3	3	2	2	1	1	1	0	1	1	2
10	5	4	4	3	3	2	2	2	1	0	1	2
11	4	3	3	2	2	1	2	2	1	1	0	1
12	4	3	3	2	2	1	2	3	2	2	1	0

Table 11.2: entries of table 11.1

1	Groningen
2	Friesland
3	Drenthe
4	Overijssel
5	Flevoland
6	Gelderland
7	Utrecht
8	Noord-Holland
9	Zuid-Holland
10	Zeeland
11	Noord-Brabant
12	Limburg

Figure 11.1: Map of Dutch provinces.

11.4.1.1 Linear assignment

The first result is that two circuits maximise total connectropy; they present themselves as follows:

(a,i): Gelderland ⇒ Utrecht ⇒ Noord-Holland ⇒ Zuid-Holland ⇒ Zeeland ⇒ Noord-Brabant ⇒ Limburg ⇒ Gelderland;
(a,ii): Groningen ⇒ Drenthe ⇒ Overijssel ⇒ Flevoland ⇒ Friesland ⇒ Groningen.
(b,i): Overijssel ⇒ Gelderland ⇒ Limburg ⇒ Noord-Brabant ⇒ Zeeland ⇒ Zuid-Holland ⇒ Noord-Holland ⇒ Utrecht ⇒ Flevoland ⇒ Overijssel;
(b,ii): Groningen ⇒ Friesland ⇒ Drenthe ⇒ Groningen.

The symbol ⇒ represents the sequencing of the spatial elements in the formation of a group.

Furthermore, twelve other groups were generated, four with 12 members, four with 11 members, the remaining groups having respectively 8, 5, 2, and 1 member(s); total connectropy was systematically declining.

The main point is that the connectropy analysis separates neatly the North (ii) from the Center-South (i) of the Netherlands, be it in two different ways; total connectivity on the resulting solution is indeed not imposed by conditions (11.8) and (11.10). As already said, this would require quadratic assignment, which will be the object of short expositions farther down.

It should be mentioned that the precision of the numerical procedure applied has an influence on the outcome; increased precision allowed of improving the value of the objective function but only by one out of one million, resulting in reproducing case *b*, albeit with some shifts in the individual links.

In order to test the relevance of the above findings, economic linkages between the Dutch provinces were investigated. Table 11.3 hereafter reproduces the data on gross regional products (GRP's) at market prices in the Dutch provinces (1997,billions of guilders at current prices; source : Central Statistical Office, CBS, The Hague, www.cbs.nl, 'overzicht kerncijfers').

Table 11.3: GRP's of the Dutch provinces

Provinces	GRP's
1	30.7
2	22.0
3	16.8
4	41.4
5	9.1
6	74.7
7	52.2
8	119.3
9	164.2
10	16.9
11	105.0
12	46.1

From these figures monetary flows between the provinces (call them 'potential flows'; one will come back to this point in the conclusions) can be computed as the product of the corresponding GRP's (see Leontief and Strout, 1966), and, in order to generate magnitudes formally similar to those of equations (11.5) and (11.6), they were transformed to:

$$x_{ij} = (\max_{ij} f_{ij})(f_{ij})^{-1} \tag{11.12}$$

where f_{ij} represents a computed potential flow, x_{ij} now playing the role of d_{ij}, and again, for the sake of clarity, $i \neq j$.

The result of applying the methodology of section 11.3 has again been to generate two groups of equal maximal total connectropy, but composed of only 9 members, the provinces of Drenthe, Flevoland and Zeeland being systematically absent and only a single Groningen-Friesland flow being present; this is due in fact to the presence of negligeable x_{ij}'s, which has also lead to an unstructured set of 16 residual potential flows. In order to include all the flows in the maximising sets, conditions (11.8) will be rendered binding further down; but first some detailed results will be presented.

Again, in each maximising case, North and Center-South have been separated; noticeable is also the fact that each time three groups were formed:
(a,i,1): Overijssel ⇒ Limburg ⇒ Utrecht ⇒ Overijssel;
(a,i,2): Gelderland ⇒ Noord-Holland ⇒ Noord-Brabant ⇒ Zuid-Holland ⇒ Gelderland;
(a,ii): Groningen ⇒ Friesland;
(a,iii): Friesland ;
(b,i,1): Overijssel ⇒ Utrecht ⇒ Limburg ⇒ Overijssel;
(b,1,2): Gelderland ⇒ Zuid-Holland ⇒ Noord-Brabant ⇒ Noord-Holland ⇒ Gelderland;
(b,ii): Friesland ⇒ Groningen;
(b,iii); Groningen ⇒ Friesland;

One should also stress the fact that the links do not necessarily correspond to first order contiguities, Limburg ⇒ Utrecht, Noord-Holland ⇒ Noord-Brabant and Gelderland ⇒ Noord-Holland being examples of this.

As to the other groups, two of them had 10 members, four of them 9 members, three of them 8 members, the 7, 6, 4, 2 and 1-member classes being only represented by one group; as already mentioned, a residual set of 16 flows was also present.

In order to recover the absentee provinces in the maximising groups, the analysis was rerun with constraints (11.8) binding; the results were the following.

The value of the objective function did not change for the two maximising groups, a fact due again to the weakness of the previous absentee links; moreover the above group structure was also retained. The latter presented itself as follows:
(a,i,1): Overijssel ⇒ Limburg ⇒ Utrecht ⇒ Overijssel;
(a,i,2): Gelderland ⇒ Zuid-Holland ⇒ Noord-Brabant ⇒ Noord-Holland ⇒ Gelderland;
(a,ii): Groningen ⇒ Friesland ⇒ Zeeland ⇒ Flevoland ⇒ Drenthe ⇒ Groningen.

(b,i,1): Overijssel ⇒ Utrecht ⇒ Limburg ⇒ Overijssel;
(b,i,2): Gelderland ⇒ Noord-Holland ⇒ Noord-Brabant ⇒ Zuid-Holland ⇒ Gelderland;
(b,ii) : Groningen ⇒ Drenthe ⇒ Flevoland ⇒ Zeeland ⇒ Friesland ⇒ Groningen.

The Center-South groups are now combinations of the previous ones; the previously missing links have all been integrated in the Northern group, which has also absorbed a south-western province, Zeeland, which has everything to do with the measured relative weakness of the latter's links with the rest of the Dutch economy. Again the (a)-groups were reproduced by increasing the precision of the program with the weak constraints (11.8), only Zeeland taking up an isolated position; and again the objective functions differed by less than three out of ten thousand.

It can be easily verified that the procedure outlined with respect to equation (11.11) (though leading up to an expression formally similar to equation 11.6) does not define a proximity relation that satisfies condition (11.4); in order to guarantee this the connectropy relation has been modified to:

$$c_{ij} = -ln(f_{ij}^*)f_{ij}^* \tag{11.13}$$

where the stars denote potential flows divided by the square of the domestic product, in other words the provincial domestic products have been expressed in terms of shares. In fact property (1) is no longer satisfied, but for the analysis a pre-connectivity concept is sufficient.

Applying the above program, i.e. with constraints (11.8) binding, gave the following results, the objective function having equal values:
(a,i,1): Overijssel ⇒ Utrecht ⇒ Gelderland ⇒ Limburg ⇒ Overijssel;
(a,i,2): Noord-Holland ⇒ Noord-Brabant ⇒ Zuid-Holland ⇒ Noord-Holland;
(a,ii): Groningen ⇒ Zeeland ⇒ Flevoland ⇒ Drenthe ⇒ Friesland ⇒ Groningen;
(b,i,1): Overijssel ⇒ Limburg ⇒ Gelderland ⇒ Utrecht ⇒ Overijssel;
(b,i,2): Noord-Holland ⇒ Zuid-Holland ⇒ Noord=Brabant ⇒ Noord-Holland;
(b,ii): Groningen ⇒ Friesland ⇒ Drenthe ⇒ Flevoland ⇒ Zeeland ⇒ Groningen.

The main difference with the previous results is the shift of Gelderland from (a,i,2)-(b,i,2) to (a,i,1)-(b,i,1), but the fundamental trisection remains; anyhow, correctly specifying the nearness measure (and also, of course, a different choice for that measure) influences the results, in this case accentuating the group structure of three south-western provinces.

11.4.1.2 Quadratic assignment

As said before, applying quadratic assignment in stead of linear assignment would generate completely connected groups.

Research is underway to derive an original algorithm to treat this problem (Kaashoek and Paelinck, 2000), in particular the connectropy one which has the structure of a traveling salesman problem, however with an objective function to be maximised.

In the course of this research, the following solution has been obtained:

Groningen ⇒ Zuid-Holland ⇒ Noord-Brabant ⇒ Noord-Holland ⇒ Limburg ⇒ Gelderland ⇒ Friesland ⇒ Overijssel ⇒ Zeeland ⇒ Drenthe ⇒ Utrecht ⇒ Groningen

The striking fact is the the presence still of the grouping Zuid-Holland ⇒ Noord-Brabant ⇒ Noord-Holland and of the partial groupings Limburg ⇒ Gelderland and Drenthe ⇒ Utrecht, and this despite the fact that quadratic assignment forces entities to join a large group, much like a traveling salesman would prefer cutting his one-tour journey into separate, but less tiring, parts! It will be interesting to compare this result with the one that will be computed for Belgium hereafter.

11.4.2 Belgium

The analysis presented in section 11.2 will first be applied to the Belgian matrix of regional contiguity degrees; table 11.4 hereafter pictures that matrix, table 11.5 defining the entries and map in Figure 11.2 visualising the data.

Table 11.4: contiguity degrees of the Belgian provinces

Regions	**1**	**2**	**3**	**4**	**5**	**6**	**7**	**8**	**9**	**10**	**11**	**12**
1	0	2	2	2	1	3	2	2	2	3	3	4
2	2	0	1	1	1	2	2	2	2	3	3	4
3	2	1	0	2	1	3	2	2	1	2	2	4
4	2	1	2	0	1	1	2	1	2	3	2	4
5	1	1	1	1	0	2	1	1	1	2	2	4
6	3	2	3	1	2	0	2	1	3	3	2	4
7	2	2	2	2	1	2	0	1	1	2	1	4
8	2	2	2	1	1	1	1	0	2	2	1	4
9	2	2	1	2	1	3	1	2	0	1	1	4
10	3	3	2	3	2	3	2	2	1	0	1	4
11	3	3	2	2	2	2	1	1	1	1	0	4
12	4	4	4	4	4	4	4	4	4	4	4	0

Table 11.5: entries of table 11.4

1	Region of Bruxelles-Capitale
2	Antwerpen
3	Limburg
4	Oost-Vlaanderen
5	Vlaams Brabant
6	West-Vlaanderen
7	Brabant wallon
8	Hainaut
9	Liège
10	Luxembourg
11	Namur
12	Extra-territorial units

The Belgian regions are in principle provinces, except entries 1 and 12, the last one grouping the European Union, embassies and NATO, these entities being located inside the region 'Bruxelles-Capitale'; to this group a fictitious maximal degree of contiguity 4 has been assigned.

In order to test the relevance of those findings, economic linkages between the Belgian provinces will also be investigated. Table 11.6 hereafter reproduces the share data on gross regional products (GRP's) at market prices in the Belgian provinces (1997; source: Institute of National Accounts, *Regional Accounts*).

In order to derive (potential) flows, transformation (11.13) of section 11.4.1 was again introduced.

11.4.2.1 Linear assignment

First applying expression (11.6) to contiguity matrix table 11.4 there results from program (11.7) through (11.9) a large number of solutions with declining objective functions, function (11.7) having equal values only for solutions 1 and 2, and for solutions 3 and 4; moreover the resulting first two groups are just permutations of each other. Hereafter the results will only be detailed for the maximising groups.

Taking up the first maximising value (3.9495), one observes the following cyclical groupings:

Figure 11.2: Map of Belgian provinces.

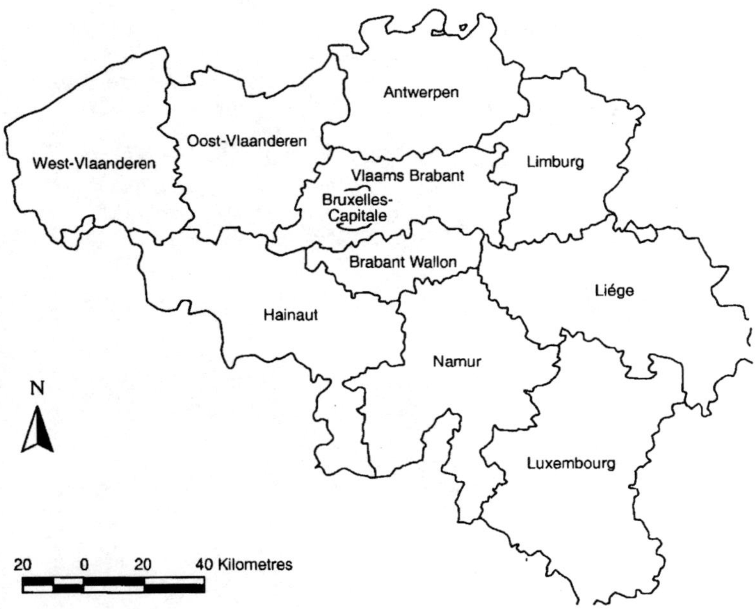

(a,1): Bruxelles-Capitale ⇒ Vlaams Brabant ⇒ Limburg ⇒ Antwerpen ⇒ Bruxelles-Capitale;
(a,2): Oost-Vlaanderen ⇒ Hainaut ⇒ West-Vlaanderen ⇒ Oost-Vlaanderen;
(a,3): Brabant Wallon ⇒ Liège ⇒ Luxembourg ⇒ Namur ⇒ Brabant Wallon.

The Extra-territorial units are completely absent.

The three groupings are typically supra-regional geographic entities, in plain words the North-East, the South-West and the South-East.

The next two solutions (objective function value 3.46340) generate only two cyclical clubs, but of different composition, this time a large one and a smaller one; the smaller one in one case is a Northern region, in the other one a North-Eastern one.

As a curiosum, one should mention the fact that in a lower value solution (2.4680) all regions, except the extra-territorial one, formed one group.

As to the application of expression (11.11) to table 11.6, the main result is that in all the cycles that could be derived (9 exactly), Bruxelles-Capitale always joined one cycle; only the maximal connectropy value (.4770) generated three groups, otherwise only one or two groups were observed.

Table 11.6: GRP shares of the Belgian provinces

Regions	GRP shares
1	.143
2	.214
3	073
4	.123
5	.084
6	.113
7	.026
8	.085
9	.085
10	.020
11	.033
12	.001

Here follow the three maximising cycles:
(a,i,1): Bruxelles-Capitale ⇒ Antwerpen ⇒ Oost-Vlaanderen ⇒ West-Vlaanderen ⇒ Bruxelles-Capitale;
(a,i,2): Limburg ⇒ Liège ⇒ Hainaut ⇒ Vlaams Brabant ⇒ Limburg;
(a,i,3): Brabant wallon ⇒ Namur ⇒ Luxembourg ⇒ Brabant Wallon;
(a,ii): Extra-territorial units.

The three cycles could be characterised as Central-North-West, Central-North-East and South-East; curiously enough three contiguous super-regions, but not completely overlapping with the political-administrative division in a Flemish and a Walloon region, the main outsider group being the second one.

Be it also mentioned that the value of the objective function drops smoothly in the first six groupings generated, then decreases rapidly.

If one now compares the two groups of solutions, one notices that the flow solutions do not necessarily follow the contiguity solutions. In group 1, Vlaams Brabant and Limburg are absent in the flow version, but Oost-Vlaanderen and West-Vlaanderen are added; in group 2 where Oost- and West-Vlaanderen are present in the contiguity version, Limburg, Liège and Vlaams Brabant join the flow version; groups 3 have the most resemblance, only Liège making them differ.

11.4.2.2 Quadratic assignment

The outcome has been the following:
Bruxelles-Capitale ⇒ Antwerpen ⇒ Hainaut ⇒ Limburg ⇒ Brabant wallon ⇒ Oost-Vlaanderen ⇒ Luxembourg ⇒ Vlaams Brabant ⇒ West-Vlaanderen ⇒ Namur ⇒ Liège ⇒ Extra-territorial units ⇒ Bruxelles-Capitale.

Curiously enough, only small bits and pieces of the 'linear' groups above show up in the solution (e.g. Bruxelles-Capitale ⇒ Antwerpen and Hainaut ⇒ Limburg); a ready explanation is not available at this point of the analysis.

11.5 Conclusion

The concept and model exposed in this chapter did allow to derive formally the connections between spatial entities. Though much of the findings correspond to prior knowledge about the spatial economic structure of the Netherlands and Belgium, links with the spatial positioning of its constituent parts and between them and the estimated economic interrelations have now been derived in a more axiomatic way. The comparison between position and interaction, the latter being studied here by means of 'potential' flows, should be sharpened and made specific, by using real flows (commuting, migration, interregional trade in goods and services...). Still it has been shown here that economically logical connections do not necessarily completely match purely geographical ones; the reason is that positions and distances alone are not fully explanatory, economic impulses (demand and supply) have to be added to get a complete picture.

As the method is still in its infancy as far as its applications are concerned, other extensions have to be envisaged: alternative structures of the **J**-matrix of equations (11.8), application to more numerous and smaller spatial units; the latter extension would not be impeded by computing problems, as linear programming programs are nowadays extremely efficient, and can handle a very large number of variables. It has shown anyway that it can be operationalised and lead to acceptable spatial analytical results.

12 Locally Weighted Regression and Time-Varying Distance Gradients

Daniel P. McMillen
University of Illinois at Chicago

Abstract

The nonparametric estimator used in spatial analysis is a version of locally weighted regression in which observations closer to a target point receive greater weight. I show how locally weighted regression can be used to estimate housing price indexes and time-varying distance gradients. The empirical results suggest that the City of Chicago has become more centralized during the past decade. During much of the 1980s, housing prices were higher farther from the city center. A negative distance gradient re-emerged during the 1990s, and housing prices in the City of Chicago once again peak near the traditional city center.

12.1 Introduction

Standard econometric models are based on an assumption of a known functional form. Either errors are presumed to be independently and identically distributed or autocorrelation and heteroskedasticity are assumed to have known and convenient functional forms. Although these assumptions are often appropriate for standard cross sectional data sets, spatial relationships are typically more complicated. Statistical tests based on simple functional forms often reveal that coefficients vary over space, implying that more complicated functional forms are necessary to represent the data adequately. Incorrect functional form assumptions and missing variables that are correlated over space can lead to complicated autocorrelation and heteroskedasticity processes that cannot be represented adequately by simple functions.

Nonparametric estimation procedures are useful alternatives to standard regression analysis when functional forms are unknown. In spatial modeling, the most common nonparametric estimator is locally weighted regression (LWR), which is a type of nearest neighbor estimator. As its name implies, LWR uses linear regressions to approximate a function locally by placing more weight on nearby observations. In spatial analysis, 'nearby' has a straightforward interpretation as simple Euclidean distance between sites, but more complicated metrics can be used. Instead of estimating a single regression for the entire sample, locally weighted regressions are estimated for a set of target points. The regressions imply a series of predicted values of the dependent variable and estimates of the marginal effect of each explanatory variable. The target points may be actual observations or arbitrarily defined points.

Nonparametric estimates of the marginal effects of an explanatory variable are difficult to summarize when there are many target points. One alternative is to present means and standard deviations, or other summary statistics such as the median and other percentiles. Another common procedure is to calculate the estimates for 'representative' data points.

In this chapter, I show how locally weighted regression can be used to construct an index of distance gradients that vary over time. I first show that the simple geographic weighting scheme used by geographers and urban economists is a special case of the more general local linear estimator. I then describe how indexes can be constructed readily using an analog to the average derivative estimator of Härdle-Stoker (1989) and Powell, Stock, and Stoker (1989). The average derivative estimator uses estimated densities to construct a weighted average of the estimates. The procedures used here are readily extended to other situations, such as constructing estimates for a variety of geographic sites within the data set.

I use the LWR estimator to show that house prices in the City of Chicago became much more centralized during the 1990s. During the 1980s, house prices rose with distance to the city center – the distance gradient was positive. There was a significant change in the distance gradient during the 1990s. A sharp increase in house prices near the city center led to a reemergence of the negative distance gradient by the end of the decade.

12.2 Locally Weighted Regression

Standard regression procedures start with an assumed parametric functional form, which is usually written $y_i = \beta' X_i + u_i$, where X_i is a kx1 vector of explanatory variables, u_i is an error term, and β is a vector of coefficients. In contrast, nonparametric estimators begin with the more general function:

$$y_i = g(X_i) + u_i \qquad (12.1)$$

The only restrictions on the function $g(\bullet)$ are that it is smooth and has continuous first derivatives. In this section, I outline some alternative nonparametric estimators for $g(X_i)$ and $\partial g(X_i)/\partial X_i$. I show that the estimator most commonly used in spatial data analysis is a special case of a more general locally weighted regression procedure[1].

12.2.1 Kernel Regression

Nadaraya (1964) and Watson (1964) derive the standard kernel estimator by minimizing the following expression with respect to m:

$$\sum_{i=1}^{n}(y_i - m)^2 K\left(\frac{X_i - X}{h}\right) \qquad (12.2)$$

In equation (12.2), $K(\bullet)$ is a weighting function that determines the weight given to observation i when constructing an estimate for the target point, X. Common choices include the bi-square, tri-cube, and gaussian weighting functions, each of which is a smooth, single-peaked function[2]. The parameter h is the bandwidth. Large values of h increase the weight placed on distant observations. Maximizing equation (12.2) with respect to m leads to the standard kernel estimator:

$$\hat{y}(X) = \frac{\frac{1}{hn}\sum_{i=1}^{n} K\left(\frac{X_i - X}{h}\right) y_i}{\frac{1}{hn}\sum_{i=1}^{n} K\left(\frac{X_i - X}{h}\right)} \qquad (12.3)$$

Consistent estimates of marginal effects are given by $\partial \hat{y}(X)/\partial X$.

12.2.2 Locally Weighted Regression

Locally weighted regression is a simple extension of the standard kernel estimator in which a linear function of the explanatory variables is added to m in equation (12.2). One advantage of LWR over kernel regression is that the additional explanatory variables help to improve the fit of the equation. Another advantage is that LWR can be analyzed with standard regression techniques, which makes the statistical properties easier to derive[3].

The estimator is derived by minimizing equation (12.4) with respect to m and β:

$$\sum_{i=1}^{n}(y_i - m - \beta'(X_i - X))^2 K\left(\frac{X_i - X}{h}\right) \qquad (12.4)$$

After writing $Z_i = (1\ X_i - X)'$ and $\theta = (m\ \beta)'$, the resulting estimator is given by:

$$\hat{\theta}(X) = \left(\sum_{i=1}^{n} K\left(\frac{X_i - X}{h}\right) Z_i Z_i'\right)^{-1} \sum_{i=1}^{n} K\left(\frac{X_i - X}{h}\right) Z_i y_i \qquad (12.5)$$

The expression in equation (12.5) is a standard weighted least squares (WLS) estimator. It is constructed by regressing $K(P_i)^{1/2} y_i$ on $K(P_i)^{1/2}$ and $K(P_i)^{1/2}(X_i - X)$, where $P_i = (X_i - X)/h$. The coefficient on $K(P_i)^{1/2}$ is the estimated value of y at X, while the coefficients on $K(P_i)^{1/2}(X_i - X)$ are the estimates of the marginal effects of the explanatory variables on y at the target point.

Standard errors are easy to calculate. Following Pagan and Ullah (1999), the covariance matrix for $\hat{\theta}(X)$ is

$$V(\hat{\theta}(X)) = \sigma^2 \left(\sum_{i=1}^{n} Z_i K(\psi_i) Z_i'\right)^{-1} \left(\sum_{i=1}^{n} Z_i K(\psi_i)^2 Z_i'\right) \left(\sum_{i=1}^{n} Z_i K(\psi_i) Z_i'\right)^{-1} \qquad (12.6)$$

Equation (12.6) requires information on one new parameter, Φ^2, which is the variance of the error term, u. The standard way to estimate this parameter is to calculate $\hat{y}(X)$ using each observation as the target point. The estimated variance is

$$\hat{\sigma}^2 = \frac{1}{n} \sum_{i=1}^{n} (y_i - \hat{y}(X_i))^2 \qquad (12.7)$$

This estimator can be modified to take into account the heteroskedasticity that is endemic to spatial models. Rather than estimating the variance as a simple average of the squared residuals, a kernel or LWR estimate of $\Phi^2(X)$ can be estimated for each observation, where X is a set of explanatory variables for the variance. The vector X can differ from the original set of explanatory variables if desired.

12.2.3 Spatial LWR

In the standard LWR model, the same variables appear as explanatory variables for y and as arguments in the weighting function. Suppose, for example, that we have two explanatory variables, x_1 and x_2. Using a simple multiplicative gaussian weighting function, we have $K(P_i) = N((x_{1i}-x_1)/h_1)N((x_{2i}-x_2)/h_2)$, where N(.) is the standard normal density function. Standard LWR estimates are obtained by regressing $K(P_i)^{1/2} y_i$ on $K(P_i)^{1/2}$, $K(P_i)^{1/2}(x_{1i} - x_1)$, and $K(P_i)^{1/2}(x_{2i} - x_2)$. In contrast, the standard spatial nonparametric estimator uses a different set of variables in the weighting function from those used as explanatory variables for y. Weights are a function of *geographic* distance, i.e., the Euclidean distance between locations rather than a more general metric of distance between the explanatory variables. A common weighting scheme is the tri-cube. Letting

d_i be the distance between the target point and observation i, the weight given to observation i is:

$$K(\psi_i) = \left(1 - \left(\frac{d_i}{d_i^*}\right)^3\right)^3 I\left(d_i < d_i^*\right) \tag{12.8}$$

In equation (12.8), d_i^* is referred to as the *window size*. It is the distance between the target point and the most distant observation given weight in estimation[4]. This procedure is referred to as the 'nearest neighbor' approach in the nonparametric literature.

The window size plays an analogous role to the bandwidth. The difference is significant, however. The bandwidth divides the data into increments. In areas where data are sparse, few observations are given positive weight in estimation, whereas many observations receive weight in regions with ample data. Estimates are less precise in regions where data are sparse. The nearest neighbor approach fixes the number of observations used in estimation while letting the implied bandwidth vary. Making the bandwidth larger in areas where data are sparse increases the precision of the estimates and ameliorates some of the 'boundary effects' that are prevalent in spatial analysis.

Two approaches can be used to demonstrate that the nonparametric spatial estimation procedure is a special case of standard locally weighted regression. First, note that most explanatory variables in spatial models are themselves functions of distance. For example, in McMillen (1996), I use distances from downtown Chicago, Lake Michigan, and O'Hare and Midway Airports as explanatory variables for land values in the City of Chicago. Each of these variables is itself a function of the geographic coordinates of the site. Let W represent this list of explanatory variables, and let G represent the geographic coordinates. Since W is a function of G, we can write the expression for observation i in equation (12.4) as $\left(y_i - m - \beta_G'(G_i - G)\right)K\left((G_i - G)/h\right)$, or as $\left(y_i - m - \beta_w'(W_i - W)\right)K\left((G_i - G)/h\right)$. The only difference is the point of the linear expansion used to approximate the function. Using W rather than G has two advantages: (1) it results in a better fitting equation for a given bandwidth or window size; and (2) it greatly simplifies the calculation of the marginal effects of interest, β_w.

The second approach for demonstrating that spatial LWR is a special case of the more general approach is based on specific forms of the weighting function. Simple product kernels often are used to form weighting functions for multivariate models. As in the first approach, let W represent the explanatory variables and let G represent the geographic coordinates. In addition, let W^* represent the union of the two sets of variables. The standard LWR estimator then is based on the expression $\left(y_i - m - \beta'^*\left(W_i^* - W^*\right)\right)K\left(\left(W_i^* - W^*\right)/h\right)$. Let us divide the weighting function into two parts: $K\left(\left(W_i^* - W^*\right)/h\right) = K_G\left(\left(G_i - G\right)/h_G\right)K_w\left(\left(W_i - W\right)/h_W\right)$. If we use a nearest neighbor estimator for $K_G(\ )$ then only the nearest B observations are given positive weight in estimation, where B is the window width. If we use a uniform kernel for $K_w(\ )$ with equal weight given to the B available observations, then $K_w(\ ) = ½$ for each value of W_i used in estimation. This combination of a nearest neighbor estimator for $K_G(\ )$ and a uniform kernel for $K_w(\ )$ leads directly to the spatial LWR estimator.

This version of the LWR estimator has important advantages for spatial modeling. In many spatial models, a simple parametric function fits the data fairly well, but there may be geographic heterogeneity that is omitted from the base model. The LWR estimator readily handles modest departures from the base model. The uniform kernel for the other variables is appropriate if the underlying functional form is close to being correct apart from minor variations that are explained by the geographic coordinates. Using a kernel that puts less weight on observations that are far away in terms of W increases the variance of the LWR estimates. The spatial LWR model is appropriate when geographic effects are the primary source of error in the base econometric specification.

12.3 Summarizing Nonparametric Estimates

Nonparametric estimators produce a full set of coefficients for each target point. The estimates become difficult to summarize when there are many target points, as in the common case where separate estimates are produced for each observation. One possibility is to present simple summary statistics for the full set of estimates, such as means and standard deviations. The mean places equal weight on all estimates, including those for boundary observations with few neighbors. The average derivative estimator of Härdle and Stoker (1989) and Powell, Stock and Stoker (1989) also uses a simple weighted average to summarize the full set of estimates, but more weight is placed on observations that are drawn from regions with many neighbors[5]. Average derivative estimates are more reliable than simple averages or point estimates because they place heavier weight on regions where nonparametric procedures estimate the function more accurately. In this section, I show how this approach can be used to summarize LWR estimates and to calculate time indexes of coefficients.

The critical device for estimating average derivatives is the nonparametric density function, which is a direct counterpart to a standard statistical density function[6]. The density estimate at the target point X is simply:

$$f(X) = \frac{1}{hn}\sum_{i=1}^{n} K\left(\frac{X_i - X}{h}\right) \tag{12.9}$$

As with our earlier notation, h represents the bandwidth and $K(\ )$ is a kernel weighting function. The same weighting functions can be used for LWR and density estimation, but different functions can be used also. The density at observation i is given by $f(X_i)$ and the LWR coefficient estimates for this target point are $\hat{\theta}(X_i)$. The average derivative estimates are:

$$\hat{\lambda} = \frac{\sum_{i=1}^{n} f(X_i)\hat{\theta}(X_i)}{\sum_{i=1}^{n} f(X_i)} \qquad (12.10)$$

The term in the denominator assures that the weights sum to one.

The average derivatives are a simple measure of the explanatory variables' marginal effects, but unlike a simple average, they put more weight on estimates obtained from regions of the data set with ample numbers of observations. Estimating the densities imposes little additional cost because they can be calculated in the same pass through the data set used to estimate the LWR coefficients. However, standard errors are much more difficult to calculate for average derivatives than for simple point estimates. The estimates of $\lambda(X_i)$ are not independent across observations because the same data are used to estimate values at each target point. Thus, the variance of the average derivative estimate is not a simple weighted average of the variances of each $\hat{\theta}(X_i)$. More complicated variance estimators, such as the wild bootstrap of Härdle (1990) are necessary to calculate variance estimates for $\hat{\theta}(X_i)$.

In the empirical section of this chapter, my objective is to calculate an index of distance gradients over 1983-98, a period of 64 quarters. The data set includes 44190 observations of house prices in the City of Chicago. Two strategies are available for LWR estimation of a time index of gradients for distance from the city center. First, we can form an estimator that is designed to estimate a distance gradient for each of 64 target time periods. Alternatively, we can calculate LWR estimates for all 44190 data points (or a subset), and summarize them using density estimates that are centered on the target times.

Time is a continuous variable in this application, representing the number of days since 1 January 1983. In a slight modification of the notation used previously, let T_i represent the time of sale for observation i, and let X_i be the vector of other explanatory variables. For the first estimation strategy, the simple weighting function for target time T and explanatory variable set X is $K_T((T_i - T)/h_T)K_X((X_i - X)/h_X)$, which is a product of two standard kernels such as the tri-cube function shown in equation (12.8). Suppose that we use a nearest neighbor estimator for time. For example, we may place positive weight on the 1,000 most recent sales for time T. These 1,000 sales will include houses with varying characteristics at different locations, i.e., X_i will vary across the 1000 most recent sales.

For the second part of the weighting function, $K_X(\ )$, we might calculate LWR estimates using each of 1000 values of X_i as target points, which leads to 1000 estimates of the distance gradient for each of the 64 target times. The 1000 estimates then could be collapsed into a single estimate by calculating either a simple average or a density-weighted average. Alternatively, we could estimate the index for a representative set of values of X, e.g., a 2000 square foot, three-bedroom house at a distance of 10 miles from the city center. A third strategy, which is the one I employ, is to use a uniform kernel for X for these 1000 values. This strategy leads to an estimator in which the observations are weighted only by the time between an observation's sale and the target time (the variables in X enter as explanatory variables for the locally weighted regression, but uniform weights are placed on all values of X).

Using a uniform kernel for X greatly reduces the cost of estimation because we need only estimate one regression per target time. Standard errors are easy to calculate for each target time because we are not using multiple estimates to construct the index[7]. Equation (12.6) provides standard error estimates. This estimation strategy is appropriate if a simple functional relationship between X and y is adequate over short time periods. It would not be appropriate if the functional form does not account adequately for spatial effects.

The second strategy for LWR estimation of a time index of distance gradients uses nonparametric density estimates to collapse the $n = 44190$ coefficient estimates into an index for 64 target times. The combination T_i, X_i serves as the target point for $i=1,\ldots,n$. The density for observation i at target time T is then

$$\frac{1}{nh_T h_X}\sum_{j=1}^{n} K_T\left((T_j - T)/h_T\right)K_X\left((X_j - X_i)/h_X\right) \tag{12.11}$$

The key thing to note about equation (12.11) is the arguments for $K_T()$ and $K_X()$: T does not have a subscript i because it is set at a pre-defined target value. After normalization, equation (12.11) provides the weights for 64 time-varying average derivatives. This approach is more flexible than the previous one because it is easy to calculate alternative indexes for a variety of times, locations, and coefficients once LWR estimates have been obtained for all observations. However, it typically will be much more computer intensive than the first strategy, which is designed to form specialized estimators for the index that is of direct interest.

12.4 Data and Descriptive Statistics

The data set includes all sales of single-family homes in the City of Chicago from 1 January 1983 to 31 December 1998, subject to the restriction that the homes sold at least twice during the sample period. The data, which are obtained from tax records, reflect actual transaction prices. A small number of observations without addresses or sales dates were eliminated from the sample.

I restrict the analysis to single-story, one-family residences. Nonparametric estimators are better suited to analyzing continuous data than dummy variables. Though Horowitz and Härdle (1996) show that it is possible to recover the effects of discrete covariates in average derivative estimators, dummy variables are usually analyzed by one of two methods. The first method, which I follow, is to estimate separate nonparametric models for subsets of the data with common values of the discrete variables. An alternative method is to estimate semiparametric models in which the dummy variables are modeled as parametric intercept shifts (Robinson, 1988; Stock 1991). I choose the first method because my data set is already sufficiently large to make nonparametric estimation costly, and the distance gradient (which is my primary interest) is unlikely to be altered by omitting a dummy variable for multiple-story homes. The final data set has 44,190 observations.

Table 12.1 presents descriptive statistics on the explanatory variables used in the analysis. Building area and lot size enter in logarithmic form to be consistent with the Cobb-Douglas housing production function, which has ample empirical support (Thorsnes, 1997). In levels, building areas average 1,115 square feet, and lot sizes average 4043 square feet. The homes are nearly 60 years old on average. Several homes are within half a mile of the traditional city center at the intersection of State and Madison Streets, while others are nearly 17 miles away. From 1983 to 1998, house sales prices averaged $97513, with a range of $10010 to $12749977.

Table 12.1: Descriptive Statistics and Base Hedonic Regression

Variable	Mean (Standard Deviation)	Minimum (Maximum)	Coefficient (Standard Error)
ln Building Area (square feet)	6.9837 (0.2524)	5.9915 (9.6178)	0.4894 (0.0086)
ln Lot Size (square feet)	8.2755 (0.2294)	5.9915 (11.1019)	0.3599 (0.0106)
Age (10's of years)	5.9854 (2.2712)	0.0000 (12.9000)	-0.0511 (0.0011)
Distance from the City Center (miles)	8.8010 (2.5058)	0.2912 (16.6345)	-0.0289 (0.0011)

Notes. The dependent variable for the hedonic regression is the natural logarithm of house value, which has a mean of 11.3465 and a standard deviation of 0.5378 over the 44,190 observations. The regression also includes an intercept and 63 quarterly dummies. The R^2 is 0.313.

Few explanatory variables are available for hedonic estimation because the Illinois Department of Revenue does not collect much information for reviewing assessments. Through differencing, Bailey, Muth, and Nourse's (1963) repeat sales estimator controls for the effects of missing variables that do not change over time. Using the same data set, I find in McMillen (2001) that price indexes and an index of distance gradients are very similar whether constructed by hedonic or repeat sales methods. This result suggests that using a limited set of explanatory variables does not lead to significant bias. The coefficients and standard errors for building area, lot size, building age, and distance from the city center are presented in the last column of Table 12.1. The coefficients have the expected signs and are highly significant. The estimated distance gradient indicates that house prices fell by about 2.9 percent per mile over the full 1983-98 time period.

Accurate estimation of price and distance gradient indexes requires ample observations during each time period. I analyze quarters because this unit of time is small enough that the indexes are informative, while not being so small as to produce time periods during which there are very few sales. Figure 12.1 shows the number of sales by quarter. No quarter has fewer than 200 sales, while several quarters have more than 1000. Despite these large numbers, I combine the last two quarters into one for hedonic estimation because high standard errors for the estimates imply that there is not enough variation in the explanatory variables during this quarter to produce accurate estimates.

Figure 12.1: Number of Sales by Quarter

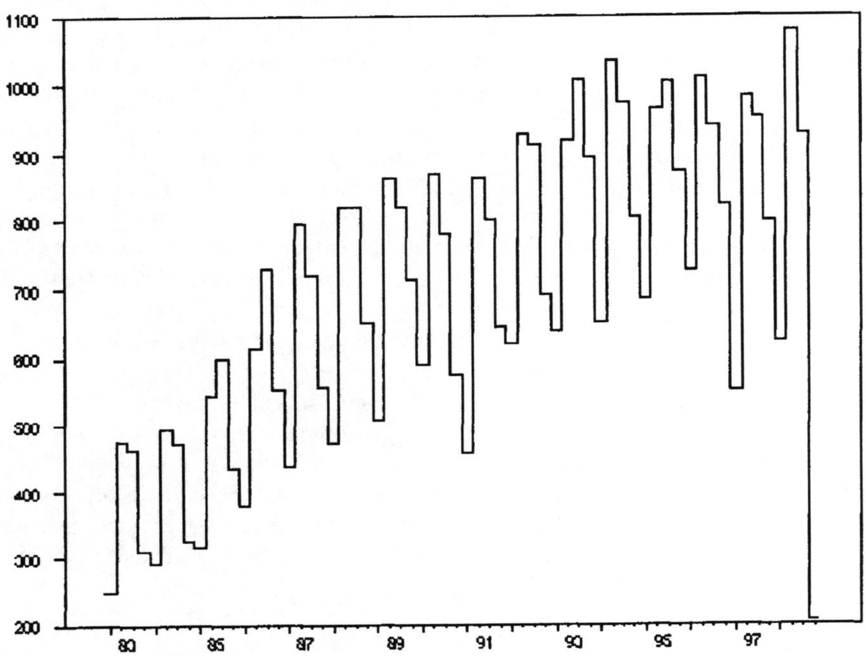

12.5 Estimated Price and Gradient Indexes – Implementation

In the empirical section of the chapter, I present estimates of housing price and distance gradient indexes for the City of Chicago. I compare the results of two estimators. The first is a standard hedonic estimator. These estimates are obtained by 63 quarterly regressions of the natural logarithm of housing sales prices on an intercept, the natural logarithms of building area and lot size, building age, and distance from the city center – one regression for each quarter from 1983 to 1998, with the last two quarters of 1998 combined. The estimated hedonic price index is the series of predictions for a standard 1100 square foot, 60-year old house on a 4000 square foot lot that is 9 miles from the city center. The distance gradient index is the series of coefficients for distance from the city center. Confidence intervals for the distance gradients are calculated using the estimated standard errors for the coefficients.

The second estimator is locally weighted regression[8]. I use the tri-cube weighting function given by equation (12.8) for the continuous time variable. The target points for estimation are the midpoints of each of the 64 quarters (15 February 1983 through 15 November 1998). The weight given to an observation is a decreasing function of the time between the sale date and the target time. I vary the window size, using the closest 1, 5, 10, 25, and 50 percent of the 44,190 observations. I focus on the results for the 10 percent window size because these results appear to provide the right amount of smoothing when compared to the simple hedonic results.

I use a uniform kernel for the weighting function for the other explanatory variables, with the window size set at 100 percent of the available observations. Under this specification, the LWR index requires only 64 weighted least squares regressions, which are no more difficult to calculate than the simple hedonic model. The weights are a function of time only. The explanatory variables include an intercept, x_{1i}-1100, x_{2i}-4000, x_{3i}-60, x_{4i}-9, and T_i-T, where x_1 is building area, x_2 is lot size, x_3 is age, x_4 is distance from the city center, T_i is the time of sale for observation i, and T is the target time. The rescaling of the explanatory variables implies that the price index is simply the series of 64 intercept estimates. The distance gradient index is the series of coefficients for distance from the city center.

Standard errors are more difficult to calculate for the LWR distance gradient estimates than for the hedonic estimates because of the error variance estimate. I use equation (12.7) to estimate the error variance. I use a tri-cube weighting function for time, with a window size of 10 percent of the 4419 observations. I again use a simple uniform kernel for the other variables, with a window size of 100 percent of the available observations. The resulting estimator requires 44190 weighted least squares regressions, one for each observation[9]. After calculating the estimated error variance, I use equation (12.6) to form standard error estimates for the estimated distance gradients.

12.6 Estimated Price and Gradient Indexes – Results

The estimated hedonic and LWR price indexes are presented in Figure 12.2. Both indexes suggest that housing prices rose rapidly throughout the 1980s, leveled off during the early 1990s' recession, and then rose rapidly again to the end of the sample period. The difference between the estimators is superficial on the one hand yet critical on the other. The LWR estimates appear to be a simple smoothing of the hedonic index. Smoothing is attractive theoretically. There is no reason why prices should stay the same from 1 January to 31 March, followed by a sharp change on 1 April, yet this is precisely the pattern implicit in the hedonic estimator. The LWR estimator incorporates a more plausible pattern in which prices change smoothly over time.

An obvious alternative to LWR estimation is to use the hedonic approach to estimate the index, and then to smooth the index. The LWR estimator carries out the smoothing in one step, and does it at no additional computational cost when simple kernel weighting functions are used.

Figure 12.2: Estimated Price Indexes

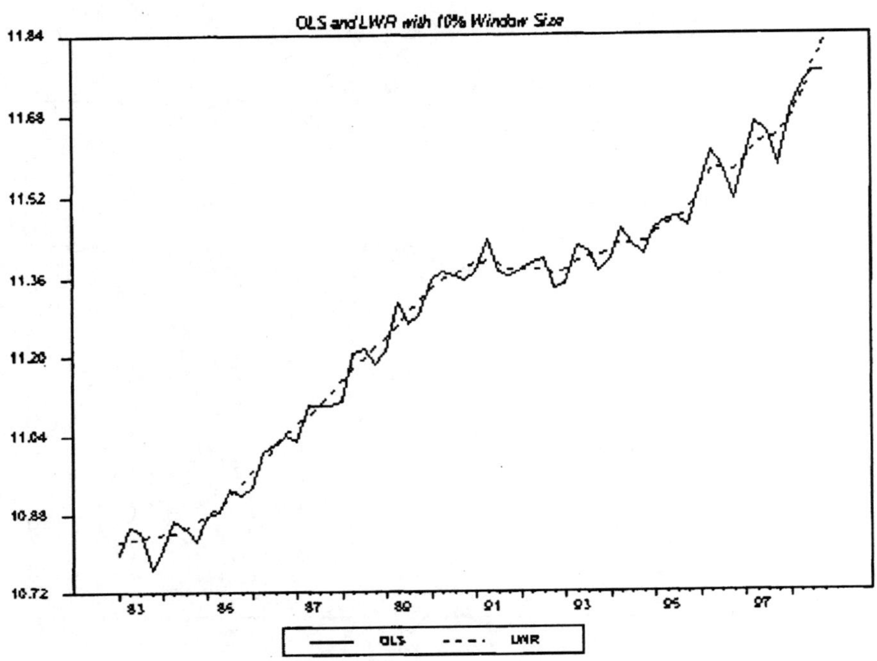

The estimated distance gradients are displayed in Figure 12.3. Here the undesirable discontinuities of simple hedonic estimates are more pronounced. Both estimators suggest that housing prices rose by nearly 2 percent per mile during the 1980s. A sharp change in structure occurred during the early 1990s as the estimated gradient turned negative. Prices fell by as much as 7 percent per mile with distance to the city center during the 1990s. Figures 12.4 and 12.5 present 95 percent confidence intervals for the estimated distance gradient indexes. Both hedonic and LWR estimates have tight confidence intervals, and the figures suggest that there are no statistical advantages to the hedonic estimates.

All of the LWR results discussed so far are based on a window size of 10 percent, which was chosen by a simple comparison of hedonic and LWR estimates. The estimated distance gradients under various window sizes are presented in Figure 12.6. The 1 percent window size produces results that are very similar to the standard hedonic estimates. Indeed, the hedonic estimator is simply a special case of the LWR estimator. The hedonic estimator implicitly uses a uniform kernel for time, with the calendar determining the window size: all sales that take place within the target quarter are given uniform weight, and sales from other quarters receive zero weight. Visual inspection of Figure 12.6 suggests that LWR estimates based on a 10 percent window size provide an appropriate amount of smoothing.

Figure 12.3: Estimated Distance Gradients

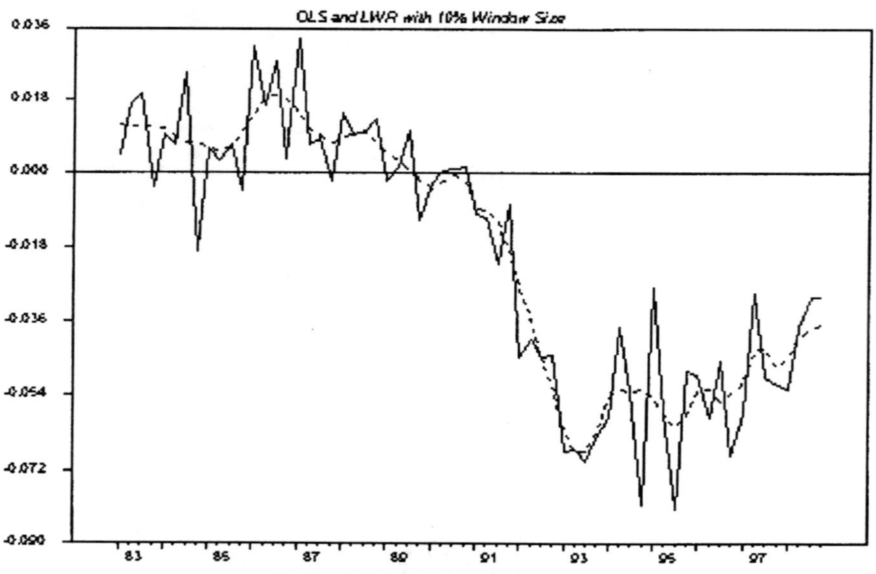

Figure 12.4: OLS 95% Confidence Intervals for Distance Gradients

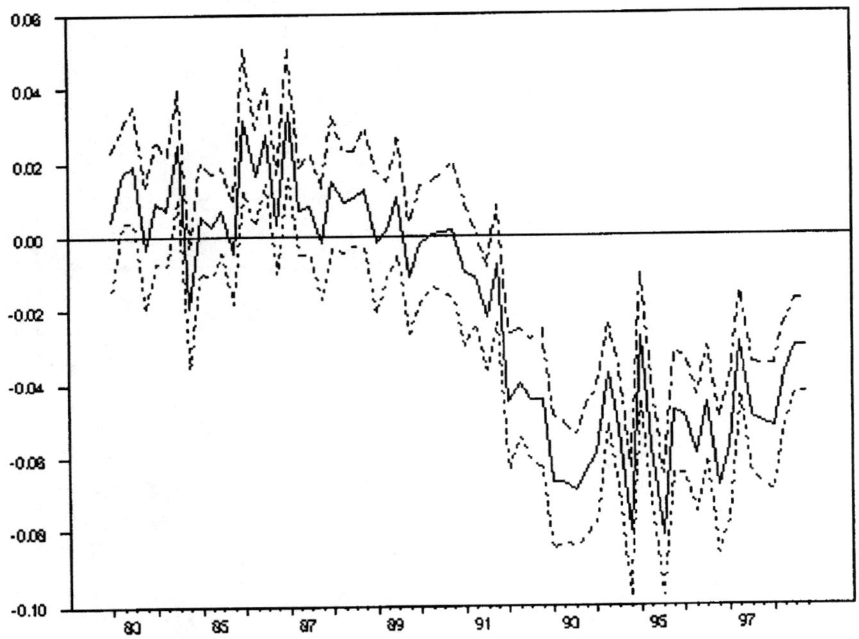

The estimated distance gradient index suggests the reemergence of centralization in the City of Chicago. Standard urban models such as Mills (1972) and Muth (1969) predict that the *unit* price of housing declines with distance from the city center. Many researchers have attempted to test this prediction using data on house prices, with mixed results[10]. Findings of positive distance gradients have led some researchers to argue that the standard urban model is no longer relevant. One critical point is that the standard models do not produce any predictions regarding house sales prices, which are the product of the unit price and the quantity of housing. If wealthier homeowners live farther from the city center, then house sales prices will increase with distance from the city center even if the unit price of housing falls. However, the results here suggest sales prices of housing decrease with distance along with the unit price. The reemergence of a negative sales price gradient in Chicago suggests that the traditional urban model is not ready for wholesale abandonment. The center of the City of Chicago again has extremely high house values.

Figure 12.5: LWR 95% Confidence Intervals for Distance Gradients

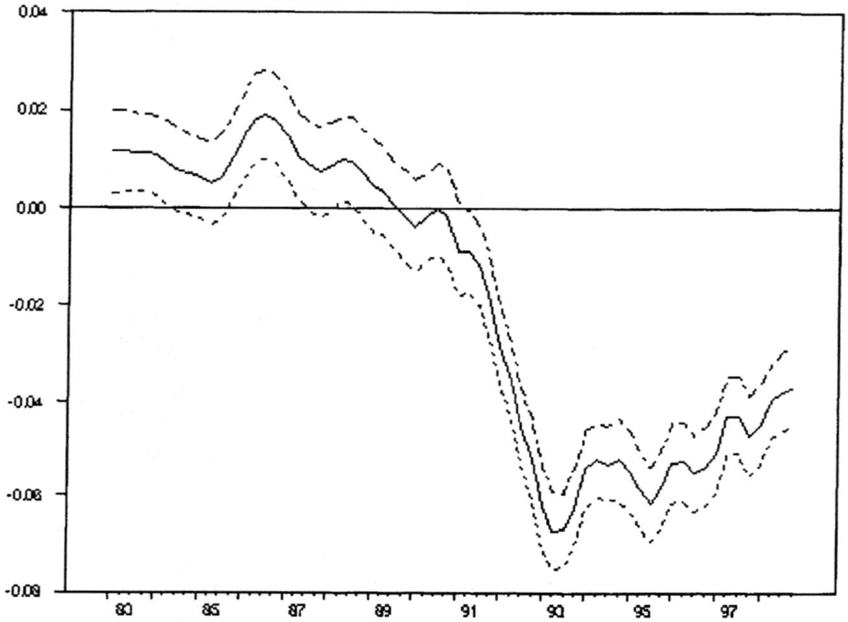

12.7 Conclusion

Nonparametric estimators are well suited to spatial data analysis. Geographic variation in the effects of explanatory variables, which can be handled readily by nonparametric estimators, leads to complicated and potentially intractable parametric functions. Using a subset of the data to estimate econometric relationships helps to reduce the severity of heteroskedasticity and spatial autocorrelation problems, which are often caused by misspecified functional forms and missing variables that are correlated over space. Programming is not difficult for nonparametric estimators, requiring only repeated use of standard weight least squares estimation procedures.

Nonparametric estimation is becoming more common in spatial analysis. Not surprisingly, some confusion has arisen over the relationship between various estimators. The nonparametric estimator most commonly employed by geographers and urban economists is referred to as 'locally weighted regression' or as a 'local linear smoother' or a 'local regression smoother' by statisticians. Spatial data analysts have used nearest neighbor versions of these estimators, taking the word 'nearest' literally to mean geographic distance. Nonetheless, the resulting estimator remains a special case of the more general LWR estimator.

Figure 12.6: LWR Distance Gradients

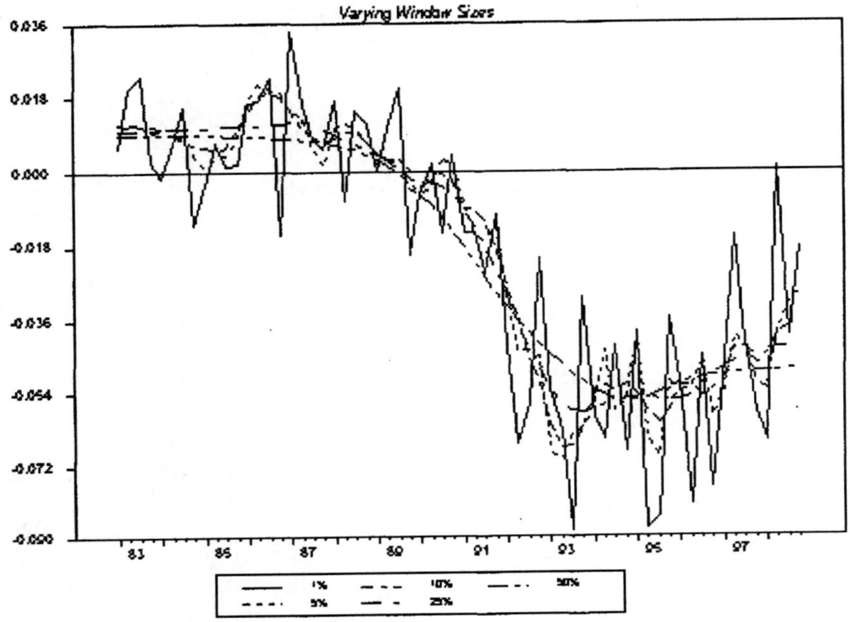

This chapter demonstrates how to use locally weighted regression to obtain estimates of housing price indexes and to estimate time-varying derivatives. I use a simple specification of the kernel weighting function that places more weight on recent sales when estimating the housing price function for a target time. The specification allows a smooth index to be estimated for a 16-year time period with only 64 weighted least squares regressions. Another advantage of the specification is it makes hypothesis testing nearly as easy to conduct as with standard parametric modeling.

The results have important implications for urban spatial modeling. The standard theoretical urban model sometimes is dismissed because it predicts that the unit price of housing falls with distance to the city center while empirical results suggest that house values are an increasing function of distance. This rejection of the model is based in part on a fundamental misunderstanding: if higher-income households live farther from the city center, then the model predicts this very result. Nevertheless, the evidence presented here suggests the reemergence of a downward sloping house price function of the City of Chicago. Among others, high income, childless families with two earners are moving to the city center and driving up house values. Locations near the city center are again hotbeds of real estate activity. As assumed in theoretical models, homeowners are willing to pay a significant premium to live near downtown Chicago in order to reduce their commuting time and expense. Although the rise of suburban employment centers means that Chicago can no longer be modeled adequately as a monocentric city, the traditional central business district still dominates the metropolitan area. The city-center distance gradient again has become pronounced.

Endnotes

[1]The best introduction to nonparametric estimation is Pagan and Ullah (1999), which is the basis for much of this section. Other useful surveys are Härdle (1990), Härdle and Linton (1994), and Yatchew (1998).

[2]The bi-square function is given by $K(P) = (15/16)(1-P^2)^2 I(|P|<1)$, where $I(\ )$ is an indicator function that equals one when the condition is true and zero otherwise. The tri-cube function is written $K(P) = (70/81)(1-|P|^3)^3 I(|P|<1)$, and the gaussian is given by $K(P) = (2\pi)^{-1/2}\exp(-P^2/2)$. The leading constants in these expressions play no practical role in estimation.

[3]Locally weighted regression was proposed by Stone (1977) and Cleveland (1979). It has been developed further by Cleveland and Devlin (1988), Fan (1992, 1993), Fan and Gijbels (1992), and Ruppert and Wand (1994). Pagan and Ullah (1999) provide an excellent review.

[4]This approach was first used by Meese and Wallace (1991). Recent applications include Brunsdon *et al.* (1996, 1999), Clapp (2001), McMillen (1996), McMillen and McDonald (1997), Meese and Wallace (1997), Pavlov (2000), and Yuming and Somerville (2001). In the geography literature, the estimator is sometimes referred to as 'geographically weighted regression,' and presented as though the approach is unique to the field. However, it should be clear from the discussion here that the approach taken by geographers is neither original nor unique.

[5]Applications include Huang and Fu (1999) and Lee, Kwak, and List (2000).

[6]In spatial analysis, the term 'density' can be confusing because it more often refers to population or employment density. For average derivative estimation, a region with high density is a set of data points that are close to other observations, where to be 'close' means to have similar values. Typically, regions with high population or employment density also have high values for the nonparametric density estimates because they are drawn from areas with many observations, but the two concepts are not identical. Excellent reviews of nonparametric estimation include Silverman (1986) and Pagan and Ullah (1999).

[7]Note, however, that we still must estimate many regressions to calculate $\hat{\sigma}^2$, as given by equation (12.7). It is possible to obtain an accurate estimate of the error variance using a subset of the observations as target points, but a larger number than 64 is desirable.

[8]Some applications of nonparametric estimators to housing price functions include Anglin and Gencay (1996); Clapp (2001); Gencay and Yang (1996); Iwata, Murao, and

Wang (2000); McMillen and Thorsnes (2000); Meese and Wallace (1991, 1997); Pace (1995); Pavlov (2000); and Stock (1991).

[9]The calculations are simplified somewhat because several houses sometimes sell on the same day, which reduces the number of required regressions.

[10]Examples include Coulson (1991); Heikkila *et al.* (1989); Hoch and Waddell; and Waddell, Berry, and Hoch (1993a, 1993b).

13 The Statistical Significance of the Maximum Local Statistic

Peter A. Rogerson
University at Buffalo

Abstract

Statistics aimed at testing for local spatial association are now well developed and enjoy widespread use. In many applications, there is often interest in testing the statistical significance of more than one local spatial statistic. For instance, analysts may compute local statistics for each subregion in a study area. The fact that multiple testing is being carried out calls for an adjustment to the significance level of each test, so that the overall significance level can be maintained at a desired level. The commonly employed Bonferroni adjustment is conservative, since it assumes that individual tests are independent; the correlation between the local test statistics for nearby areas requires a less conservative adjustment. In this chapter I suggest how the conservative Bonferroni upper bound for the critical value of a local statistic may be improved upon by accounting for the correlation between tests. I also demonstrate how a liberal lower bound for the critical value may be obtained.

13.1 Introduction

The application of local spatial statistics (Anselin 1995; Getis and Ord 1992; Ord and Getis 1995) is rapidly becoming more widespread in the field of spatial analysis. An open problem in the use of local spatial statistics is how to properly account for multiple tests of significance. Often there is interest in testing for significant spatial association around more than one local site, and one possibility is to test the local statistics around all n sites. However, if a large number of local tests are carried out, one will almost certainly obtain some 'significant' results by chance alone. Thus if n=1000 local tests are made, each using $\alpha = 0.05$, one would expect 50 of the 1000 test statistics to lie beyond their critical values. Clearly an adjustment to the usual method is needed if one is to avoid finding so many spurious 'significant' findings.

In this chapter, I suggest a number of alternatives for assessing the statistical significance of the maximum local statistic. In the second section, I show how common upper bounds for the critical value may be improved upon, by accounting for the correlation between tests in adjacent regions. I also demonstrate how a liberal, lower bound for the critical value may be obtained. The third section contains several illustrations, and the final section provides a summary.

13.1.1 Bonferroni and Sidak Adjustments

The most common method of correcting for multiple testing is the Bonferroni adjustment. If k separate, independent tests are made, then instead of choosing a critical value of the test statistic using α as the Type I error probability for each test, one uses α/k for each test. A Type I error will be committed when at least one of the k observed local statistics exceeds its critical value (say b). The overall Type I error is then equal to the probability that the maximum local statistic is greater than b, which in turn is equal to one minus the probability that all k tests are accepted, i.e.

$$pr(\max_{i;i=1..,k} z_i > b) = 1-(1-\frac{\alpha}{k})^k,$$

where z_i is the local statistic for region i. This probability is just slightly less than the desired overall Type I error of α, whenever $k \geq 2$ David (1956) provides an early description and application of the Bonferroni adjustment.

A small improvement is possible by using the Sidak adjustment (see, e.g., Sidak 1968), which replaces the α/k used for the Bonferroni correction with $1-(1-\alpha)^{1/k}$. Then, when tests are independent, $pr(\max_{i;i=1..,k} z_i > b) = \alpha.$

The Bonferroni and Sidak adjustments are accurate when multiple tests are independent, but when the outcomes of individual tests are positively correlated, the adjustment will be conservative. With k positively correlated tests, the 'effective' number of independent tests is less than k. For example, if the outcomes of all k tests were perfectly correlated with one another, there would effectively be only one test (since knowing the outcome of one test, you would know the outcomes of all other tests). Hence in this case no adjustment for multiple testing would be necessary; if a Bonferroni or Sidak adjustment was made, the Type I error would be much less than α, and the power of the test would be reduced. To say that using a Bonferroni or Sidak adjustment for k tests is conservative means that the adjusted critical value is too extreme, significant results may be missed, and the actual value of $\geq\alpha$is less than the nominal value.

Local tests carried out for neighboring geographic areas are positively correlated since some of the same data is used in the computation of their statistics. It is of interest to know just how conservative the Bonferroni and Sidak adjustments are, and whether there are better, alternative adjustments that can be made. A principal objective of the next section is to provide bounds for the probability that the maximum local statistic exceeds some specified critical value, *b*.

13. 2 Bounds

In this section, I focus upon deriving upper and lower bounds for the critical value associated with the maximum local statistic. Fuchs and Kenett (1980) discuss a similar problem in an aspatial context, where interest is in outlying cells in multinomial distributions and contingency tables. For example, for the multinomial distribution, they define local (aspatial) statistics for each cell:

$$z_i = \frac{n_i - Np_i^{(0)}}{\sqrt{Np_i^{(0)}(1-p_i^{(0)})}},$$

where n_i is the observed frequency in cell *i*, $p_i^{(0)}$ is the hypothesized probability that an observation falls in cell *i*, and where there are *N* observations. They reject the null hypothesis of no outliers when the largest local statistic ($\max_i z_i$) is greater than some critical value, *b*. They find bounds on the probability that $\max_i z_i > b$ under the null hypothesis (and, equivalently, bounds for the critical value *b*) from the general inequality

$$\sum_{i=1}^{k} \Pr(E_i) - \sum_{i<j} \Pr(E_i \cap E_j) \le \Pr(\bigcup_{i=1}^{k} E_i) \le \sum_{i=1}^{k} \Pr(E_i) \tag{13.1}$$

where the E_i are arbitrary events. By taking E_i to be the event $z_i>b$, the middle term in the inequality is the desired probability that $\max_i z_i > b$. I next describe how (13.1) may be improved and used in the context of local spatial statistics. The development relies upon the assumption that the distribution of the local statistic is approximately normal, though this assumption could be relaxed.

13.2.1 The Joint Probability that Two Local Statistics Exceed the Critical Value

The bounds in (13.1) require evaluation of $\Pr(E_i \cap E_j)$, which is the joint probability that two regions exceed the critical value. If the local statistics in regions *i* and *j* (denoted z_i and z_j) are independent, and the probability of an event is $pr(E_i) = \theta_i = \theta$, then the joint probability is simply $\Pr(E_i \cap E_j) = \Pr(E_i)\Pr(E_j) = \theta^2$. More generally, when individual local statistics have a normal distribution, $\Pr(E_i \cap E_j) = \Pr(z_i > b, z_j > b;\ r_{ij}) = L(b, b, r_{ij})$ may be evaluated using the known correlation between tests (denoted r_{ij}) and tables of the bivariate normal distribution (see, e.g., National Bureau of Standards 1959). Alternatively, one can use mathematical software that is able to either (a) evaluate the bivariate normal function directly, or (b) carry out the appropriate numerical integration. These latter options are discussed in the appendix.

13.2.2 Upper Bounds

With $\Pr(z_i > b) = \Pr(E_i) = \theta_i = \theta$, it is easy to see that the right-hand side of the inequality (1) is associated with the Bonferroni adjustment. Set the right-hand side of (13.1) equal to α

$$\Pr(\max_i z_i > b) \le k\theta = \alpha \tag{13.2}$$

This implies that for an individual test one should find an upper bound (b_U) for the critical value such that $\Pr(z_i > b_U) = \theta = \alpha / k$.

The slightly better Sidak adjustment results from using a tighter inequality for the right-hand side of (13.1)

$$\Pr(\bigcup_{i=1}^{k} E_i) \le 1 - \prod_{i=1}^{k} [1 - \Pr(E_i)]. \tag{13.3}$$

Based on (13.3), the Type I error to be used for each of the *k* independent tests (θ) is found via the solution of

$$\alpha = 1 - (1 - \theta)^k$$

for θ:

$$\theta = 1-(1-\alpha)^{1/k}. \tag{13.4}$$

For example, with α = 0.05 and k = 100 and a test statistic with a standard normal distribution, the critical value would be b = 1.645 if no adjustment for multiple testing was made. The result would be that at least one of the 100 tests would be significant far more often than the nominal value of 5% of the time. A Bonferroni adjustment would use θ=0.05/100=0.0005 for each test. This corresponds to a critical value of $b_U = \Phi^{-1}(1-\theta) = 3.2905$, where $\Phi^{-1}(x)$ denotes the standard normal deviate associated with an area of x in the left tail of the cumulative standard normal distribution. The Sidak adjustment implies

$$\theta = (1-0.05)^{1/100} = 0.0005128,$$

corresponding to a critical value of $\Phi^{-1}(1-.0005128) = b_U = 3.2834$.

For situations where tests are correlated, Worsley (1982) improved upon the upper bounds given in (13.1) and (13.3). In particular,

$$\Pr(\bigcup_{i=1}^{k} E_i) \le \sum_{i=1}^{k} \Pr(E_i) - \sum_{i=1}^{k-1} \Pr(E_i \cap E_{i+1}). \tag{13.5}$$

This was a rediscovery of an earlier result proved by Hunter (1976). To make optimal use of (13.5) in applications to local spatial statistics, one would take $\Pr(E_i \cap E_{i+1})$ to be the joint probability that two adjacent regions both had local statistics that exceeded the critical value, b. A desirable feature of (13.5) is that implementation only requires calculation of the correlation for adjacent regions; the correlations of a local statistic with other, nonadjacent cells, are not required.

To use (13.5) to find an upper bound for the critical value b that is associated with a Type I error of α, the following equation is solved for b_U:

$$\alpha = k\theta - \sum_{i=1}^{k} L(b_U, b_U, r_{ij}) \tag{13.6}$$

where r_{ij} is the correlation between the local statistic in cell i and the local statistic in adjacent cell j, and as a reminder, $L(b,b,r_{ij}) = \Pr(z_i > b, z_j > b)$. The subscript 'U' indicates that the resulting value is an upper bound for the actual value of b. When data are on a regular lattice, this simplifies to

$$\alpha = k\theta - (k-1)L(b_U, b_U, r)$$

where r is the correlation between adjacent cells in the lattice.

13.2.3 Lower Bounds

Let us now wish to focus on the left-hand side of the inequality in (13.1), used previously by Fuchs and Kenett (1980). This same inequality was used earlier by David and Newell (1965) to find lower bounds associated with tests for raised disease incidence during more than one time period.

Fuchs and Kenett were able to use the negative correlations that exist between cell frequencies to derive reasonably tight bounds on the desired probability. However, in the present case, positive correlations exist between neighboring local tests, and this ultimately limits the usefulness of these bounds. However, it is still possible to make use of them in some cases. First note a necessary condition that must hold to employ the lower bound in (13.1). Since this condition was easily satisfied in the previous applications cited above, it was not mentioned explicitly. It will, however, be important in applications to local spatial statistics.

A Necessary Condition for Calculating the Lower Bound from (13.1)
When tests are independent, the function

$$f_1(\theta) = k\theta - k(k-1)\theta^2 / 2$$

may be set equal to an overall Type I error probability (say $\alpha = 0.05$), and then solved for $\theta = 1 - \Phi(b_L)$, where Φ is the cumulative distribution function for the normal distribution. This yields a lower bound for the critical value to be used when performing k independent tests.

More generally, with correlated tests, one can attempt to find $\theta = 1 - \Phi(b_L)$ such that

$$f_2(\theta) = k\theta - \sum_{i=1}^{k}(k - m_i - 1)\theta^2 / 2 - \sum_{j=1}^{m} L(b_L, b_L, r_{ij}))/2 = \alpha \quad (13.7)$$

where m_i is the number of cells that have nonzero correlations with a given cell (i), and where r_{ij} is a nonzero correlation between cell i and another cell j. For a regular lattice,

$$f_2(\theta) = k\theta - k(k-m-1)\theta^2/2 - k\sum_{j=1}^{m} L(b_L, b_L, r_j))/2 = \alpha$$

where $m_i = m$, and r_j is the correlation that all cells i have with their adjacent cells j.

The function f_2 is quadratic in θ; it has a maximum at

$$\theta_{\max} = \frac{1 - \sum_{i=1}^{m} d_j/2}{k-m-1}$$

where d_j is the conditional probability that cell j is significant, given that cell i is significant. The quantity d_j is the solution to

$$d_j = \int_b^{\infty} \left\{ 1 - \Phi\left[\frac{(b - r_{ij} z_i)}{\sqrt{1 - r_{ij}^2}} \right] \right\} \frac{f(z_i)}{1 - \Phi(z_i)} dz_i$$

At the maximum, f_2 is equal to

$$f_2(\theta) = \frac{k(1 - \sum d_j/2)^2}{2(k-m-1)}.$$

A necessary condition for calculating the lower bound is that $f_2(\theta) \geq \alpha$. Since the d_j terms are positively related to the correlations between cells, if the correlations between cells are sufficiently high, $f_2(\theta_{\max})$ may be less than α. In these cases, there is no solution for the lower bound, b_L.

An Alternative Lower Bound

An alternative lower bound that can improve upon (12.7) is

$$\frac{2}{h+1} \sum_{i=1}^{k} \Pr(E_i) - \frac{2}{h(h+1)} \sum_{i<j}^{k} \Pr(E_i \cap E_j) \leq \Pr(\bigcup_{i=1}^{k} E_i) \qquad (13.8)$$

where

$$h = 1 + [\frac{2\sum_{i<j}^{k} \Pr(E_i \cap E_j)}{\sum_{i=1}^{k} \Pr(E_i)}]$$

and the square brackets indicate that the integer part of the quantity inside of the brackets should be taken (see, for example Sobel and Uppuluri 1972; Kwerel 1975; Galambos 1975, 1978). Galambos (1978) notes that this is the best linear lower bound. When h=1, (13.8) reduces to (13.7) (and to the left-hand side of (13.1)). The lower bound determined from the inequality in (13.8) could be improved further with information on the joint probabilities of three or more events (e.g., with information on the quantities $\Pr(E_i \cap E_j \cap E_k)$.

13.3 Illustration

The foregoing is now illustrated by using a study region consisting of a 30x30 lattice of square cells. Suppose that there is interest in searching for local pockets of spatial association, and consequently, local statistics are derived for each of the cells. The objective is to compare the maximum local statistic with an appropriate critical value to test the null hypothesis that there is no spatial association. How can one find an appropriate critical value (or more specifically, bounds for the critical value)?

Suppose further that the local statistic is a $G_i^*(d)$ statistic, following Ord and Getis (1995):

$$G_i^*(d) = \frac{\sum_j w_{ij}(d)x_j - \sum_j w_{ij}(d)\bar{x}}{s\sqrt{(n\sum_j w_{ij}(d)^2)/(n-1)}} \tag{13.9}$$

where n is the number of cells, and $\bar{x}$ and s represent the mean and standard deviation of the x values, respectively. The statistic has, approximately, a normal distribution with mean zero and variance one, under the null hypothesis of no local spatial association. The quantity $w_{ij}(d)$ is a weight equal to one if cell j lies within a distance d of cell i. Thus the statistics have a 'range of influence' of d, where the distance between adjacent cell centroids is taken to be unity. A value of d=1 implies that m=5 cells are used in the calculation of a local statistic for a given cell (i.e., the given cell, plus those four lying within d=1 via rook's adjaceny). For d=2, m=13 cells would be used in calculating the local statistic, since in addition to the given cell there are 12 other cells (8 via queen's adjaceny, and 4 via rook's adjacency at a distance of 2 units). When d=3, m=29.

For the case d=1, I used the n=26x26=676 cells in the interior to avoid edge effects. Similarly, for d=2, I used n=22x22=484 cells in the interior of the grid, and for d=3, I used the 18x18=324 central cells.

To find appropriate bounds, it is necessary to find the correlations between nearby local statistics. The correlations between local statistics may be derived as follows (Ord and Getis 1995). Let c_{ij} represent the number of cells that the local statistics in cells i and j have in common, divided by m. This value of c_{ij} is also equal to the correlation coefficient, r_{ij}.

The d=1 case

When d=1 on a square lattice, m=5, and the correlations of a given cell with other cells are as follows:

c_{ij} = 2/5 =0.4 for the 8 cells adjacent by queen's case adjacency

c_{ij} = 1/5 =0.2 for the 4 cells that are two cells away in either the horizontal or vertical direction

c_{ij} = 0 for the remaining n-8-4-1 = n-13 cells

The d=2 case

When d=2, the correlations of a given cell with other cells are:

c_{ij} = 8/13=0.615 for the 8 cells adjacent by queen's case adjacency

c_{ij} = 5/13=0.385 for the 4 cells that are two cells away in a horizontal or vertical direction

c_{ij} = 4/13 = 0.308 for the 8 cells that are two cells away in one direction, and one cell away in the other direction.

c_{ij} = 3/13=0.231 for the 4 cells that are two cells away in the horizontal direction, and two cells away in the vertical direction,

c_{ij} = 2/13=0.154 for the 8 cells that are three cells away in either the horizontal or vertical direction, and also for the 4 cells that three cells away in one direction and one cell away in the other.

c_{ij} = 1/13 =0.077 for the 4 cells that are four cells away in either the horizontal or vertical direction.

c_{ij} = 0 for the remaining n-8-4-8-4-8-4-4-1 = n-41 cells.

The d=3 case

When d=3, the correlations of a given cell with other cells are:

c_{ij} = 22/29 for the 4 cells adjacent by rook's case adjacency

c_{ij} = 20/29 for the 4 cells that are one cell away in both horizontal and vertical direction

c_{ij} = 17/29 for the 4 cells that are two cells away in a horizontal or vertical direction

c_{ij} = 16/29 for the 8 cells that are two cells away in one direction, one cell away in the other direction

c_{ij} = 13/29 for the 4 cells that are two cells away in each direction

c_{ij} = 12/29 for the 4 cells that are three cells away in a horizontal or vertical direction

c_{ij} = 10/29 for the 8 cells that three cells away in one direction an one cell away in the other direction

c_{ij} = 8/29 for the 8 cells that are three cells away in one direction and two cells away in the other direction

c_{ij} = 7/29 for the 4 cells that are four cells away in either direction

c_{ij} = 6/29 for the 8 cells that are four cells away in one direction and one cell away in the other direction

c_{ij} = 5/29 for the 8 cells that are four cells away in one direction and two cells away in the other direction

c_{ij} = 2/29 for the (a) 4 cells that are five cells away in either the vertical or horizontal direction, (b) 8 cells that are four cells away in one direction and three cells away in the other, (c) 8 cells that are five cells away in one direction and one cell away in the other, and (d) 8 cells that are five cells away in one direction and two cells away in the other.

c_{ij} = 1/29 for the 4 cells that are six cells away in either direction, and for the four cells that are four cells away in each direction

Results

In the d=1 case, n=676 and m=5. The Sidak bound (13.4) yields b_U = 3.788 and the Bonferroni bound from (13.2) is b_U = 3.795. The improved Bonferroni bound from (13.6) is b_U = 3.792. In this case, where the influence of local statistics is spatially limited relative to the size of the grid, the Bonferroni and Sidak adjustments will serve quite well, and the bounds around the appropriate b are very tight. Note that here the improved Bonferroni bound does not result in too much improvement, and in fact the Sidak bound is better. This will not be the case for d=2 and d=3, where the local statistics are not as 'local', and the correlations between adjacent cells are higher. The correlation values imply a lower bound, using (13.7), of b_L = 3.778.

For the d=2 case, n=484 and m=13. The Sidak bound (13.4) is b_U = 3.704 and the Bonferroni bound (13.2) is b_U = 3.711. The improved Bonferroni bound is b_U = 3.696, a slight improvement on both the Bonferroni and Sidak bounds. The lower bound, derived from (13.7), is b_L = 3.61. The difference between lower and upper bounds is a bit larger, in comparison with the d=1 case.

For d=3, the lower bound for the critical value can not be determined from (13.7), since the necessary condition described above is not satisfied. However, (13.8) was used to derive a lower bound of b_L =3.25. The upper Bonferroni and Sidak bounds (with n=324) yield b=3.608 and b=3.601, respectively. The improved upper bound (13.6) is b=3.563.

For each of the three cases, Monte Carlo simulations of the null hypothesis of no local clustering confirmed that the actual value of b consistent with $\alpha = 0.05$ falls within these bounds. The simulations were carried out by first filling each cell of the grid with a normally distributed random variable having mean zero and variance one. The local statistics were then found using (13.9).

For d=1, two Monte Carlo trials of 10,000 simulations each yielded 3.774 and 3.776 for the 95th percentiles of the maximum local Getis-Ord statistic -- both very near the lower bound of 3.778 (note that the simulations can actually yield values slightly outside of the bounds). In the d=2 case, two separate Monte Carlo experiments of 10,000 trials each resulted yielded 3.663 and 3.680 for the 95th percentiles of the maximum local Getis-Ord statistic. These values are slightly closer to the best upper bound than to the lower bound.

With d=3, the 95th percentile of the distribution of the maximum local Getis-Ord statistic among the 18x18=324 central cells was found to be 3.474, which is not far from the improved upper bound of 3.563. Use of the improved upper bound, Sidak, or Bonferroni bounds in this d=3 example would have led to rejection of the null hypothesis in 4%, 3.2%, and 3.1% of the simulations, respectively. These are all slightly conservative, with the least upper bound being the least conservative.

These results are summarized in Table 13.1.

Table 13.1. Upper and Lower Bounds, and Comparison to Simulated

		Upper Bounds			Lower Bounds		Simulated 95th percentile with 10000 Monte Carlo repetitions	
d	# of cells	Bonferroni	Sidak	Improved Bonferroni	(Eq. 13.7)	(Eq. 13.8)		
		(Eq. 13.2)	(Eq. 13.4)	(Eq. 13.6)				
1	676	3.795	3.788	3.792	3.778	3.778	3.774	3.776
2	484	3.711	3.704	3.696	3.61	3.61	3.663	3.680
3	324	3.608	3.601	3.563	--	3.25	3.474	

13.4 Summary and Discussion

When the number of surrounding regions used in the computation of local statistics is small, the Bonferroni and Sidak adjustments, though slightly conservative, will be quite accurate. When the local statistics are computed by using a large number of surrounding regions, these adjustments will be overly conservative, and the improved Bonferroni inequality (13.6) is recommended. The amount of improvement achieved by (13.6) is directly related to the strength of the correlation between adjacent local statistics. For low values of d, the lower bound (13.7) is accurate and confirms the adequacy of the Bonferroni and Sidak adjustments. For higher d, (13.7) must be abandoned in favor of (13.8). Due to both the complexity of the calculations, and the fact that it is a fairly liberal lower bound, it is unlikely that routine computation of (13.8) will be particularly useful in the application of local statistics. In contrast, the improved Bonferroni inequality in (13.6) should prove useful in providing a less conservative critical value, especially in those instances where local tests are highly correlated.

Finally, there are also other ways to assess the significance of the maximum local statistic. When the local statistic is defined using a Gaussian kernel, the critical value associated with the null hypothesis that there is no local clustering may be closely approximated with a simple analytical expression (Rogerson 2001). For normally distributed data (standardized to have zero mean and unit variance) on a regular lattice, the local statistic

$$z_i = \sum_j w_{ij} y_j$$

has a standard normal distribution, where

$$w_{ij} = (\pi\sigma)^{-1} \exp(\frac{-d_{ij}^2}{2\sigma^2})$$

and where d is the distance between cells and σ is the width of the kernel. The maximum local statistic, $M = \max_i z_i$, is then compared with the critical value

$$M^* \approx \sqrt{-\sqrt{\pi} \ln[4\alpha(1+0.81\sigma^2)/A]}$$

where A is the area of the study region, and where the distance between adjacent cell centroids is scaled to equal one. For irregular lattices, it is recommended that the new, scaled weights be used; these are defined by:

$$\tilde{w}_{ij} = \frac{w_{ij}}{\sqrt{w_{ij}^2}}.$$

There are alternative procedures for deciding whether to accept or reject hypotheses, when many are being tested. Holm (1979) suggested the following sequential procedure. Rank the T p-values, with p_1 being the smallest and p_t the largest. Compare p_1 with α/n; if $p_1 > \alpha/n$ then none of the tests should be rejected. If $p_1 < \alpha/n$, reject that hypothesis and then compare p_2 with $\alpha/(n-1)$ (since there are at most n-1 true null hypotheses). Continue in this manner until for some j, $p_j > \alpha/(n-j+1)$. The first j-1 hypotheses are rejected and one fails to reject the others. As Aickin and Gensler (1996) note, this is guaranteed to (a) bound the Type I error at α, and (b) reject at least as many hypotheses as the Bonferroni procedure.

There are also graphical approaches to finding the appropriate critical value when multiple testing carried out -- an example is the *p*-value plot of Schweder and Spjotvoll (1982). Here the *p*-values of all individual tests are first ordered (p_i, i=1,2,..., n), and the number of observed p values greater than p_i is plotted on the vertical axis, against 1-p_i on the horizontal axis. The slope is equal to the estimated number of true null hypotheses (T), and the authors recommend using a Bonferroni adjustment for T tests. Haybittle et al. (1995) find this still too conservative, and recommend rejecting those hypotheses corresponding to all of the points that deviate from the straight line in the right-hand side of the plot. While this may be of value in uncovering more false null hypotheses, one should be cautioned that the Type I error is no longer bounded by α and hence this latter strategy should be considered more exploratory.

Acknowledgements

This chapter is based upon work supported by Grant No.98-IJ-CX-K008 awarded by the National Institute of Justice, Office of Justice Programs, U.S. Department of Justice. It was also supported by the National Science Foundation under Award No. BCS-9905900. Its contents are solely the responsibility of the author and do not necessarily represent the official views of the sponsoring agencies. Points of view in this document are those of the author and do not necessarily represent the official position or policies of the U.S. Department of Justice or the National Science Foundation. The support of a Guggenheim Fellowship is also gratefully acknowledged.

Appendix

Write the probability of the joint event $(z_i > b, z_j > b)$. as

$$\Pr(z_j > b, z_i > b) = \Pr(z_j > b) \mid z_i > b) \Pr(z_i > b).$$

The conditional probability on the right-hand side may be determined by considering a regression of z_j on z_i ; the slope of this regression is equal to r. Thus

$$\hat{z}_j = r z_i + \varepsilon ,$$

where $\varepsilon \sim N(0,1\text{-}r^2)$ The desired conditional probability is therefore

$$\Pr(r z_i + \varepsilon > b \mid z_i > b) = \Pr(\varepsilon > b - r z_i \mid z_i > b);$$

The joint probability $(z_i > b, z_j > b)$ may then be written as

$$L(b,b,r) = \Pr(z_j > b, z_i > b) = \int_b^{\infty} \Pr(\varepsilon > b - rz_i)\Pr(z_i \mid z_i > b)\Pr(z_i > b)dz_i$$

Using the fact that the variable ε may be standardized by dividing by its standard deviation, this may be rewritten as

$$L(b,b,r) = \Pr(z_j > b, z_i > b) = \int_b^{\infty} \left\{ 1 - \Phi\left[\frac{(b - r_{ij} z_i)}{\sqrt{1 - r_{ij}^2}} \right] \right\} \frac{f(z_i)}{1 - \Phi(z_i)} [1 - \Phi(z_i)] dz_i \quad \text{(A.1)}$$

where f and Φ represent the density and cumulative distribution function of a standard normal variable, and where the subscript ij has been added to the correlation term to remind us that the magnitude of the correlation is a function of the two locations. Canceling terms yields

$$L(b,b,r) = \Pr(z_j > b, z_i > b) = \int_b^{\infty} \left\{ 1 - \Phi\left[\frac{(b - r_{ij} z_i)}{\sqrt{1 - r_{ij}^2}} \right] \right\} f(z_i) dz_i$$

As with the right-hand side of the inequality in (A.1), the next step is to set the left-hand-side of (A.1) equal to 0.05

$$k\theta - \sum_{i<j} \Pr(E_i E_j) = 0.05, \quad \text{or}$$

$$k[1 - \Phi(b)] - \sum_{i<j} \int_b^{\infty} \left\{ 1 - \Phi\left[\frac{(b - r_{ij} z_i)}{\sqrt{1 - r_{ij}^2}} \right] \right\} f(z_i) dz_i = 0.05 \quad \text{(A.2)}$$

Equation (A.1) may be solved by choosing an arbitrary value of b, finding the left-hand side, and then determining whether it should be increased or decreased in order to make it equivalent to the desired value of 0.05.

Though Equation (A.1) involves numerical integration, carrying the integration out requires only a few lines of code in an analytical mathematical package. For example, the code required for the integration in Maple, for the case r=0.396 and b=3.776 is:

```
with(stats): r:=.396; b:=3.778;
int((1-statevalf[cdf,normald] ((b-r*zx)/(sqrt(1-r^2))))*(statevalf[pdf,normald] (zx))/(1-
statevalf[cdf,normald](b)),zx=b..infinity);
evalf(%);
```

The language in SPLUS 2000 for *L(b,b,r)* is even more direct:

```
L <- 1+pmvnorm(c(b,b,rho=r)-2*pmvnorm(c(b,10,rho=r)
```

14 Detecting Spatial Clusters: a Bayesian Approach with Application to Identifying Disability Patterns in Mississippi and Alabama

Li-an Xu
Pharmaceutical Research Institute

Shuming Bao
University of Michigan

Abstract

In this chapter, we discuss some methods for identifying spatial clusters with a local focus. First, we provide a review on some recent developments in global spatial statistics for identifying spatial clusters and local spatial statistics. We then propose a spatial mixture model by applying a Bayesian approach for identifying spatial clusters of local interest. Gibbs sampling is used to make inferences about parameters in our model. The method is demonstrated using county disability rate data from Mississippi and Alabama.. The final estimate of the disability rate map with possible clusters is displayed and compared with the results from other models for categorical data.

14.1 Introduction

The study of spatial clustering of objects has been a subject of considerable interest in many fields, such as clusters of galaxies in astronomy, plants and animal species in ecology, prevalence of diseases in epidemiology, and social habitat of human beings in social studies (Cliff and Ord, 1981; Ripley, 1981; Diggle, 1983; Cressie, 1991). The common statistical approach for detecting the existence of spatial clusters is to test if spatial observations follow a spatial distribution of completely spatial randomness (CSR). CSR assumes that events are likely to occur equally over a space and that events do not interact with each other, either repulsively (regular events) or attractively (clustering events). A departure from CSR may suggest the presence of clusters.

To identify spatial clusters, many methods have been developed in the literature. They can be mainly classified as three groups: a) methods based on cell counts (quadrat procedure); b) adjacencies of cells with 'high' counts; and c) distance methods (Cuzick and Edwards, 1990; Cressie, 1991). Some spatial statistics have been proposed for exploratory spatial data analysis which include some global spatial statistics and local spatial statistics. Under the null hypothesis of spatial randomness, the global statistics (such as Moran's I, Geary's c, Tango's Cg, and Rogerson's R) can be used to test if there exist a significant spatial autocorrelation or clusters over space (Goodchild, 1986; Tango, 1995; Rogerson, 1998; Lin and Zeng, 1999). For local tests, some local spatial statistics (such as the G statistics, LISA and GLISA) can be used to identify spatial associations or clusters in local regions (Ord and Getis, 1992; Getis and Ord, 1994; Anselin, 1996; Bao and Henry, 1996). Those methodologies have been widely applied to identify the existence of spatial clusters or autocorrelation. However, they cannot define the spatial boundary of clusters. For quantifying the boundary of spatial clusters, see Fortin et al. (1996).

In this chapter, we propose a spatial mixture model for detecting spatial clusters. A Bayesian approach is applied for the estimation. This unified approach can be applied to both categorical and continuous data with a focus on the boundary of clusters. The methodology we propose here is also applicable to many other studies.

In the next section, we introduce our methodology and a general spatial mixture model for identifying spatial clusters. In section 14.3, we propose the spatial mixture models for Binomial processes and discuss the use of Gibbs sampling to estimate the posterior distributions of interest. In section 14.4, we conduct an exploratory data analysis by applying some traditional approaches to the county disability rate data from Mississippi and Alabama. The results from our spatial mixture model on the disability data will also be presented. In section 14.5, we briefly discuss the relationship between our model and the generalized linear model. Finally, some computational and practical issues for applying the methodology for detecting spatial clusters is discussed.

14.2 Methodology

Our methodology is to use a Markov random field (MRF) to model the latent process that creates the spatial clusters. The observed data are viewed as a 'corrupted' version of the latent data, corrupted by some independent 'noise' mechanism. We assume that the latent process X, which creates spatial clusters, is generated by a Markov random field (MRF), and the observable process Y has a density function $f(y|x; \alpha(x))$ for given $X = x$, where $\alpha(x)$ is an unknown parameter. The objective is to recover the cluster x and estimate $\alpha(x)$.

We use Bayesian methods for inference. The model for X can be viewed as a prior distribution. Given the prior distribution of X and the model $f(y|x)$, we use the posterior distribution of X given $Y = y$ to recover the cluster x. To give an example, suppose the X process takes on three values 0,1, 2 (see Figure 14.1). Given $x = i$, the conditional density of Y is $f(y|x=i, \alpha_i)$; i = 0; 1; 2. Assuming a Markov random field (MRF) prior P(x) for X, we define the posterior density of x; α_i; i = 0; 1; 2 as

$$P(x)\, f(y|x=0, \alpha_0)\, f(y|x=1, \alpha_1)\, f(y|x=2, \alpha_2)$$

Based on this posterior density, we can make inference on the cluster x and α_i; i = 0; 1; 2.

Figure 14.1: An illustration of spatial mixture processes

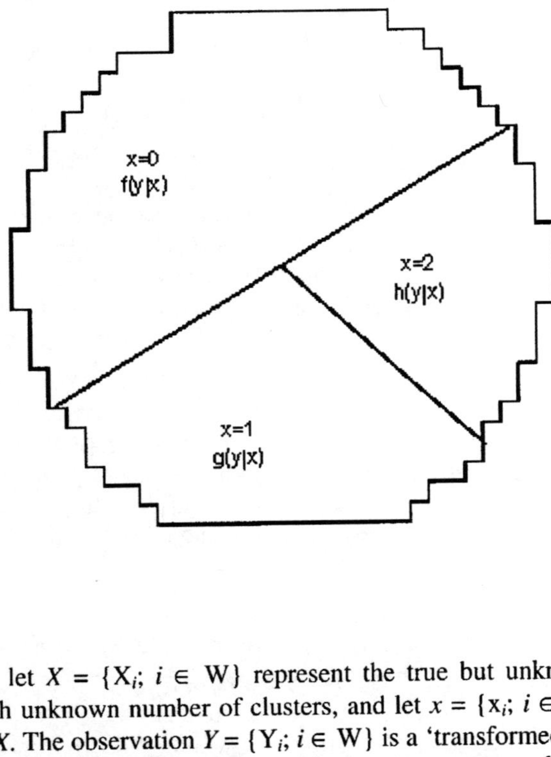

In more detail, let $X = \{X_i;\ i \in W\}$ represent the true but unknown spatial clusters or an image with unknown number of clusters, and let $x = \{x_i;\ i \in W\}$ be one particular realization of X. The observation $Y = \{Y_i;\ i \in W\}$ is a 'transformed' version of X with known mechanism transforming X to Y and unknown parameters to be estimated. The observed data y is one particular realization of the random vector Y. Y has a conditional density function $f(y|x)$ for given $X = x$. To recover X from Y , we use Bayesian inference. Assuming P(x) as a prior distribution on X. We can obtain a maximum a posteriori (MAP) estimate for recovered cluster $\hat{x}$ by maximizing the posterior density of X:

$$f(x|y) = P(x)f(y|x)$$

In the case that $P(x)$ is complex, it would be difficult to find the MAP estimate. Besag (1986) proposed the Iterated Conditional Mode (ICM) algorithm for finding a local maximum. The idea of the ICM algorithm is as follows: supposing we have an initial estimate $\hat{x}$ of the image, a new one is re-estimated by visiting each pixel in turn. The current value at pixel i is replaced by the value that maximizes the conditional density

$$f(x_i|\hat{x}_{-i}, y)$$

where $\hat{x}_{-i}$ denotes all the values except $\hat{x}_i$. The posterior density $f(x|y)$ can be represented by the product of $f(x_i|x_{-i};y)$ and $f(x_{-i}|y)$. As a result, the posterior density $f(x|y)$ will be increased. In this chapter, instead of using the ICM algorithm which leads to a local maximum, we use Gibbs Sampling for the inference, which provides us with the whole posterior distributions.

The prior distribution $P(x)$ is usually assumed to be a Markov random field (Chellappa and Jain, 1993), i.e. $P(x_i = x_i|x_{-i}) = P(x_i = x_i|x_{\delta i})$, where δ_i denotes the neighbourhood of pixel i. Since a Markov random field (MRF) has the property that the probability distribution for the i-th pixel depends only on its neighbours, we can redefine $f(x_i|\hat{x}_{-i}, y)$ as $f(x_i|\hat{x}_{\delta i}, y)$, which usually has a much simpler form.

This general methodology provides a flexible framework for detecting spatial clusters. First, the local spatial correlation is specified through the prior distribution, which can be chosen differently in different situation. Secondly, the likelihood function accommodates almost all types of data in reality. Thirdly, there is no need to specify the number of clusters in the data in advance. This is a big advantage over a lot of other clustering methods such as the K-Means method. We will discuss it in section 14.5.

14.3 Spatial Mixture Model for Binomial Processes and the Inference Using Gibbs Sampling

In this section, we propose a spatial mixture model and apply Bayesian approach for the estimation. Let X be the underlying process for clusters and K be the number of possible values for X. We will discuss the choice of K in section 14.5 in more details. Here we assume that X takes on three values 0, 1, and 2. We denote that there are n_i events (such as the number of elderly people) at location i, y_i is the count of observed events (q) of interest (such as the number of elderly people in disability) among n_i events. The event (q) may be observed at different locations with different probabilities of p_0, $p1$, or p_2. Given $X = x$, we consider the following distribution for Y:

$$P(Y_i = y_i \mid X_i = 0) = \binom{n_i}{y_i} p_0^{y_i}(1-p_0)^{n_i-y_i}; P(Y_i = y_i \mid X_i = 1) = \binom{n_i}{y_i} p_1^{y_i}(1-p_1)^{n_i-y_i};$$

$$P(Y_i = y_i \mid X_i = 2) = \binom{n_i}{y_i} p_2^{y_i}(1-p_2)^{n_i-y_i}$$

Since Y has different distributions in different spatial regions, we consider our model as a spatial mixture model.

The prior distribution for the underlying cluster X is given by a Potts model (Cressie, 1991), which is a natural extension of the Ising model (Cressie, 1991) in Ferromagnetism. If x is one possible realization of the random vector X with each component being 0, 1 or 2, then the prior distribution is given by

$$P(X = x) = Z(\beta) e^{\beta} \sum_{k=0}^{2} \sum_{i \sim j} I_k(i) I_k(j)$$

where β is a parameter, $Z(\beta)$ is the normalizing constant, $I_k(i)$ is the indicator function which is 1 if $x_i = k$ and 0 otherwise. Here, $i \sim j$ means that i and j are neighbours for a given neighbourhood structure. We will discuss the neighbourhood structures in section 5 in more detail. Note that the value of β indicates the strength of the dependency among neighbouring pixels. We assume for now that the value of β is known (the issue of estimating β will be discussed in section 14.5).

Figure 14.2 shows some simulated samples from the Potts model with different β values (β =.2; .25; .3; .7 respectively). As we can see, the estimated clusters will get larger as the β value gets bigger.

The prior distributions of p_i; i = 0; 1; 2 are assumed to be uniform on unit interval or Beta(1,1),which are non-informative. Under this specification, the posterior distribution of X and p can then be defined as

$$f(x, p_0, p_1, p_2 \mid y) \propto \prod_{x_i=0} p_0^{y_i}(1-p_0)^{n_i-y_i} \prod_{x_i=1} p_1^{y_i}(1-p_1)^{n_i-y_i} \prod_{x_i=2} p_2^{y_i}(1-p_2)^{n_i-y_i}$$

Now the goal is to get the MAP (maximum a posteriori) estimate of x, which would be our estimate of the spatially clustered area.

For such a complicated distribution, it is difficult to directly obtain the MAP. As an alternative approach, Gibbs sampling have been developed to generate samples from a complicated joint distribution (Geman and Geman, 1984; Casella and George,

Figure 14.2: Simulated Samples of the Potts model with different β The upper left sample is with $\beta = .2$. The upper right, the bottom left, and the bottom right are with $\beta = .25$; .3; .7, respectively

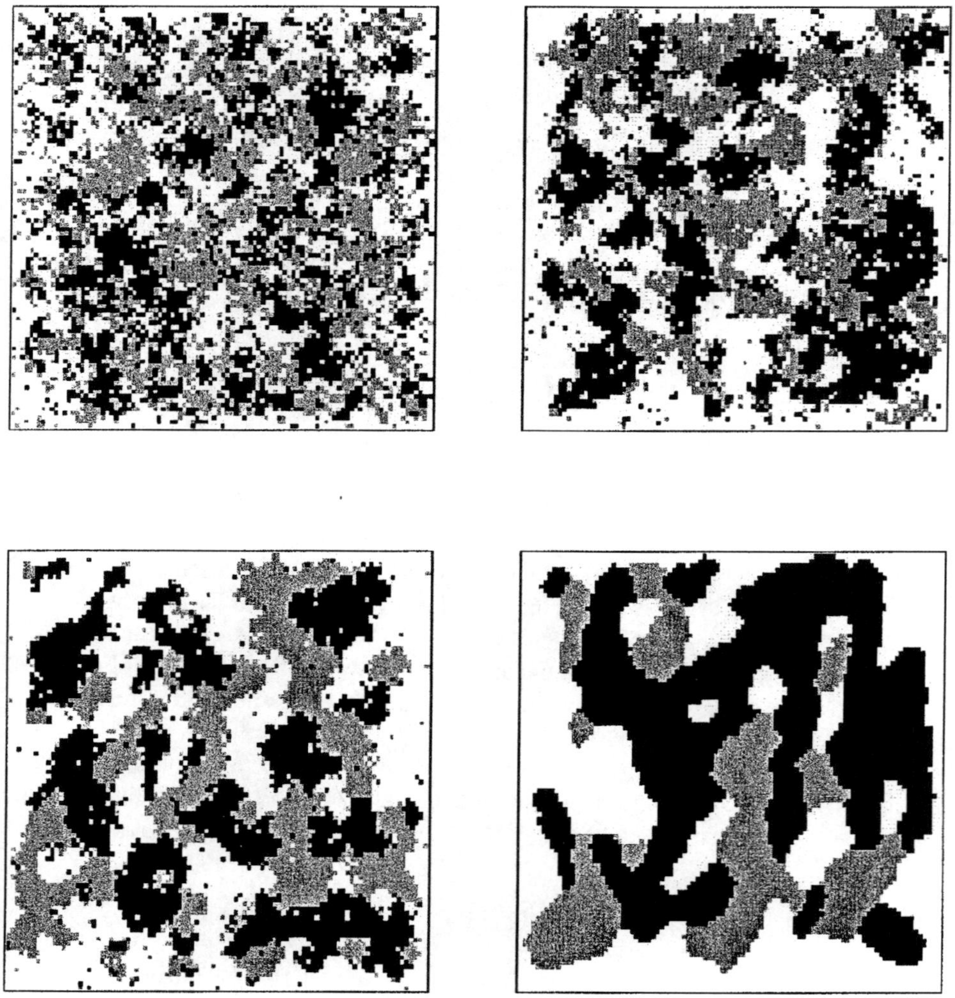

1992; and Tanner, 1996). The procedure of Gibbs sampling for sampling from the joint distribution of ($\beta_0, \beta_1, \ldots; \beta_m$) is described as follows

At the i-th iteration, we sample $\beta_0^{(i)}$ based on the conditional distribution on $\beta_1^{(i-1)}; \beta_2^{(i-1)}; \ldots, \beta_m^{(i-1)}$ which usually has a simpler form. Here $\beta_j^{(i-1)}$ denotes the sample of β_j at $(i-1)$-th iteration for $j = 0; 1; 2; \ldots; m$. Then, we sample $\beta_1^{(i)}$ conditioning on $\beta_0^{(i)}; \beta_2^{(i-1)}; \ldots; \beta_m^{(i-1)}$, $\beta_2^{(i)}$ conditioning on $\beta_0^{(i)}; \beta_1^{(i)}; \beta_3^{(i-1)}; \ldots; \beta_m^{(i-1)}$, and $\beta_l^{(i)}$ conditioning on $\beta_0^{(i)}; \beta_1^{(i)}; \ldots; \beta_{l-1}^{(i)}\, \beta_{l+1}^{(i-1)}; \ldots; \beta_m^{(i-1)}$. Thus, we have obtained the i-th iteration sample from the $(i-1)$-th iteration sample. We can obtain the final sample by repeating this procedure until convergence.Now we use Gibbs sampling to obtain the posterior distributions of $p_0; p_1; p_2$ and X. Given X = x, it is easy to show that p_0 has Beta distribution with parameters $\sum_{i:x_i=0} y_i + 1$ and $\sum_{i:x_i=0}(n_i - y_i) + 1$. The posterior distribution of p_1 and p_2 can be worked out similarly. Now, given $p_0; p_1; p_2$, and y, the posterior distribution of x is still very complicated. We consider a further layer of Gibbs sampling and sample x_i from $f(x_i \mid x_{-i}\ ;\ p_0; p_1; p_2; y)$. It can be shown that x has a multinomial distribution with parameters $(q_1/q;\ q_2/q;\ q_3/q)$, where $q_1 = e^{2\beta u_i(0)} p_0^{y_i}(1-p_0)^{n_i - y_i}$, $q_2 = e^{2\beta u_i(1)} p_1^{y_i}(1-p_1)^{n_i - y_i}$, $q_3 = e^{2\beta u_i(2)} p_2^{y_i}(1-p_2)^{n_i - y_i}$, $q = q_1 + q_2 + q_3$, and $u_i(j)$ is the number of j's in the neighbourhood of the i-th location. The algorithm is described as follows:

1-Initialize $p_0^{(0)}; p_1^{(0)}; p_2^{(0)}$, and $x^{(0)}$ ($p_0 < p1 < p_2$. The choice of $x^{(0)}$ is important but not crucial. A better choice of $x^{(0)}$ will give a faster convergence of the algorithm.)

2-Conditional on y and x, sample p_0 from Beta($\sum_{i:x_i=0} y_i + 1$, $\sum_{i:x_i=0}(n_i - y_i) + 1$).

3- Conditional on y and x, sample p_1 from Beta($\sum_{i:x_i=1} y_i + 1$, $\sum_{i:x_i=1}(n_i - y_i) + 1$).

4- Conditional on y and x, sample p_0 from Beta$(\sum_{i:x_i=2} y_i + 1$, $\sum_{i:x_i=2}(n_i - y_i) + 1)$.

5-Given p_0; p_1; p_2, and y, sample x.

-Given x_{-i}; p_0; p_1; p_2, and y, sample x_i from Multinomial distribution with parameter $(q_1/q;\ q_2/q;\ q_3/q)$, where $q_1 = e^{2\beta u_i(0)} p_0^{y_i}(1-p_0)^{n_i-y_i}$, $q_2 = e^{2\beta u_i(1)} p_1^{y_i}(1-p_1)^{n_i-y_i}$, $q_3 = e^{2\beta u_i(2)} p_2^{y_i}(1-p_2)^{n_i-y_i}$, $q = q_1 + q_2 + q_3$

-Continue this loop till convergence.

6-Go back to step 2 till convergence.

In this illustration, we assume K=3. The algorithm for K = n can be generalized in a straightforward manner.

14.4 Empirical Studies with the Disability Rate Data from Mississippi and Alabama

It has been observed that there are distinct patterns in spatial distributions of elderly disabilities across the United States according to the 1990 census data. There is a high concentration of elderly disabilities in the South and a low concentration of elderly disabilities the Midwest (Lin, 2000). Alabama and Mississippi have the highest elderly disability rates among all states. It is of interest to know if there are local spatial clusters of elderly disability rates among these two states. If it is not the case, one may conclude that the high disabilities are due to the common environmental factor in the South. Otherwise, one may suspect some local environmental or social factors contribute to the high disabilities in those two states.

The elderly disability rates aggregated at the county level are based on the long-form of the1990 population Census data. The data were processed in S-PLUS (MathSoft, 2000a). The elderly is defined as those who are 65 or over. A person is considered as a disability if one has a mobility or self-care limitation. Figure 14.3 shows a map of disability rates with counties of Mississippi and Alabama States. The histogram of the disability rates suggests a normal distribution of disability rates in general. The map reveals several possible clusters locally.

As a comparison, we did some exploratory data analysis in the disability data by applying some other approaches for detecting the spatial clusters. Figure 14.4 shows the result of the K-Means method (K = 3) by in-cooperating the spatial coordinates. It partitions the two states into three regions smoothly.

Figure 14.3. Disability rates with counties in Mississippi and Alabama States

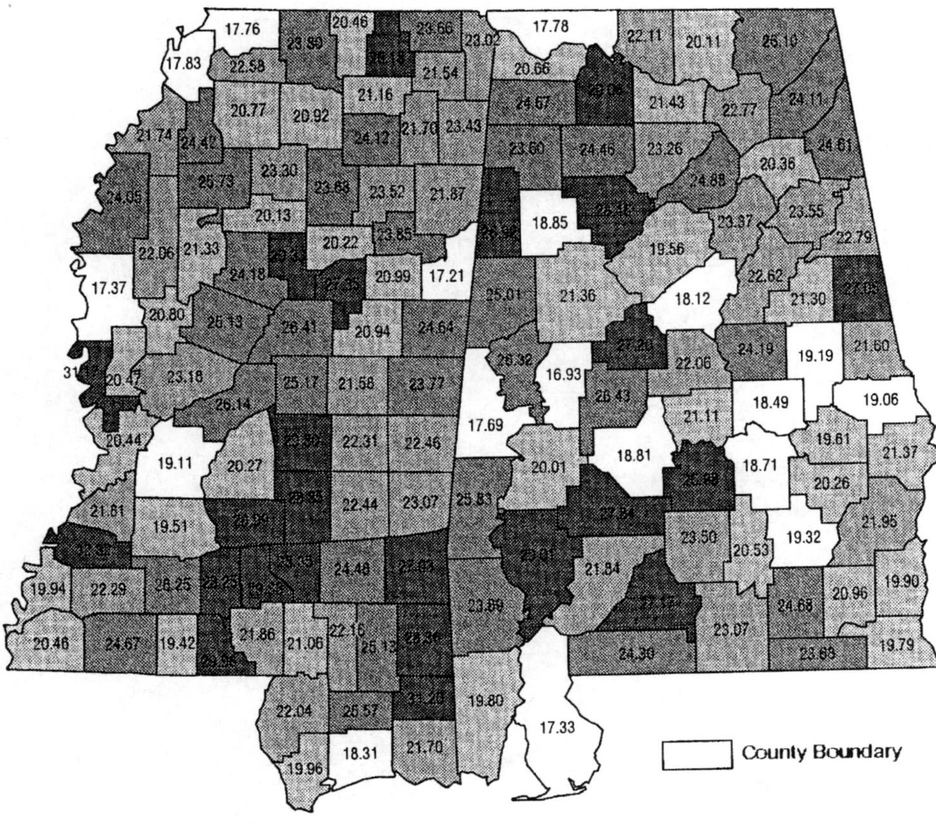

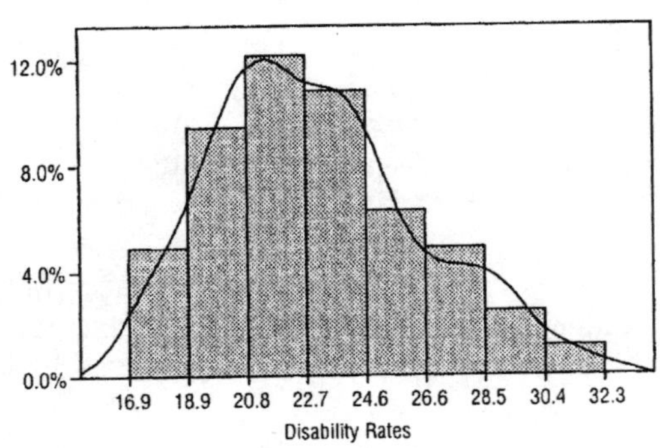

Figure 14.4. Clusters calculated using K-Means methods

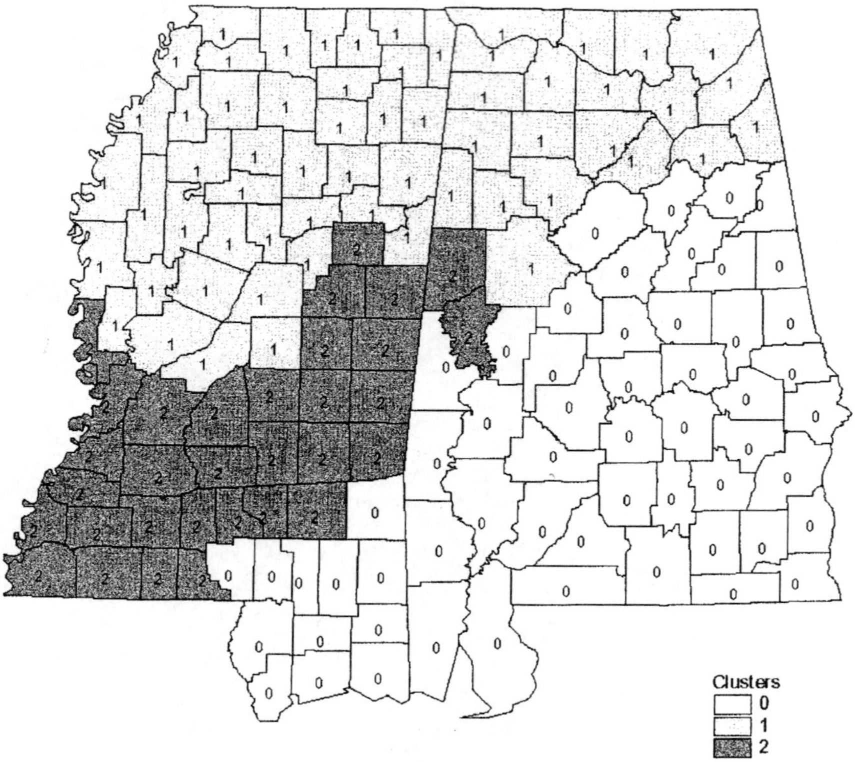

Figure 14.5 displays the results of smoothed disability rates using an interpolation method (Akima, 1978). With the smoothing and thresholding techniques, we can use two dimensional splines to fit a smooth surface and then classify clusters with high values or low values.

We can also use Krigging method (Cressie, 1991) for such an analysis. We can reconstruct the response surface using weighted each observations of neighbourhood. The weighting scheme depends on the covariance structure specified in advance. The Figure 14.5a is a smoothed image by interpolation. The Figure 14.5b is obtained by thresholding the image in Figure 14.5a. The two thresholds are chosen to be 21 and 27.

Figure 14.5. Results of the smoothing method. Figure 14.5a is the smoothed map by applying an interpolation method. Figure 14.5b shows the clusters with the thresholds of 21 and 27

Figure 14.5a Smoothed map by applying an interpolation method

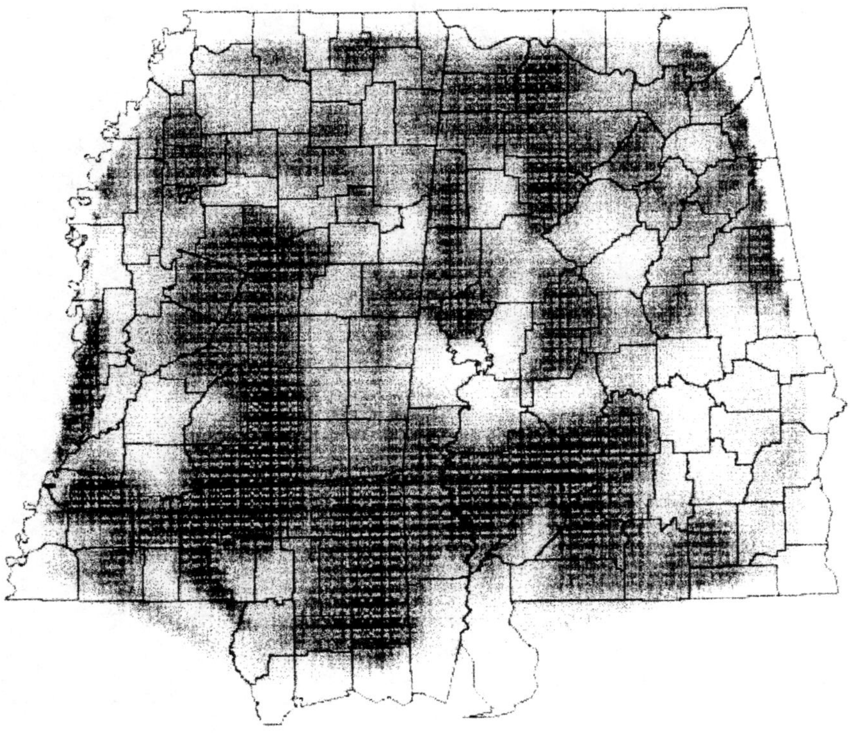

The area with disability rates less than 21 is colored white. The area with disability rates greater than 27 is colored black. The rest is colored grey. The resulted map suggests that there are some local clusters in these two states. The advantage of this method is that it is simple and effective. The shortcoming is that it is usually difficult to choose a proper threshold. For choosing the thresholds based on statistical significance, see Rogerson (2001).

Figure 14.6 displays the test results of Local Moran's on the county disability data from the Mississippi and Alabama State. Although Local Moran's suggest that there exist some local clusters, the boundaries of those clusters are not clear.

Figure 14.5b Clusters with the thresholds of 21 and 27

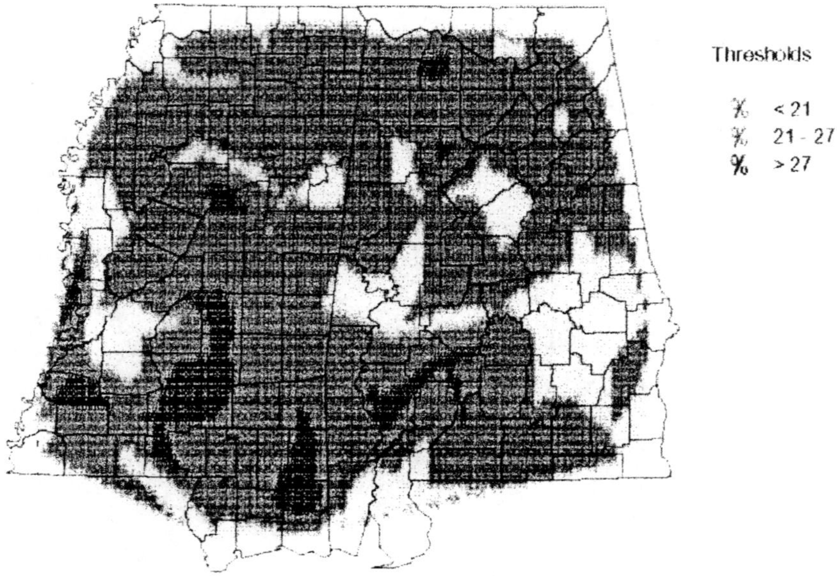

Figure 14.7 shows the results estimated from our spatial mixture model. It reveals some local clusters in presence. We assume that n_i is the total count of the elderly people in the county i, y_i is the observed count of the elderly people in disability in the county i. We set initial values $p_0^{(0)} = .1;\ p_1^{(0)} = .2\ ;\ p_2^{(0)} = .3,,\ x_i^{(0)}$ to 0 if its corresponding y_i is less than the first quartile of observed y, to 1 if its corresponding y_i is between the second and third quartiles, and to 2 if its corresponding y_i is greater than the third quartile. The darkest area with $x = 2$ represents clusters with high disability rates. The lightest area with $x = 0$ represents clusters with low disability rates. Figure 14.8 shows the time series plots of the sampled p_0; p_1; p_2. The stationarity indicates a convergence of the algorithm. One interesting finding in the plot of p_2; p_3 is that the distributions seems oscillating. One explanation is that some counties cannot be clearly classified. They are classified back and forth between 1 and 2 during the iteration. From the posterior distribution of p_0; p_1 and p_2, we obtain the 95% highest posterior density (HPD) intervals for them, which are(0.1924;0.1945); (0.2267;0.2317), and (0.2751;0.2852) respectively.

Figure 14.6. Tests of Local Morans on the disability rate data in the Mississippi and Alabama counties

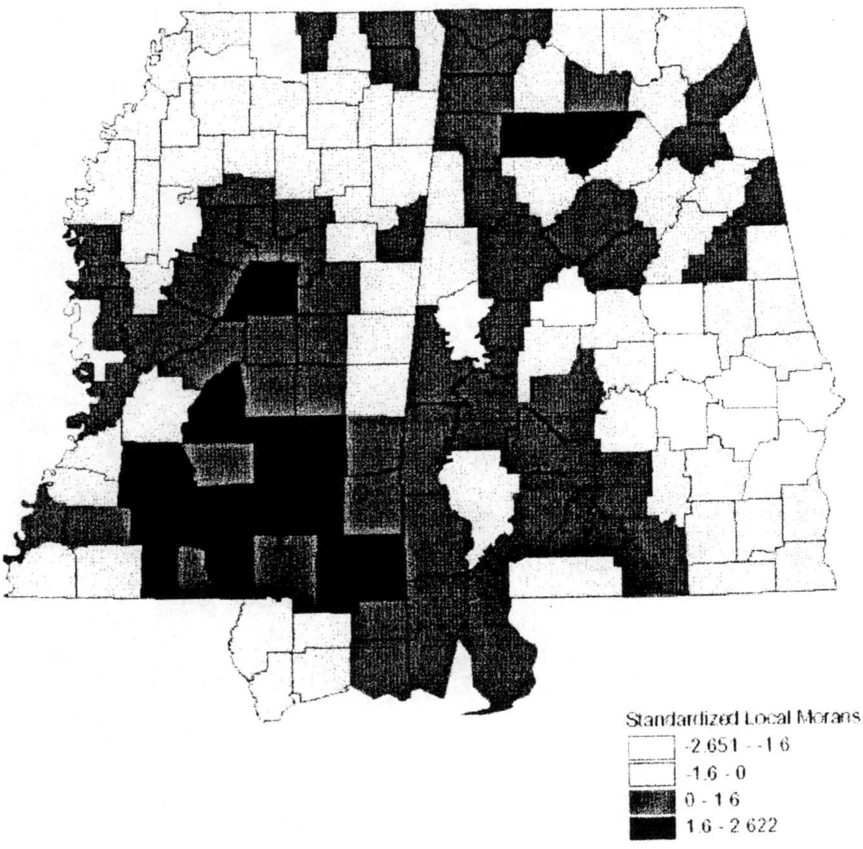

14.5 Discussion and Conclusion

In section 14.3, we have presented the Gibbs sampling algorithm to make inference for the spatial mixture model for Binomial processes. This Bayesian provides a flexible framework for detecting spatial clusters. In fact, this model can be considered as an extension to the generalized linear model.

Let $I_i(X)$ be an indicator function. It equals 1 if $X = i$ and 0 otherwise, $i = 0; 1; 2$. We can redefine our model in section 14.3 as

Figure 14.7. Results from spatial mixture model using Gibbs sampling (K=3)

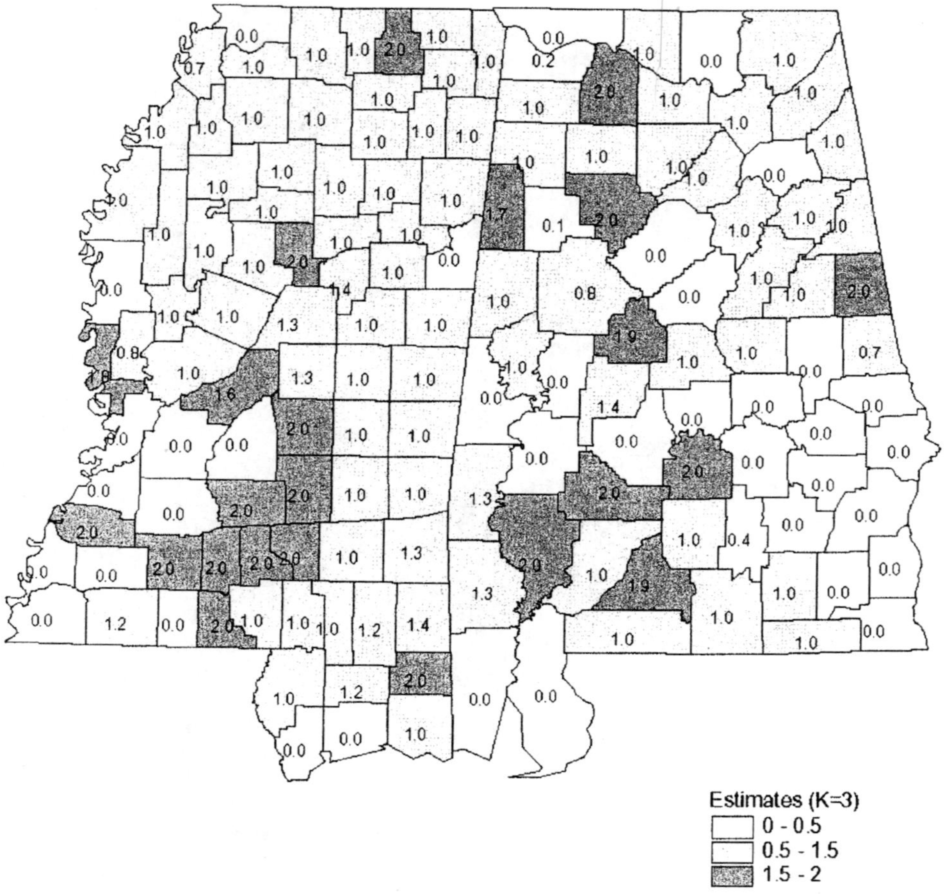

$$logit(p_i) = \alpha_0 I_0(X) + \alpha_1 I_1(X) + \alpha_2 I_2(X)$$

where X has a distribution specified as the Potts model. The only difference between the general logit model and our model is that X is random in our model. With this model, we can easily accommodate more external variables, such as income, and temperature in our example.

Figure 14.8: Time series plots of the sampled p_0; p_1; p_2

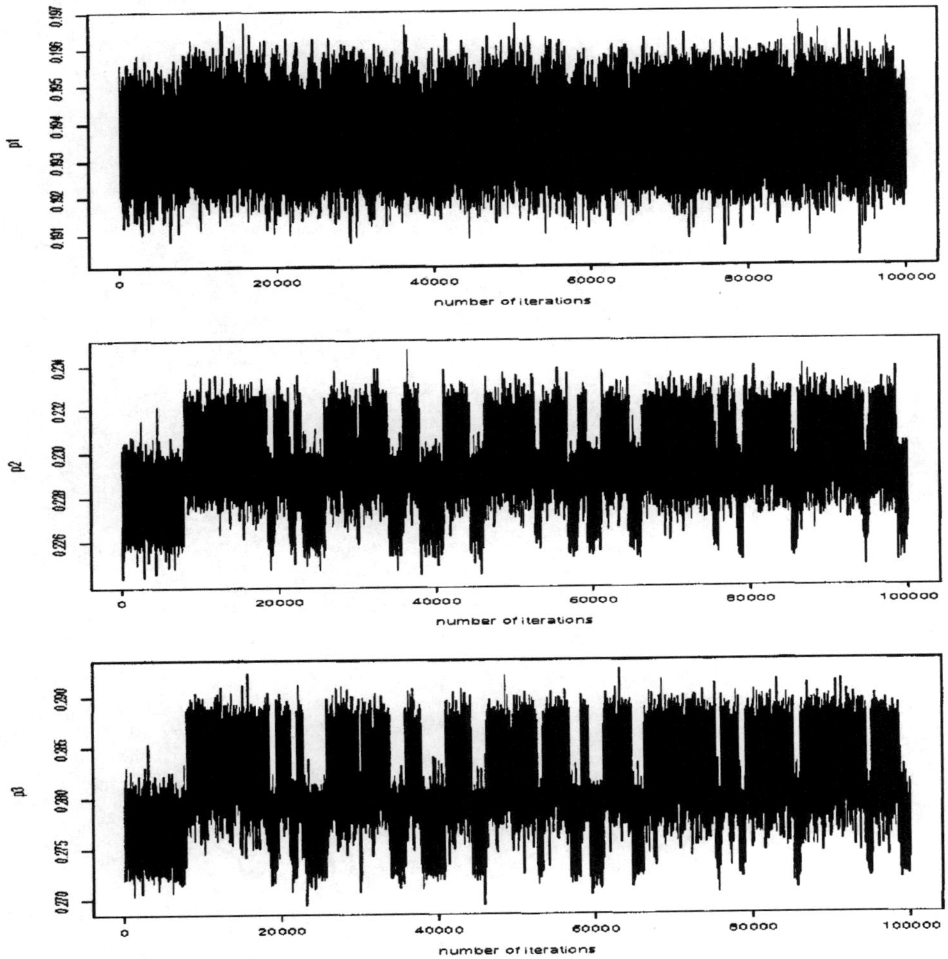

As an important parameter in the model, the choice of β value introduces the degree of local spatial dependence. If we have prior information of the spatial correlation, we can assign a prior distribution for β. The framework remains intact. It is also possible to use pseudo-likelihood method to estimate β with the observed data (Cressie, 1991). Different β value will result in different number of estimated clusters. The larger the β value, the bigger the clusters.

Figure 14.9. Results from spatial mixture model using Gibbs sampling (K=2)

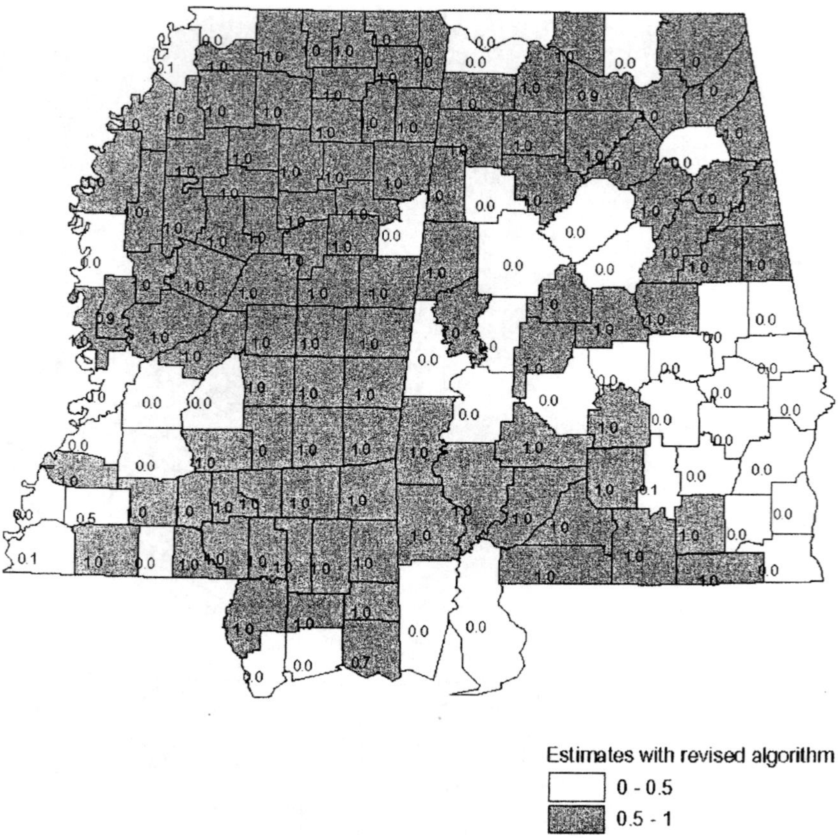

The neighbourhood structure in the model is another important issue. In our application, we used the first order approach (MathSoft, 2000c) in defining the neighbourhood structure, which means only two adjacent counties are defined as neighbours. There are also many other methods in defining the neighbourhood structure (Bao et al., 1995). For example, we can use the geographical distance between two counties to define the neighbourhood structure. The estimated results may be sensitive to different neighbourhood structures. For example, if we choose the strip type of neighbourhood structure, the algorithm will put more weights to detect clusters in strip shape. This sensitivity, however,can provide practitioners the flexibility in detecting different shapes

Figure 14.10: Time series plots of the sampled p_0; p_1

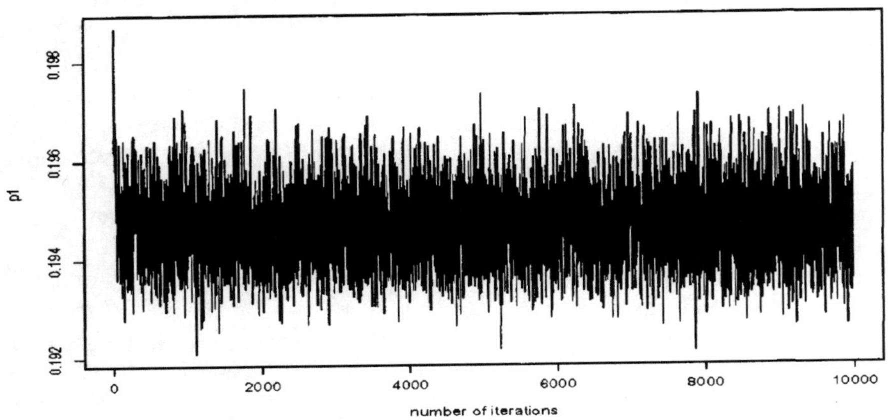

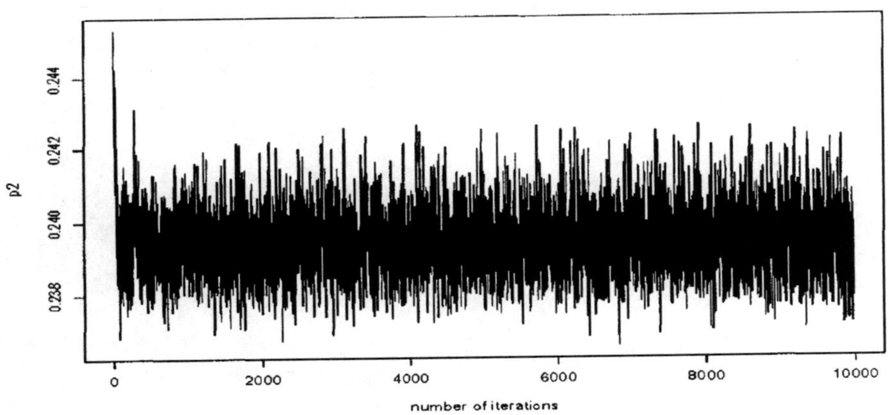

of clusters. In image processing, both the first order and second order nearest neighbourhood structures are commonly used. Theoretically, what kind of methodology for defining neighbourhood structure should depend on the theoretical framework for each particular study. A good neighbourhood structure should better represent spatial linkages or interactions among spatial observations. In practice, we should compare the results from exploratory data analysis by trying different approaches in defining spatial neighbourhood structure.

In our application, as we mentioned in section 14.4, the number (K) of possible values of X is set to 3. We should point out that this parameter, which is the number of types of clusters, is different from the number of estimated clusters, which usually has to be pre-specified in many other clustering methods. This offers more flexibility for detecting clusters. We may have many spatial clusters of the same type. For example, there are two types of clusters (K = 2) in our application (see Figure 14.9) while each type include several clusters.

There is no easy solution for the choice of K. One possible solution is to assign a discrete prior distribution for K first, and then to use the observed data to estimate K. Different K's should be tried to compare the estimated results. Generally, the model based clustering method would work well if the practitioner has already had some idea of the problem in mind. In this study, we classify all counties of Mississippi and Alabama States into three groups with different disability rates (high, intermediate and low) by assuming $K = 3$.

If we assume only two types of clusters with high disability rates and low disability rates, X will only takes on 0 or 1. The Ising model can then be used as the prior distribution of X. The results are displayed in Figure 14.9. As we can see in Figure 14.9, the low disability clusters are very similar to that we obtained when $K = 3$. The high disability cluster is the combination of the high and intermediate clusters when $K = 3$. Figure 14.10 shows that the convergence is achieved. The 95% HPD intervals for p_0 and p_1 are (.194; .1954) and (.2385; .242), respectively.

In summary, we proposed a general framework for detecting local spatial clusters in this chapter. We demonstrate that this approach works well on the disability rate example. Gibbs sampling provides a powerful computing tool for making inference of such models. Our general approach can be applicable to a wide variety of spatial clustering problems.

Acknowledgements

We would like to thank Professor Ge Lin for providing the county disability data for Mississippi and Alabama, Professor Vijay Nair for stimulating discussion on the chapter, and the valuable comments from an anonymous reviewer.

Bibliography

Abdel-Rahman, H. and M. Fujita, 'Product variety, Marshallian externalities, and city size'. *Journal of Regional Science*, 30, (1990) 165-183.

Aghion, P. and P. Howitt, *Endogenous Growth Theory*, (Cambridge: MIT Press, 1998).

Aickin, M. and H. Gensler, 'Adjusting for multiple testing when reporting research results: the Bonferroni vs. Holm methods', *American Journal of Public Health*, 86, (1996) 726-728.

Akima, H. 'A Method of Bivariate Interpolation and Smooth Surface Fitting for Irregularly Distributed Data Points', *Transactions on Mathematical Software*, 4, (1978) 148-164.

Albert, J.H. and S. Chib, 'Bayesian Analysis of Binary and Polychotomous Response Data', *Journal of the American Statistical Association*, 88, (1993) 669-679.

Amable, B. 'Catch-up and convergence: a model of cumulative growth', *International Review of Applied Economics*, 7, (1993) 1-25.

Amemiya,T. *Advanced Econometrics*, (Cambridge: Harvard University Press, 1985).

Anglin, P.M. and R. Gencay, 'Semiparametric Estimation of Hedonic Price Function', *Journal of Applied Econometrics*, 11, (1996) 633-648.

Anselin, L. 'A note on small sample properties of estimators in a first-order spatial autoregressive model', *Environment and Planning A*, 14, (1982) 1023-1030.

Anselin, L. *Spatial Econometrics: Methods and Models*, (Dordrech: Kluwer Academic Publishers,1988).

Anselin, L. 'Local indicators of spatial analysis–LISA', *Geographical Analysis*, 27, (1995) 93-115.

Anselin, L. *Spacestat version 1.9*, (Ann Arbor: BioMedware Inc., 1999).

Anselin, L. and R.J.G.M. Florax, (eds), *New directions in spatial econometrics*, (Berlin: Springer-Verlag, 1995).

Anselin, L. and R.J.G.M. Florax,. 'Small Sample Properties of Tests for Spatial Dependence in Regression Models: Some Further Results', in *New Directions in Spatial Econometrics*, L. Anselin and R.J.G.M Florax (eds.) (Berlin: Springer-Verlag, 1995).

Anselin, L. and S. Hudak, 'Spatial econometrics in practice; A review of software options', *Regional Science and Urban Economics*, 22, (1992) 509-536.

Anselin, L. and R. Moreno, 'Properties of Tests for Spatial Error Components', paper presented at the Applied Econometric Association Conference held in Conjunction with the WRSA Special 40th Anniversary Annual Meeting. Palm Springs, (2000).

Arbia, G. *Spatial Data Configuration in Statistical Analysis of Regional Economic and Related Problems*, (Dordrecht: Kluwer Academic Publishers, 1989)

Bailey, M.J., R.F. Muth, and H.O. Nourse, 'A Regression Method for Real Estate Price Index Construction', *Journal of the American Statistical Association*, 58, (1963) 933-942.

Bao, S., M.S. Henry, K. Brooks and D. Barkley. 'RAS-An Integrated Regional Analysis System with ARC/INFO', *Computers, Environment, and Urban Systems*, 19, (1995)

1.

Bao, S. and M.S. Henry, 'Heterogeneity issues in local measurements of spatial association', *Geographical Systems,* 3, (1996) 1-13.

Barro, R. and X. Sala-i-Martin, 'Convergence across states and regions' *Brookings papers on Economic Activity*, 1, (1991) 107–182.

Barro, R. and X. Sala-i-Martin, '*Economic Growth'*, (Boston: McGraw Hill, 1995).

Bartels, C.P.A. 'Operational statistical methods for analysing spatial data', in *Exploratory and explanatory statistical analysis of spatial data,* C.P.A Bartels and R.H. Ketellapper (eds.) (Boston: Martinus Nijhoff, 1979).

Baumont, C., C. Ertur, and J. Le Gallo, 'Geographic spillover and growth: a spatial econometric analysis for European regions', paper presented at the 6th RSAI World Congress, Lugano, (2000).

Bavaud, F. 'Models for Spatial Weights: A Systematic Look', *Geographical Analysis,* 30, (1998) 153-171.

Beckmann, M.J. 'On the equilibrium distribution of a population in space,' *Bulletin of Mathematical Biophysics,* 19, (1957) 81-90.

Bennett, R.J. *Spatial time series: analysis, forecasting and control.*(London: Pion, 1979).

Besag, J.E. 'Spatial Interaction and the Statistical Analysis of Lattice Systems', *Journal of the Royal Statistical Society B*, 36, (1974) 192-225.

Besag, J.E. 'Statistical Analysis of Nonlattice Data', *The Statistician,* 24, (1975) 179-195.

Besag, J.E. 'On the Statistical Analysis of Dirty Pictures. (with discussion)', *Journal of the Royal Statistical Society, B*, 48, (1986) 259-302.

Bivand, R. 'Regression Modeling with Spatial Dependence: An Application of Some Class Selection and Estimation Methods', *Geographical Analysis*, 16, (1984) 25-37.

Björck, A. *Numerical Methods for Least Squares Problems*, (Philadelphia: SIAM, 1996).

Blommestein, H. and P. Nijkamp, 'Testing the Spatial Scale and the Dynamic Structure in Regional Models (A Contribution to Spatial Econometric Specification Analysis)', *Journal of Regional Science*, 26, (1986) 1-17.

Blumenthal, L.M. *Theory and Applications of Distance Geometry*, (New York: Chelsea Publishing Company, 1970).

Breusch, T. and A. Pagan, 'The LM test and its applications to model specification in econometrics', *Review of Economic Studies*, 47, (1980) 239–254.

Bröcker, J. 'Regional policy lessons from new theories of growth and location', Manuscript, Tecnishe Universität Dresden (1997).

Bronars, S.G. and D.W. Jansen, 'The geographic distribution of unemployment rates in the U.S', *Journal of Econometrics,* 36, (1987) 251-279.

Brunsdon, C. A.S. Fotheringham and M.E. Charlton, 'Geographically Weighted Regression: A Method for Exploring Spatial Nonstationarity', *Geographical Analysis*, 28, (1996) 281-298.

Brunsdon, C., A.S. Fotheringham and M.E. Charlton, 'Spatial Nonstationarity and Autoregressive Models', *Environment and Planning A*, 30, (1998) 957-973.

Brunsdon, C., A.S. Fotheringham, and M.E. Charlton, 'Some Notes on Parametric Significance for Geographically Weighted Regression', *Journal of Regional Science*, 39, (1999) 497-524.

Buja, A., T. Hastie, and R. Tibshirani, 'Linear Smoothers and Additive Models (with Discussion)', *Annals of Statistics*, 17, (1989) 453-555.

Burke, S., L. Godfrey and R. Tremayne, 'Testing AR(1) Against MA(1) Disturbances in the Linear Regression Model: An Alternative Procedure', *Review of Economic Studies*, 57, (1990) 135-145.

Burridge, P. 'Testing for a Common Factors in a Spatial Autorregresion Model', *Environment and Planning A*, 13, (1981) 795-800.

Carlin, B.P. and T.A. Louis, *Bayes and Empirical Bayes Methods for Data Analysis*, (London: Chapman & Hall, 1998).

Casella, G. and E.I. George, 'Explaining the Gibbs Sampler', *Journal of the American Statistical Association,* 46, (1992) 167-173.

Chellappa, R. and A.K. Jain, *Markov random fields: theory and application*, (Boston: Academic Press, 1993).

Chib, S. 'Bayes Inference in the Tobit Censored Regression Model', *Journal of Econometrics*, 51, (1992) 79-99.

Ciccone, A. and R.E. Hall, 'Productivity and the density of economic activity', *American Economic Review*, 86, (1996) 54-70.

Clapp, J.M. 'A New Method for Estimating Land Value Surfaces over Time', Manuscript, University of Connecticut (2001).

Cleveland, W.S. 'Robust Locally Weighted Regression and Smoothing Scatterplots', *Journal of the American Statistical Association*, 74, (1979) 829-836.

Cleveland, W.S. and S.J. Devlin, 'Locally Weighted Regression: An Approach to Regression Analysis by Local Fitting', *Journal of the American Statistical Association*, 83, (1988) 596-610.

Cliff, A.D., P. Haggett, J.K. Ord, K.A. Bassett and R.B. Davies, *Elements of spatial structure*, (Cambridge: Cambridge University Press, 1975).

Cliff., A.D. and J.K. Ord, *Spatial Autocorrelation*, (London: Pion, 1973).

Cliff, A.D. and J.K. Ord, *Spatial Processes, Model and Applications*, (London: Pion, 1981).

Coulson, N.E. 'Really Useful Tests of the Monocentric City Model', *Land Economics*, 67, (1991) 299-308.

Cressie, N.A.C. *Statistics for Spatial Data*, (New York: Wiley, 1993)

Crown, W.H. and L.F. Wheat, 'State per capita income convergence since 1950: Sharecropping's demise and other influences', *Journal of Regional Science*, 35, (1995) 527–552.

Cuadrado-Roura, J., B. Garcia-Greciano and J. Raymond, 'Regional convergence in productivity and productivity structure: The Spanish case', *International Regional Science Review*, 22, (1999) 35–53.

Cuzick, J. and R. Edwards, 'Spatial Clustering for Inhomogenous Populations', *Journal of the Royal Statistical Society, B,* 52, (1990) 73-104.

David, H.A. 'On the application to statistics of an elementary theorem in probability', *Biometrika,* 43, (1956) 85-91.

David, H.A. and D.J. Newell, 'The identification of annual peak periods for a disease', *Biometrics,* 21, (1965) 645-650.

Davidson, R. and J.G. Mackinnon, *Estimation and inference in econometrics,* (Oxford: Oxford University Press, 1993).

Dixit, A. and J.E. Stiglitz, 'Monopolistic competition and optimum product diversity', *American Economic Review*, 67, (1977) 297-308.

Dixon, R.J. and A.P. Thirlwall 'A Model of Regional Growth Rate Differences on Kaldorian Lines', *Oxford Economic Papers*, 27, (1975) 201-14.

Donaghy, K.P. 'The 'Green Book' Twenty Years On: A New Look at the Research Program of Isard and Liossatos's 'Spatial Dynamics and Optimal Space-Time Development'' *Papers in regional science,* (forthcoming).

Donaghy, K.P. 'Solution and estimation of spatial dynamic models in continuous space and continuous time', *The journal of geographical systems*, (forthcoming) (2001).

Ekwaru, J.P. and S.D. Walter, 'An approximation for the rank adjaceny statistic for spatial clustering with sparse data', *Geographical Analysis*, 33, (2001a) 19–28.

Ekwaru, J.P. and S.D. Walter, 'Power of the rank adjaceny statistic for spatial clustering in a small number of regions', *Geographical Analysis*, 33, (2001b) 29–39.

Fan, J. 'Design-Adaptive Nonparametric Regression', *Journal of the American Statistical Association*, 87, (1992) 998-1004.

Fan, J. 'Local Linear Smoothers and their Minimax Efficiencies', *Annals of Statistics*, 21, (1993) 196-216.

Fan, J. and I. Gijbels, 'Variable Bandwidth and Local Regression Smoothers', *Annals of Statistics*, 20, (1992) 2008-2036.

Farebrother, R. 'Algorithm AS153: Pan's Procedure for the Tail Probabilities of the Durbin-Watson Statistics', *Applied Statistics*, 29, (1980) 224-227.

Fingleton, B. 'Estimates of time to economic convergence: An analysis of regions of the European Union', *International Regional Science Review*, 22, (1999) 5–35.

Fingleton, B. 'Spatial econometrics, economic geography, dynamics and equilibrium: a 'third way'?', *Environment and Planning A* , 32, (2000a) 1481-1498.

Fingleton, B. 'Convergence: international comparisons based on a simultaneous equation model with regional effects', *International Review of Applied Economics*, 14, (2000b) 285-305.

Fingleton, B. 'Equilibrium and economic growth: spatial econometric models and simulations', *Journal of Regional Science*, 41, (2001a) 117-148.

Fingleton, B. 'Theoretical economic geography and spatial econometrics: dynamic perspectives', *Journal of Economic Geography*, 1, (2001b), 201-225.

Fingleton, B. and J.S.L. McCombie, 'Increasing returns and economic growth: some evidence for manufacturing from the European Union regions', *Oxford Economic Papers*, 50, (1998) 89-105.

Florax, R.J.G.M. and H. Folmer, 'Specification and Estimation of Spatial Linear Regression Models: Monte Carlo Evaluation of Pre-Test Estimators', *Regional Science and Urban Economics*, 22, (1992) 405-432.

Florax, R.J.G.M., H. Folmer and S. Rey, 'The Relevance of Hendry's Methodology: Experimental Simulation Results for Linear Spatial Models', Discussion Paper no. 98-125/4, (Amsterdam: Tinbergen Institute, 1998).

Fortin, M.J. P. Drapeau and G.M. Jacquez, 'Quantification of the spatial co-occurrences of ecological boundaries', *Oikos* 77, (1996) 51-60.

Frisch, R. 'Editorial', *Econometrica,* 1, (1933) 2.

Fuchs, C. and M. Kenett, 'A test for detecting outlying cells in the multinomial distribution and two-way contingency tables', *Journal of the American Statistical Association,* 75, (1980) 395-98.

Fujita, M., P. Krugman, and A. Venables, *The Spatial Economy: Cities, Regions, and International Trade,* (Cambridge: MIT Press, 1999).

Fujita, M. and J-F. Thisse, 'Economics of agglomeration', *Journal of the Japanese and International Economies,* 10, (1996) 339-378.

Galambos, J. 'Methods for proving Bonferroni type inequalities', *Journal of the London Mathematical Society,* 9, (1975) 561-564.

Galambos, J. *The asymptotic theory of extreme order statistics*, (New York: Wiley, 1978).

Gandolfo, G. *Qualitative analysis and econometric estimation of continuous-time dynamic models,* (Amsterdam: North-Holland, 1981).

Gelman, A., J.B. Carlin, H.S. Stern and D.B. Rubin, *Bayesian Data Analysis*, (Chapman & Hall: London, 1995).

Geman, S. and D. Geman, 'Gibbs distributions and the Bayesian restoration of images', *IEEE Transactions on Pattern Analysis and Machine Intelligence*, (1984) 721-741.

Gencay, R. and X. Yang, 'A Forecast Comparison of Residential Housing Prices by Parametric versus Semiparametric Conditional Mean Estimators', *Economics Letters*, 52, (1996) 129-135.

Getis, A. and K. Ord, 'The analysis of spatial association by use of distance statistics', *Geographical Analysis,* 24, (1992) 189-206.

Geweke, J. 'Bayesian Treatment of the Independent Student t Linear Model,' *Journal of Applied Econometrics*, 8, (1993) 19-40.

Gilbert, C.L. 'Professor Hendry's econometric methodology', *Oxford Bulletin of Economics and Statistics,* 48, (1986) 283-307.

Goodchild, M.F. *Spatial Autocorrelation*, (Norwich: Geobooks, 1986).

Greene, W.H. *Econometric analysis*, 2nd edition, (New York: Macmillan, 1993).

Greene, W.H. *Econometrics* (Prentice Hall: New Jersey, 1997).

Griffith, D.A. *Advanced spatial statistics*, (Dordrech: Kluwer Academic Publishers, 1988).

Griffith, D.A. and F. Lagona, 'On the quality of likelihood-based estimators in spatial autoregressive models when the data dependence structure is misspecified', *Journal of Statistical Planning and Inference,* 69, (1998) 153-174.

Griffiths, W.E., R.C. Hill and G.G. Judge, *Learning and practising econometrics*, (New York: Wiley, 1993).

Haining, R. 'Statistical Tests and Process Generators for Random Fields Models', *Geographical Analysis*, 11, (1979) 45-64.

Haining, R. *Spatial data analysis in the social and environmental sciences*, (Cambridge: Cambridge University Press, 1990).

Härdle, W. *Applied Nonparametric Regression* (Cambridge: Cambridge University Press, 1990).

Härdle, W. and O. Linton, 'Applied Econometric Methods,' in *Handbook of Econometrics, Vol 4,* R.F. Engle and D.L. McFadden (eds.), (New York: Elsevier, 1994).

Härdle, W. and T.M. Stoker, 'Investigating Smooth Multiple Regression by the Method of Average Derivatives', *Journal of the American Statistical Association*, 84, (1989) 986-995.

Haybittle, J., P. Yuen and D. Machin, 'Multiple comparisons in disease mapping', *Statistics in Medicine,* 14, (1995) 2503-2505.

Heikkila, E., P. Gordon, J.I. Kim, R.B. Peiser, H.W. Richardson and D. Dale-Johnson, 'What Happened to the CBD-Distance Gradient?: Land Values in a Policentric City', *Environment and Planning A*, 21, (1989) 221-232.

Hendry, D.F. and G.E. Mizon, 'Serial correlation as a convenient simplication, not a nuisance: a comment on a study of the demand for money by the Bank of England', *Economic Journal,* 88, (1978) 549-563.

Hepple, L. 'Bayesian Techniques in Spatial and Network Econometrics', *Environment and Planning A*, 27, (1995) 447-469.

Hoch, I. and P. Waddell, 'Apartment Rents: Another Challenge to the Monocentric City Model', *Geographical Analysis*, 25, (1993) 20-34.

Holm, S. 'A simple sequentially rejective multiple test procedure', *Scandanavian Journal of Statistics,* 6, (1979) 65-70.

Holt, D., D.G. Steel, M. Tranmer and N. Wrigley, 'Aggregation and ecological effects in geographically based data', *Geographical Analysis*, 28, (1996) 244-262.

Hooper, P. and G. Hewings, 'Some Properties of Space-Time Processes', *Geographical Analysis*, 13, (1981) 203-223.

Hordijk, L. and P. Nijkamp, 'Dynamic models of spatial autocorrelation', *Environment and Planning A,* 9, (1977) 505-519.

Horowitz, J.L. and W. Härdle, 'Direct Semiparametric Estimation of Single-Index Models with Discrete Covariates', *Journal of the American Statistical Association*, 91, (1996) 1632-1640.

Huang, C.J. and T. Fu, 'An Average Derivative Estimation of Stochastic Frontiers', *Journal of Productivity Analysis*, 12, (1999) 45-53.

Hunter, D. 'An upper bound for the probability of a union', *Journal of Applied Probability,* 13, (1976) 597-603.

Huriot, J-M. and J-F. Thisse, *The Economics of Cities*, (Cambridge: Cambridge University Press, 2000).

Ioannides, Y.M. and H.G. Overmann, 'Spatial evolution of the US urban system', Manuscript, Department of Economics, Tufts University (2000).

Isard, W. and P. Liossatos, *Spatial dynamics and optimal space-time development*, (Amsterdam: North-Holland, 1979).

Iwata, S., H. Murao, and Q. Wang, 'Nonparametric Assessment of the Effects of Neighborhood Land Uses on Residential House Values', *Advances in Econometrics*, 14, (2000) 229-257.

Judd, K. *Numerical methods in economics*, (Cambridge: MIT Press, 1998).

Kaashoek, J.F. and J.H.P. Paelinck, 'A Bilinear Programming Solution to the Quadratic Assignment Problem', Working Paper, (2000).

Kaldor, N. 'A Model of Economic Growth', *Economic Journal*, 67, (1957) 591-624.

Kelejian, H.H. and D.P. Robinson, 'Spatial Autocorrelation: A New Computationally Simple Test with an Application to Per Capita Country Police Expenditures', *Regional Science and Urban Economics,* 22, (1992) 317-331.

Kelejian, H.H. and D.P. Robinson, 'Spatial correlation; a suggested alternative to the autoregressive model', in *New Directions in Spatial Econometrics,* L. Anselin and R.J.G.M Florax (eds.) (Berlin: Springer-Verlag, 1995).

Kelejian, H.H., D.P. Robinson and R.L. Vigil, 'Convergence and divergence of US cities:1970-90', Manuscript, Department of Economics, University of Maryland, (1997).

Kelejian, H.H. and I.R. Prucha, 'A generalized spatial two-stage least squares procedure for estimating a spatial autoregressive model with autoregressive disturbances', *Journal of Real Estate Finance and Economics,* 17, (1998) 99-121.

Kelejian, H.H. and I.R. Prucha, 'A Generalized Moments Estimator for the Autoregressive Parameter in a Spatial Model', *International Economic Review,* 17, (1998) 99-121.

Kim, S. 'Economic integration and convergence: U.S. regions, 1840-1987', *Journal of Economic History*, 58, (1998) 659–683.

King, M. 'Robusts Tests for Spherical Symmetry and their Application to Least Squares Regression', *The Annals of Statistics*, 8, (1980) 1265-1271.

King, M. 'Testing for Autoregressive Against Moving Average Errors in the Linear Regression Model', *Journal of Econometrics*, 21, (1983) 35-51.

King, M. and M. McAleer, 'Further Results on Testing AR(1) Against MA(1) Disturbances in the Linear Regression Model', *Review of Economic Studies*, 54, (1987) 649-663.

Krugman, P. *Geography and Trade*, (Cambridge: Cambridge University Press, 1991).

Kwerel, S.W. 'Most stringent bounds on aggregated probabilities of partially specified dependent systems', *Journal of the American Statistical Association,* 70, (1975) 472-479.

Lahiri, S. 'On inconsistency of estimators under infill asymptotics for spatial data', *Sankhya A*, 58, (1996) 403-417.

Lee, J., S. Kwak, and J.A. List, 'Average Derivative Estimation of Hedonic Price Models', *Environmental and Resource Economics*, 16, (2000) 81-91.

Leontief, W. and A. Strout, 'Multiregional Input-Output Analysis', in *Input-Output Economics* (pp.223-257), W. Leontief (ed.), (Oxford: Oxford University Press, 1966).

LeSage, J.P. 'Bayesian estimation of spatial autoregressive models', *International Regional Science Review*, 20, (1997) 113-129.

LeSage, J.P. 'Bayesian Estimation of Limited Dependent Variable Spatial Autoregressive Models,' *Geographical Analysis*, 32, (2000) 19-35.

LeSage, J.P. 'A Family of Geographically Weighted Regression Models', forthcoming in *Advances in Spatial Econometrics*, L. Anselin and R.J.G.M. Florax (eds.), (Berlin: Springer-Verlag, 2001).

LeSage, J.P. and R.K. Pace, 'Using Matrix Exponentials to Explore Spatial Structure in Regression Relationships', unpublished Manuscript available at http://www-spatial-statistics.com, (2000).

Lesse, P.F. 'Economics in time and space' *Papers of the Regional Science Association*, 51, (1983) 81-116.

Lin, G. and D. Zeng, 'Spatial clusters of diseases: Remodeling the concept', *Geographical Information Sciences*, 5, (1999) 175-180.

López-Bazo, E., E. Vaya, A.J. Mora and J. Suriñach, 'Regional economic dynamics and convergence in the European Union', *Annals of Regional Science*, 36, (1998) 1–28.

Maddala , G.S. (ed.), *The econometrics of panel data Vol 1*, (England: Edward Elgar, 1995).

Magalhães, A., G.J. Hewings and C.R. Azzoni, 'Spatial dependence and regional convergence in Brazil', Regional Economics Application Laboratory, Technical Paper 00-T-11, (2000).

Magnus, J.R. 'Maximum likelihood estimation of the GLS model with unknown parameters in the disturbance covariance matrix', *Journal of Econometrics,* 7, (1978) 281-312.

Magrini, S. 'The evolution of income disparities among the regions of the European Union', *Regional Science and Urban Economics*, 29, (1999) 257–281.

Mandy, D.M. and C. Martins-Filho, 'A unified approach to asymptotic equivalence of Aitken and feasible Aitken instrumental variable estimators', *International Economic Review,* 35, (1994) 957-979.

Marcus, M. and H. Minc, *A Survey of Matrix Theory and Matrix Inequalities*, (New York: Dover, 1992).

Mardia, K.V. and R.J. Marshall, 'Maximum likelihood estimation of models for residual covariance in spatial regression', *Biometrika*, 71, (1984) 135-146.

Marshall, A. *Principles of Economics*, (London: MacMillan, 1920).

Martin, R. 'The new 'geographical turn' in economics: some critical reflections', *Cambridge Journal of Economics*, 23, (1999) 65-91.

MathSoft, *S-PLUS 200,* (Seattle: MathSoft, Inc., 1999).

MathSoft, *S-PLUS 2000 User's Guide Version 4.0 for Windows*, (Seattle: MathSoft Inc, 2000a).

MathSoft, *S+SpatialStats User's Manual for Windows and Unix*, (Seattle: MathSoft Inc, 2000b).

MathSoft, *S-PLUS for ArcView GIS User' Guide*, (Seattle: MathSoft Inc, 2000c).

McMillen, D.P. 'Probit with spatial autocorrelation', *Journal of Regional Science*, 32, (1992), 335-348.

McMillen, D.P. 'Spatial effects in limited dependent variable models', in *New directions in spatial econometrics*, (pp189-228), L. Anselin and R.J.G.M Florax (eds.), (Berlin: Springer-Verlag, 1995).

McMillen, D.P. 'One Hundred Fifty Years of Land Values in Chicago: A Nonparametric Approach', *Journal of Urban Economics*, 40, (1996) 100-124.

McMillen, D.P. 'The Return of Centralization to Chicago: Using Repeat Sales to Identify Changes in House Price Distance Gradients', Manuscript, University of Illinois at Chicago, (2001).

McMillen, D.P. and J.F. McDonald, 'A Nonparametric Analysis of Employment Density in a Polycentric City', *Journal of Regional Science*, 37, (1997) 591-612.

McMillen, D.P. and P. Thorsnes, 'The Reaction of Housing Prices to Information on Superfund Sites: A Semiparametric Analysis of the Tacoma, Washington Market', *Advances in Econometrics*, 14, (2000) 201-228.

Meese, R. and N. Wallace, 'Nonparametric Estimation of Dynamic Hedonic Price Models and the Construction of Residential Housing Price Indices', *Journal of the American Real Estate and Urban Economics Association*, 19, (1991) 308-332.

Meese, R. and N. Wallace, 'The Construction of Residential Housing Price Indexes: A Comparison of Repeat Sales, Hedonic Regression, and Hybrid Approaches', *Journal of Real Estate Finance and Economics*, 14, (1997) 51-73.

Mills, E.S. *Studies in the Structure of the Urban Economy,* (Baltimore: Johns Hopkins University Press, 1972).

Mizon, G.E. 'A simple message for autocorrelation correctors: Don't', *Journal of Econometrics*, 69, (1995) 267-288.

Mur, J. 'Testing for Spatial Autocorrelation: Moving Average versus Autoregressive Processes', *Environment and Planning A*, 31, (1999) 1371-1382.

Mur, J. and F.J. Trivez, 'Scale Elements in Spatial Autocorrelation Tests', paper presented at the 40th Congress of the European Regional Science Association. Barcelona (2000).

Murty, K.G. *Linear and Combinatorial Programming*, (New York: Wiley, 1976).

Muth, R.F. *Cities and Housing,* (Chicago: University of Chicago Press, 1969).

Nadaraya, E.A. 'On Estimating Regression', *Theory of Probability and its Applictions*, 9, (1964) 141-142.

National Bureau of Standards, 'Tables of the bivariate normal distribution and related functions', *Applied Mathematics Series*, (1959) 50.

Nerlove, M. 'Properties of alternative estimators of dynamic panel models: An empirical analysis of cross-country data for the study of economic growth', in *Analysis of panels and limited dependent variable models,* C. Hsiao, K. Lahiri, L.F. Lee and M.H. Pesaran (eds), (Cambridge: Cambridge University Press, 1999).

Nijkamp, P. and J. Poot, 'Spatial perspectives on new theories of economic growth', *Annals of Regional Science*, 32, (1998) 407–437.

Odland, J. *Spatial autocorrelation*, (Newbury Park: Sage Publications, 1988).

Ord, J.K. 'Estimation Methods for Models of Spatial Interaction', *Journal of the American Statistical Association*, 70, (1975) 120-126.

Ord, J.K. 'Towards a theory of spatial statistics: a comment', *Geographical Analysis,* 13, (1981) 91-93.

Ord, J.K. and A. Getis, 'Distributional issues concerning distance statistics', paper presented at the Annual Meeting of the North American Regional Science Association International, Houston, (1993).

Ord, J.K. and A. Getis, 'Local spatial autocorrelation statistics: distributional issues and an application', *Geographical Analysis,* 27, (1995) 286-306.

Oosterhaven, J. 'Regional convergence and/or concentration: an overview', Manuscript, Department of Economics, University of Groningen, (1997).

Overman, H.G. and Y.M. Ioannides, 'Cross-sectional evolution of the US city size distribution', Manuscript, Department of Geography and Environment, London School of Economics, (1999).

Pace, R.K. 'Parametric, Semiparametric, Nonparametric Estimation of Characteristic Values with Mass Assessment and Hedonic Pricing Models', *Journal of Real Estate Finance and Economics*, 11, (1995) 195-217.

Pace, R.K. and R. Barry, 'Quick Computation of Regressions with a Spatially Autoregressive Dependent Variable', *Geographical Analysis*, 29, (1997) 232-247.

Pace, R.K. and J.P. LeSage, 'Closed-Form Maximum Likelihood Estimates for Spatial Problems,' unpublished Manuscript available at http://www-spatial-statistics.com (2000).

Paelinck, J.H.P. (with the assistance of J.-P. Ancot and J.H. Kuiper), *Formal Spatial Economic Analysis*, (Aldershot: Gower Press, 1983).

Paelinck, J.H.P. 'Modélisation réticulaire par affectation linéaire', paper presented at the annual congress of the ASRDLF, Lugano, (2000a).

Paelinck, J.H.P. 'Médicométrie relationnelle: application aux Pays-Bas', paper presented at the annual congress of the ASRDLF, Lugano, (2000b).

Pagan, A. and A. Ullah, *Nonparametric Econometrics,* (Cambridge: Cambridge University Press, 1999).

Parker, P.M. *Physioeconomics: the basis for long-run economic growth,* (Cambridge: MIT Press, 2000).

Pavlov, A.D. 'Space-Varying Regression Coefficients: A Semiparametric Approach Applied to Real Estate Markets', *Real Estate Economics*, 28, (2000) 249-283.

Pfeifer, P.F. and S.J. Deutsch, 'A three-stage iterative procedure for space-time modeling', *Technometrics,* 22, (1980) 35-47.

Phillips, P.C.B. 'Error Correction and Long-Run Equilibrium in Continuous Time,' *Econometrica*, 59, (1991) 967-980.

Plummer, P. and M. Taylor, 'Theories of local economic growth (part 1): concepts, models and measurement'. *Environment and Planning A*, 33, (2001) 219-236.

Powell, J.L., J.H. Stock, and T.M. Stoker, 'Semiparametric Estimation of Index Coefficients', *Econometrica*, 57, (1989) 1403-1430.

Puu, T. 'Outline of a trade cycle model in continuous space and time', *Geographical Analysis*, 14, (1982) 1-9.

Puu, T. 'Continuous spatial modeling in economics', *Papers of the Regional Science Association*, 56, (1985) 21-36.

Quah, D.T. 'Empirical cross-section dynamics in economic growth', *European Economic Review*, 37, (1993), 426–434.

Quah, D.T. 'Aggregate and regional disaggregate fluctuations', *Empirical Economics*, 21, (1996a) 137–159.

Quah, D.T. 'Twin peaks: Growth and convergence in models of distribution dynamics', *The Economic Journal*, 106, (1996b) 1045–1055.

Rey, S.J. 'Spatial empirics for regional economic growth and convergence', *Geographical Analysis*, forthcoming (2001).

Rey, S.J. and B.D. Montouri, 'U.S. regional income convergence: A spatial econometric perspective', *Regional Studies*, 33, (1999) 143–156.

Ripley, B. *Statistical Inference for Spatial Processes*, (Cambridge: Cambridge University Press, 1988).

Robinson, P.M. 'Root-*N*-Consistent Semiparametric Regression', *Econometrica*, 56, (1988) 931-954.

Rogerson, P.A. 'A spatial version of the chi-square goodness-of-ft test and its application to tests for spatial clustering' in *Econometric Advances in Spatial Modeling and Methodology: Essay in Honour of Jean Paelinck*, (pp. 71-84), C. Amrhein, D. Griffith and J.M. Huriot (eds.), (Dordrecht: Kluwer Academic Publishers, 1998).

Rogerson, P.A. 'A statistical method for the detection of geographic clustering', *Geographical Analysis*, 33, (2001) 215-227.

Roy, B. *Algèbre moderne et théorie des graphes*, (Paris: Dunod, 1969).

Ruppert, D. and M.P. Wand, 'Multivariate Locally Weighted Least Squares', *Annals of Statistics*, 22, (1994) 1346-1370.

Sala-i-Martin, X. 'Regional cohesion: Evidence and theories of regional growth and convergence', *European Economic Review*, 40, (1996) 1325–1352.

Schweder, T. and E. Spjotvoll, 'Plots of *p*-values to evaluate many tests simultaneously', *Biometrika*, 69, (1982) 493-502.

Shorrock, A. 'The measurement of mobility', *Econometrica*, 46, (1978) 1013–1024.

Sidak, A. 'On multivariate normal probabilities of rectangles: their dependence on correlations', *Annals of Mathematical Statistics*, 39, (1968) 1425-1434.

Silverman, A.W. *Density Estimation for Statistics and Data Analysis*, (New York: Chapman and Hall, 1986).

Smith, T. 'A Central Limit Theorem for Spatial Samples', *Geographical Analysis*, 12, (1980) 299-324.

Smith A.F.M. and G.O. Roberts, 'Bayesian computation via Gibbs and related Markov Chain Monte Carlo methods', *Journal of the Royal Statistical Society, B*, 55, (1993) 3-24.

Sobel, M. and V.R.R.R. Uppuluri, 'On Bonferroni-type inequalities of the same degree for the probability of unions and intersections', *Annals of Mathematical Statistics,* 43, (1972) 1549-1558.

Stock, J.H. 'Nonparametric Policy Analysis: An Appliction to Estimating Hazardous Waste Cleanup Benefits,' in *Nonparametric and Semiparametric Methods in Econometrics and Statistics,* W.A. Bennett, J. Powell, and G.E. Tauchen (eds.), (Cambridge: Cambridge University Press, 1991).

Stone, A.J. 'Consistent Nonparametric Regression', *Annals of Statistics*, 5, (1977) 595-645.

Strang, G. *Linear Algebra and its Applications*, (New York: Academic Press, 1976).

Swamy, P.A.V.B. 'Efficient inference in a random coefficient regression model', *Econometrica,* 38, (1970) 311-323.

Tango, T. 'A class of tests for detecting 'general' and 'focused' clustering of rare diseases', *Statistics in Medicine,* 14, (1995) 2323-2334.

Tanner, M.A. *Tools for Statistical Inference*, 3rd ed. (Berlin: Springer-Verlag, 1996).

Theil, H. *Principles of Econometrics*, (Amsterdam: North-Holland, 1971).

Thorsnes, P. 'Consistent Estimates of the Elasticity of Substitution between Land and Non-Land Inputs in the Production of Housing', *Journal of Urban Economics*, 42, (1997) 98-108.

Tsionas, E.G. 'Regional growth and convergence: evidence from the United States', *Regional Studies*, 34, (2000) 231–238.

van Gastel, R.A.J.J. and J.H.P Paelinck, 'Computation of Box-Cox Transform Parameters. A New Method and its Application to Spatial Econometrics', in *New Directions in Spatial Econometrics*, (pp 136-155), L. Anselin and R.J.G.M. Florax (eds.), (Berlin: Springer-Verlag, 1995).

Waddell, P., B.J.L. Berry, and I. Hoch, 'Residential Property Values in a Multinodal Urban Area: Evidence on the Implicit Price of Location', *Journal of Real Estate Finance and Economics*, 7, (1993a) 117-141.

Waddell, P., B.J.L. Berry, and I. Hoch, 'Housing Price Gradients: The Intersection of Space and Built Form', *Geographical Analysis*, 25, (1993b) 5-19.

Wald, A. 'Note on the consistency of the maximum likelihood estimate', *Annals of Mathematical Statistics*, 20, (1949) 595-601.

Watson, G.S. 'Smooth Regression Analysis', *Sankhya, Series A*, 26, (1964) 359-372.

White, H. 'A Heteroskedasticity-Consistent Covariance Matrix Estimator and a Direct Test for Heteroskedasticity', *Econometrica*, 48, (1980) 817-838.

White, H. 'Maximum Likelihood Estimation of Misspecified Models', *Econometrica,* 50, (1982) 1-25.

Wilks, S.S. 'Certain generalizations in the analysis of variance', *Biometrika*, 24, (1932) 471-494.

Wilson, A.G. *Papers in urban and regional analysis*, (London: Pion, 1972).

Worsley, K.G. 'An improved Bonferroni inequality and applications', *Biometrika,* 69, (1982) 297-302.

Wrigley, N. and R.J. Bennett (eds), *Quantitative geography*, (London: Routledge & Kegan Paul, 1981).

Wymer, C.R. 'Econometric estimation of stochastic differential equation systems', *Econometrica*, 40, (1972) 566-577.

Wymer, C.R. 'Estimation of non-linear continuous-time models from discrete data', in *Models, methods, and applications of econometrics*, P. Phillips and V. Hall (eds), (Oxford: Basil Blackwell, 1993).

Wymer, C.R. 'Structural non-linear continuous-time models in econometrics', *Macroeconomic dynamics*, 1, (1997) 518-548.

Yatchew, A. 'Nonparametric Regression Techniques in Economics', *Journal of Economic Literature*, 36, (1998) 669-721.

Yuming, F. and S.T. Somerville, 'Site Density Restrictions: Measurement and Empirical Analysis', *Journal of Urban Economics*, 49, (2001) 404-423.

Zhu, L., A.E. Gelfand and B.P. Carlin, 'Hierarchical regression with misaligned spatial data: Relating ambient ozone and pediatric asthma ER visits in Atlanta'. Research Report 2001-010, Division of Biostatistics, University of Minnesota, 2000.

Subject Index

Average derivative estimator	232
Bayesian Inference	265
Bayesian Markov Chain Monte Carlo Estimation	105
Binary Dependent Variables	105
Binomial Process	265
Chicago	232
Connectropy	217
Consistency	150
Contiguity	217
Continuous space	89
Continuous time	89
Correlated tests	250
Decentralization	265
Discrimination	150
Distance measure	217
Entropy	217
Error heterogeneity	8
Filtering	150
First observation(s)	176
Geographically Weighted Regression	31, 232
Gibbs Sampling	265
Group	217
Heteroskedasticity	135, 150
House prices	232
Improved Bonferroni adjustments	250
Income convergence	194
Inequality	194
Kelejian-Robinson test	135
King Point optimal test	150
Labor force participation	176
Lagrangian Multiplier	135
Linear assignment	217
Local linear regression	232
Local statistics	250
Markov Random Field	265
Mathematical program	217
Maximum likelihood	31, 176
Metric	217
Monte Carlo	135, 176
Moran's I	135

Multiple tests 250
Nearest neighbour 105
New economic geography 8
Non-muted and Muted heteroskedasticity 135
Nonparametric estimation 232
Parameter dynamics 8
Phillips-curve 176
Potential flow 217
Projection method 89
Proximity 217
Quadratic assignment 217
Quasi-FIML estimation 89
Recursive Estimation 8, 31
Repeat sales 232
Returns to Scale 8
Sample Censoring 105
Serial autocorrelation 176
Sidak adjustment 250
Simulation 150
Space-time data 176
Spatial Autocorrelation 150, 176, 194
Spatial Autogressive Processes 135
Spatial Autoregression 31
Spatial Autoregressive models 105
Spatial Clustering 265
Spatial Dynamics 89
Spatial Econometrics 31, 194
Spatial Error components 135
Spatial Mixture Models 265
Spatial models 232
Spatial Moving Average 150
Spatial statistics 31, 105, 265
Stationarity 176
Technical progress rate 8
Technological externalities 8
Technology diffusion 8
Urban Economics 8
Variance criterion 150
Verdoorn law 8
Wald test 150